ESO ASTROPHYSICS SYMPOSIA
European Southern Observatory

Series Editor: Jacqueline Bergeron

Springer
Berlin
Heidelberg
New York
Barcelona
Hong Kong
London
Milan
Paris
Singapore
Tokyo

Raffaella Morganti   Warrick J. Couch (Eds.)

Looking Deep in the Southern Sky

Proceedings of the ESO/Australia Workshop
Held at Sydney, Australia,
10–12 December 1997

Springer

Volume Editors

Raffaella Morganti
Istituto di Radioastronomia
Via Gobetti 101
I-40129 Bologna, Italy

Warrick J. Couch
School of Physics
University of New South Wales
Sydney NSW 2052, Australia

Series Editor

Jacqueline Bergeron
European Southern Observatory
Karl-Schwarzschild-Strasse 2
D-85748 Garching, Germany

Cataloging-in-Publication data applied for

Die Deutsche Bibliothek - CIP-Einheitsaufnahme

Looking deep in the southern sky : proceedings of the ESO/ Australia workshop, held at Sydney, Australia, 10 - 12 December 1997 / Raffaella Morganti ; Warrick J. Couch (ed.). - Berlin ; Heidelberg ; New York ; Barcelona ; Hong Kong ; London ; Milan ; Paris ; Singapore ; Tokyo : Springer, 1999
 (ESO astrophysics symposia)
 ISBN 3-540-65286-8

ISBN 3-540-65286-8 Springer-Verlag Berlin Heidelberg New York

This work is subject to copyright. All rights are reserved, whether the whole or part of the material is concerned, specifically the rights of translation, reprinting, re-use of illustrations, recitation, broadcasting, reproduction on microfilms or in any other way, and storage in data banks. Duplication of this publication or parts thereof is permitted only under the provisions of the German Copyright Law of September 9, 1965, in its current version, and permission for use must always be obtained from Springer-Verlag. Violations are liable for prosecution under the German Copyright Law.

© Springer-Verlag Berlin Heidelberg 1999
Printed in Germany

The use of general descriptive names, registered names, trademarks, etc. in this publication does not imply, even in the absence of a specific statement, that such names are exempt from the relevant protective laws and regulations and therefore free for general use.

Typesetting: Camera ready by authors/editors
Cover design: *design & production* GmbH, Heidelberg
SPIN: 10651950   55/3144-543210 - Printed on acid-free paper

Preface

The idea of a joint ESO/Australia meeting on the large number of exciting new facilities that are, or will soon be, available in the southern hemisphere arose quite naturally. In the optical and the near-infrared, the Very Large Telescope (VLT) will soon be operational. In the radio, the Australia Telescope Compact Array is going to be upgraded to higher frequencies (20 and 100 GHz), together with an improvement in very long baseline interferometry (VLBI) facilities. Other major facilities, such as the Large Millimetre Array and the 1kT are being planned. Moreover, new deep surveys are underway in the southern hemisphere: the southern Hubble Deep Field, the ESO Imaging Survey (EIS), panoramic deep surveys with the UK Schmidt telescope, and the Anglo-Australian Telescope (AAT) 2dF galaxy/QSO redshift survey in the optical; and the Parkes multibeam HI survey and Molonglo Observatory Synthesis Telescope (MOST) Wide Field continuum survey at radio wavelengths. With all these new facilities, important progress will be made regarding important issues such as the large-scale structure of the universe, the very early universe and the associated first epoch of galaxy formation. The generation of large databases, and the opportunity for sensitive follow-up observations in complementary wavebands, mean that coordinated radio, infrared and optical projects in the southern hemisphere are likely to become increasingly attractive and important.

The aim of the workshop was to bring together a diverse range of astronomers to discuss these developments and, in particular, to define areas of collaboration, and to provide a focus for stimulating new coordinated projects covering most of the electromagnetic spectrum. The meeting addressed a wide range of scientific topics relevant to deep surveys and particularly how these surveys will feed the new generation of 8-m telescopes being constructed at ESO and other observatories around the world. Of note was an impromptu lunchtime discussion (organized by Dr. B.J. Boyle of the AAO) on the coordination of observations of the southern Hubble Deep Field.

The meeting was held at the State Library of New South Wales which provided a pleasant and informal atmosphere and a great location in the heart of Sydney. It gathered approximately 110 astronomers, of whom about 50 came from overseas. To encourage an atmosphere of informality and ensure that participants were "well-oiled" for discussion during the meeting, a "Wine tasting" was held at the time of registration. The harbour cruise conference dinner was also a great opportunity for carrying on the discussions while enjoying the wonderful sights of Sydney harbour!

We are grateful to a number of sponsors who made this workshop possible. The funding received from the Australian Government Department of Industry, Science and Tourism (DIST) under the Major National Research Facilities International Collaboration program was used to support travel to Australia for some overseas participants. Support was also received from the European Southern Observatory, the Australia Telescope National Facility, the Anglo-Australian Observatory, the Special Research Centre for Theoretical Astrophysics (Sydney University) and the Mount Stromlo and Siding Spring Observatory.

Also crucial to the success of the meeting were the members of the Local Organizing Committee, all of whom gave great support. In particular, Elaine Pacey put in a major effort in taking care of most of the logistics and social events while Helen Woods played a vital role with the administration. Helen Sim was a key person in finding the venue, creating the logo and arranging the public relations. Vincent McIntyre took care of the installation of the computers at the Library: without this (i.e. without email) many astronomers would not have survived the meeting! Thanks also go to all the students and post-docs who helped during the meeting, in particular, to Katrina Sealey who collected and ordered the questions and answers from the discussion at the end of each contribution. We are also grateful to Diana O'Mally of the State Library for being so helpful.

Finally, we are greatly indebted to Pamela Bristow from the European Southern Observatory for her great care and valuable editing skills in preparing these Proceedings for publication.

Sydney, October 1998 *Raffaella Morganti and Warrick Couch*

Scientific Organizing Committee:

Brian Boyle (AAO), Russell Cannon (AAO), Matthew Colless (MSSSO), Luiz da Costa (ESO), Warrick Couch (University of NSW), Sandro D'Odorico (ESO), John Danziger (OAT), Ron Ekers (ATNF), Bob Fosbury (ST-ECF, co-Chair), Wolfram Freudling (ST-ECF), Isobel Hook (ESO), Raffaella Morganti (ATNF/ Ist. Radioastronomia, Bologna, co-Chair), Elaine Sadler (Sydney Univ.)

Local Organizing Committee:

V. McIntyre (Sydney Univ.), R. Morganti (ATNF/IRA), E. Pacey (ATNF), H. Sim (ATNF), H. Woods (AAO)

Workshop participants on the forecourt steps of Sydney Opera House prior to the conference dinner/cruise on Sydney harbour.

Contents

PART 1. INTRODUCTION

Opening Remarks
J. Mould 3

PART 2. DEEP IMAGING AND REDSHIFT SURVEYS

The 2dF Galaxy Redshift Survey
M. Colless 9

The 2dF QSO Redshift Survey
B.J. Boyle, S.M. Croom, R.J. Smith, T. Shanks, L. Miller,
N. Loaring 16

A Complete 2dF Survey of Fornax
M.J. Drinkwater, E.M. Sadler, J.I. Davies, R.J. Dickens,
M.D. Gregg, Q.A. Parker, S. Phillipps, R.M. Smith 21

Galaxies Behind the Deepest Extinction Layer of the Southern Milky Way
R.C. Kraan-Korteweg, B. Koribalski, S. Juraszek 23

Surveys of Peculiar Velocities of Galaxies
W. Freudling 29

ESO Imaging Survey
L. da Costa, A. Renzini 34

Cosmological Tests from the New Surveys
O. Lahav 42

Mapping the Dark Matter with Weak Gravitational Lensing
P. Schneider 51

Weak Lensing with MEGACAM and the VLT
      Y. Mellier, L. van Waerbeke, F. Bernardeau, O. Le Fèvre 59

The Cosmological Uncertainty Principle
      C. Fluke, R. Webster 66

Gravitational Lensing in the 2dF Galaxy Redshift Survey
      D.J. Mortlock, R.L. Webster 68

A Weak Gravitational Lensing Cluster Survey
      G. Squires, P. Rosati, J. Silk, T. Broadhurst 70

Surveys with the BTC Mosaic CCD Camera at the Blanco 4m Telescope
      M.G. Smith 72

Deep Images of Bright Galaxies
      D. Malin, B. Hadley 78

A Deep Tech Pan Survey of Dwarf Spheroidal Galaxies in Virgo
      Q.A. Parker, S. Phillipps 83

Surveys with a 4 m Liquid Mirror Telescope
      C. Jean, J.-F. Claeskens, J. Surdej 89

The Metagalactic Ionizing Field in the Local Group
      J. Bland-Hawthorn 91

PART 3. COORDINATING MULTI-WAVELENGTH OBSERVATIONS

The New Molonglo Radio Survey
      E.M. Sadler 103

Testing Models of Radio Source Space Density Evolution with the SUMSS Survey
      C.A. Jackson, J.V. Wall 110

Variance and Skewness in the FIRST Survey
      M. Magliocchetti, S.J. Maddox, O. Lahav, J.V. Wall 112

The AT–ESP Radio Survey: Goals, Description, First Results
      I. Prandoni, L. Gregorini, P. Parma, R.H. de Ruiter,
      G. Vettolani, M.H. Wieringa, R.D. Ekers 114

The Phoenix Deep Survey
       A. Hopkins, L. Cram, B. Mobasher, A. Georgakakis 120

All-Sky Radio Surveys
       R. Wielebinski 125

A Study of Low/Intermediate-Redshift Radio Sources and Clues to the Nature of High-Redshift Objects
       R. Morganti, C. Tadhunter, M. Villar-Martin, R. Dickson 130

The Local Galaxy Population from the HIPASS Survey
       L. Staveley-Smith, R.L. Webster, G. Banks, V. Kilborn, B. Koribalski, M. Putman 132

The MNRF Upgrade to the Australia Telescope
       R.P. Norris 140

The Square Kilometer Array Radio Telescope
       R. Ekers 146

Prospects with Large Millimeter Arrays
       P. Shaver 153

Galileo & d.o.lo.res.
       E. Molinari, P. Conconi, M. Pucillo, S. Monai 157

The Future of Observations of the Sunyaev-Zel'dovich Effect
       H. Liang, M. Birkinshaw 159

Cosmology in a Nutshell + an Argument Against $\Omega_\Lambda = 0$ Based on the Inconsistency of the CMB and Supernovae Results
       C.H. Lineweaver 167

ASCA Results and Future Japanese X-Ray Missions
       K. Yamashita 174

High-Redshift X-Ray Clusters
       I.M. Gioia 181

The Deep X-Ray Radio Blazar Survey (DXRBS)
       P. Padovani, E. Perlman, P. Giommi, R. Sambruna, L.R. Jones, A. Tzioumis, J. Reynolds 187

X-Ray Surveys and Their Follow-Up
       I.J. Danziger 194

Cosmic Rays and the Structure of the Local Universe
  R.W. Clay, B.R. Dawson 199

Looking Deep from the South Pole:
Star Formation in the Thermal Infrared
  M.G. Burton, J.W.V. Storey, M.C.B. Ashley 201

PART 4. THE HIGH-REDSHIFT UNIVERSE

Surveys for High Redshift QSOs
  I.M. Hook, R.G. McMahon, P.A. Shaver 211

The Evolution of the Clustering of QSOs
  F. La Franca, P. Andreani, S. Cristiani 216

Local Population and Evolution of Optically Bright QSOs
  L. Wisotzki 221

Gas and Dust at High Redshift
  R.W. Hunstead 226

Evidence of Structure in the Lyman-α Forest
  J. Liske, J.K. Webb 234

Very High Redshift Radio Galaxies
  W. van Breugel, C. De Breuck, H. Röttgering, G. Miley,
  A. Stanford 236

The Highest Redshift Radio Galaxy Known
in the Southern Hemisphere
  C. De Breuck, W. van Breugel, H. Röttgering, G. Miley,
  C. Carilli 246

Radio Continuum and Emission Line Morphologies
of Southern Seyfert Galaxies
  Z. Tsvetanov, R. Morganti, R.A.E. Fosbury, M.G. Allen,
  J. Gallimore 248

Surveying High z Galaxies with HST and Keck
  G. Illingworth 250

AUSTRALIS: A Multi-Fibre Near-IR Spectrograph
for the VLT
  K. Taylor 257

Surveys for Galaxies at $z > 2$,
and an Introduction to the HDF–South
  M. Dickinson 262

High-Redshift Galaxies: The HDF and More
  A. Fernández-Soto, K.M. Lanzetta, A. Yahil 270

AAO Support Observations for the Hubble Deep Field South
  B.J. Boyle 275

The Hubble Deep Field-South QSO
  K. Sealey, M. Drinkwater, J. Webb, B. Boyle 278

Hubble Deep Fever: A Faint Galaxy Diagnosis
  S.P. Driver 280

An Extremely Blue Population
in Multispectral Galaxy Surveys
  B. Rocca-Volmerange, M. Fioc 289

Spectro-Photometric Constraints on Distant Galaxies
  S. Charlot 294

Semi-Analytic Models
and Background Hydrogen-Ionizing Flux
  J.E.G. Devriendt, B. Guiderdoni, S.K. Sethi 301

Detection and Evolution of High-z Galaxies
  T. Broadhurst, R. Bouwens, B. Frye 303

On the Nature of Red Galaxies in the Early Universe
  P.J. Francis, B.E. Woodgate, A.C. Danks 309

On the Evolution of X-Ray Clusters
  P. Rosati 311

Rich Clusters of Galaxies at Low to Intermediate Redshift
  E. O'Hely, W.J. Couch, I. Smail, A. Zabludoff, A. Edge 318

Tunable Filter-Selected Hα Emission
in the Rich Cluster A 3665 (AC 106)
  D.H. Jones, J. Bland-Hawthorn 320

Luminosity Function of Cluster Galaxies
  E. Molinari, A. Moretti, G. Chincarini, S. De Grandi 322

A Dwarf Galaxy Population–Density Relation
       S. Phillipps, S. Driver, W. Couch, R. Smith 324

Radio Survey of Merging Clusters
in the Core of the Shapley Concentration
       T. Venturi, S. Bardelli, R. Morganti, R.W. Hunstead 326

Cosmological Parameters
as Measured by Type Ia Supernovae
       B. Leibundgut, B. Schmidt, J. Spyromilio, M. Phillips 328

Author Index 335

List of Participants

Name	Institution
ALLEN, Mark	MSSSO mga@mso.anu.edu.au
ANDERSON, Martin	University of Western Sydney Nepean anderson@atnf.csiro.au
BEDDING, Tim	University of Sydney bedding@physics.usyd.edu.au
BESIER, Saskia	UNSW, Sydney scb@edwin.phys.unsw.edu.au
BICKNELL, Geoff	MSSSO geoff@mso.anu.edu.au
BLAND-HAWTHORN, Joss	AAO, Sydney jbh@aaossz.aao.gov.au
BOYLE, Brian	AAO, Sydney Director@aaoepp.aao.gov.au
BROADHURST, Tom	University of California, Berkeley TBroadhurst@astro.berkeley.edu
BROWN, Michael	University of Melbourne mbrown@physics.unimelb.edu.au
BURTON, Michael	UNSW, Sydney M.Burton@unsw.edu.au
CANNON, Russell	AAO, Sydney rdc@aaoepp.aao.gov.au
CHARLOT, Stephane	Institut d'Astrophysique de Paris charlot@iap.fr
CLAY, Roger	University of Adelaide rclay@physics.adelaide.edu.au
COLLESS, Matthew	MSSSO colless@mso.anu.edu.au

COUCH, Warrick	UNSW, Sydney	wjc@edwin.phys.unsw.edu.au
CRAM, Lawrence	University of Sydney	l.cram@physics.usyd.edu.au
DALCANTON, Julianne	Carnegie Observatories	jd@ociw.edu
DANZIGER, Ivan John	Osservatorio di Trieste	danziger@oat.ts.astro.it
DE BLOK, Erwin	University of Melbourne	edeblok@isis.ph.unimelb.edu.au
DE BREUCK, Carlos	IGPP/LLNL, USA	debreuck@igpp.llnl.gov
DE LAPPARENT, Valerie	Institut d'Astrophysique de Paris	lapparen@iap.fr
DEVRIENDT, Julien	Institut d'Astrophysique de Paris	devriend@iap.fr
DICKINSON, Mark	Johns Hopkins University / STScI	med@stsci.edu
DRINKWATER, Michael	UNSW, Sydney	mjd@phys.unsw.edu.au
DRIVER, Simon	UNSW, Sydney	spd@edwin.phys.unsw.edu.au
EKERS, Ron	ATNF, Sydney	rekers@atnf.csiro.au
FERNANDEZ-SOTO, Alberto	UNSW, Sydney	fsoto@edwin.phys.unsw.edu.au
FLUKE, Christopher	University of Melbourne	cfluke@ph.unimelb.edu.au
FRANCIS, Paul	Australian National University	pfrancis@mso.anu.edu.au
FREUDLING, Wolfram	ST-ECF	wfreudli@eso.org
FRYE, Brenda	University of California, Berkeley	bfrye@astron.berkeley.edu
GIOIA, Isabella	IRA-Bologna, IfA-Hawaii	gioia@galileo.ifa.hawaii.edu
GLAZEBROOK, Karl	AAO, Sydney	kgb@aaoepp.aao.gov.au

GREEN, Anne	University of Sydney agreen@physics.usyd.edu.au
HAIGH, Andrew	University of Sydney ahaigh@Physics.usyd.edu.au
HANNIKAINEN, Diana	Observatory, University of Helsinki diana@gstar.astro.helsinki.fi
HARNETT, Julienne	University of Technology, Sydney jules@maths.uts.edu.au
HAYNES, Raymond	ATNF, Sydney rhaynes@atnf.csiro.au
HEISLER, Charlene	MSSSO heisler@mso.anu.edu.au
HILL, Tanya	University of Sydney thill@Physics.usyd.edu.au
HOOK, Isobel	ESO ihook@eso.org
HOPKINS, Andrew	University of Sydney ahopkins@physics.usyd.edu.au
HUNSTEAD, Richard	University of Sydney r.hunstead@physics.usyd.edu.au
ILLINGWORTH, Garth	Lick Obs./Univ. of California, Santa Cruz gdi@ucolick.org
JACKSON, Carole	University of Sydney cjackson@physics.usyd.edu.au
JAUNCEY, Dave	ATNF, Sydney djauncey@atnf.csiro.au
JEAN, Christophe	Institut d'Astrophysique, Université de Liège jean@astra.astro.ulg.ac.be
JONES, Heath	MSSSO dhj@mso.anu.edu.au
JONES, Lewis	UNSW, Sydney lewis@edwin.phys.unsw.edu.au
JONES, Paul	University of Western Sydney Nepean pjones@st.nepean.uws.edu.au
JURASZEK, Sebastian	University of Sydney juraszek@physics.usyd.edu.au
KEWLEY, Lisa	MSSSO lkewley@mso.anu.edu.au

KIM, Sungeun	MSSSO sek@mso.anu.edu.au	
KORIBALSKI, Baerbel	ATNF, Sydney bkoribal@atnf.csiro.au	
KRAAN-KORTEWEG, Renee	University of Guanajuato kraan@norma.astro.ugto.mx	
LA FRANCA, Fabio	Università 'Roma Tre' lafranca@amaldi.fis.uniroma3.it	
LAHAV, Ofer	Institute of Astronomy, Cambridge, UK lahav@ast.cam.ac.uk	
LANÇON, Ariane	Observatoire de Strasbourg, France lancon@wirtz.u-strasbg.fr	
LEIBUNDGUT, Bruno	ESO bleibund@eso.org	
LIANG, Haida	Bristol University, UK h.liang@bristol.ac.uk	
LINEWEAVER, Charley	UNSW, Sydney charley@edwin.phys.unsw.edu.au	
LISKE, Jochen	UNSW, Sydney JOL@newt.phys.unsw.edu.au	
MACCACARO, Tommaso	Osservatorio Astronomico di Brera tommaso@brera.mi.astro.it	
MAGLIOCCHETTI, Manuela	Institute of Astronomy, Cambridge, UK manuela@ast.cam.ac.uk	
MALASAN, Hakim	Bosscha Observatory, Indonesia malasan@ibm.net	
MALIN, David	AAO, Sydney dfm@aaoepp.aao.gov.au	
McGREGOR, Peter	MSSSO peter@mso.anu.edu.au	
McINTYRE, Vince	University of Sydney vjm@physics.usyd.edu.au	
McMAHON, John	UNSW, Sydney jfm@roen.phys.unsw.edu.au	
MELLIER, Yannick	Institut d'Astrophysique de Paris mellier@iap.fr	
MOLINARI, Emilio	Osservatorio di Brera Milano molinari@merate.mi.astro.it	

MORGANTI, Raffaella	Istituto di Radioastronomia, CNR, Bologna rmorgant@ira.bo.cnr.it
MORTLOCK, Daniel	University of Melbourne dmortloc@physics.unimelb.edu.au
MOULD, Jeremy	MSSSO director@mso.anu.edu.au
NORRIS, Ray	ATNF, Sydney rnorris@atnf.csiro.au
O'HELY, Eileen	UNSW, Sydney eoh@ham.phys.unsw.edu.au
OLIVER, Sebastian James	Imperial College, London s.oliver@ic.ac.uk
OOSTERLOO, Thomas	Istituto di Fisica Cosmica, CNR, Milano toosterl@ifctr.mi.cnr.it
PACEY, Elaine	ATNF, Sydney epacey@atnf.csiro.au
PADOVANI, Paolo	STScI padovani@stsci.edu
PARKER, Quentin	AAO, Sydney qap@aaocbn.aao.gov.au
PETERSON, Bruce	MSSSO peterson@mso.anu.edu.au
PRANDONI, Isabella	Istituto di Radioastronomia, CNR, Bologna prandoni@astbol.bo.cnr.it
QUINN, Peter	ESO pquinn@eso.org
RENZINI, Alvio	ESO arenzini@eso.org
ROBERTSON, Gordon	University of Sydney jgr@physics.usyd.edu.au
ROCCA-VOLMERANGE, Brigitte	Institut d'Astrophysique de Paris rocca@iap.fr
ROSATI, Piero	ESO prosati@eso.org
SADLER, Elaine	University of Sydney ems@physics.usyd.edu.au
SCHMIDT, Brian	MSSSO brian@mso.anu.edu.au

SCHNEIDER, Peter	Max-Planck-Institut f. Astrophysik, Garching peter@mpa-garching.mpg.de
SEALEY, Katrina	UNSW, Sydney kms@edwin.phys.unsw.edu.au
SHAVER, Peter	ESO pshaver@eso.org
SIM, Helen	ATNF, Sydney hsim@atnf.csiro.au
SMITH, Malcolm	AURA-Cerro Tololo msmith@noao.edu
SQUIRES, Gordon	University of California, Berkeley squires@astro.berkeley.edu
STAVELEY-SMITH, Lister	ATNF, Sydney lstavele@atnf.csiro.au
SWARUP, Govind	National Centre for Radio Astrophysics, TIFR, Pune gswarup@ncra.tifr.res.in
TAYLOR, Keith	AAO, Sydney kt@aaoepp.aao.gov.au
TSAREVSKY, Gregory	ATNF, Sydney gtsarevs@atnf.csiro.au
TSVETANOV, Zlatan	Johns Hopkins University zlatan@pha.jhu.edu
TURTLE, Tony	University of Sydney turtle@physics.usyd.edu.au
TZIOUMIS, Anastasios	ATNF, Sydney atzioumi@atnf.csiro.au
VAN BREUGEL, Wil	University of California, IGPP/LLNL wil@sundial.llnl.gov
VAN DER HULST, Thijs	Kapteyn Astronomical Institute, Groningen vdhulst@astro.rug.nl
VENTURI, Tiziana	Istituto di Radioastronomia, CNR, Bologna tventuri@astbol.bo.cnr.it
WAGNER, Stefan	MSSSO swagner@mso.anu.edu.au
WEBSTER, Rachel	University of Melbourne rwebster@physics.unimelb.edu.au
WIELEBINSKI, Richard	Max-Planck-Inst. f. Radioastronomie, Bonn rwielebinski@mpifr-bonn.mpg.de

WIERINGA, Mark	ATNF, Narrabri mwieringa@atnf.csiro.au
WISOTZKI, Lutz	Hamburger Sternwarte lwisotzki@hs.uni-hamburg.de
WOODS, Helen	AAO, Sydney hmw@aaoepp.aao.gov.au
YAMASHITA, Koujun	Nagoya University yamasita@satio.phys.nagoya-u.ac.jp

Part 1

INTRODUCTION

Opening Remarks

Jeremy Mould

Mount Stromlo & Siding Spring Observatories
Institute of Advanced Studies
Australian National University

1 Looking Deep in the Southern Sky

What a great title for a meeting! It focusses us on the purpose of our new astronomical facilities in the southern hemisphere. The theme of my opening remarks will be the great opportunities we have in the next decade with all these new facilities, which are collected for your consideration in Table 1.

Table 1: Facilities Timeline

1997	Parkes multibeam, Ceduna, NTT big bang
1998	2dF rapid reconfiguration, HDF South
1999	UT1, [Magellan I]
2000	mm ATCA
2001	UT4, Gemini S
?	VLTI, AUSTRALIS [1]
??	MMA/LSA, SPIRIT [2]

[1] For more details, see msowww.anu.edu.au/~colless
[2] For more about the proposed South Pole Infra Red Imaging Telescope, see www.unsw.phys.edu.au/~mgb

Reaching this point has been a tremendous struggle, both technically and in respect of securing funding. But in spite of the frustration, there has been success, and these facilities are on their way. All of them, except Magellan, will be directly available to the astronomical communities sponsoring this meeting.

With this abundance of new instruments, we are, it seems, only limited by our ideas. That's why this meeting is so timely. We have an opportunity here to tease out the best ideas, to plan some proposals, and to put together collaborations.

2 Instrumentation Collaborations

On the facilities side there are three ESO/Australia collaborations which deserve to be singled out.
- the fibre positioner for the VLT
- the VLT interferometer
- the complementary nature of SEST/ATCA

We hope to formalize the first of these collaborations with ESO during 1998. This project will employ the expertise developed in designing, building, and commissioning 2dF, to bring multifibre technology rapidly to the VLT. Similarly, the VLTI can, through collaboration, gain the benefit of the interferometry know-how of some of the field's pioneers who developed SUSI. And the large dish sensitivity of SEST will complement the upgraded AT compact array's superior resolution. In this case, formal collaboration plans were agreed in 1997.

Australian astronomers can gain guaranteed time on the VLT/VLTI, provided an Australian contribution of instrumentation is able to be made. For us, fundraising is the key here.

3 Support Facilities

New frontline facilities require new support facilities. We need to show the same foresight as led our predecessors 25 years ago to build the UK Schmidt Telescope and the ESO Schmidt. With this in mind, MSSSO, AAO, the University of Melbourne, and Auspace are building an 8000 × 8000 pixel Wide Field Imager; ESO and Naples Observatory are building a new survey telescope for Cerro Paranal; and we are considering the role the MACHO project's telescope can play, for example, in γ-ray Burster follow up.

An aggressive (but also expensive) program of refurbishment of the second ranked facilities is required for this purpose. In some places this is already happening, and AURA's facilities present a very good example.

4 Science Opportunities

To whet our appetites, let us look at some of the prime opportunities (Table 2). Many of these subjects will be covered in much more detail in the course of this meeting. Suffice it to say here, that we have a suite of facilities in the Southern Hemisphere which will allow us to advance our science on a broad front.

Table 2: Sample Programs

Multibeam	HI galaxies, residual galaxy formation at the current epoch Great Attractor, Zone of Avoidance
Ceduna	Structure of southern radiosources
2dF	evolution of large scale structure in the second cosmic semester
HDF South	galaxy evolution in the first Gyr
VLT/Gemini	pristine chemistry of southern globular clusters cosmology with distant supernovae
mm ATCA	star formation in southern Milky Way and LMC
VLTI	structure of the Galactic Centre
AUSTRALIS	evolution of large scale structure in the first cosmic semester
MMA/LSA	galaxy formation at $z > 4$
SPIRIT	pathfinder for NGST science

5 Conclusion

We are in a good position to continue the explosive growth which we have enjoyed in astronomy over the last quarter-century. In some disciplines students are forced to repeat their advisors' theses a little better. This will not be our fate in Astronomy, while we have a schedule of new facilities like those being prepared for the next decade.

DEEP IMAGING
AND REDSHIFT SURVEYS

The 2dF Galaxy Redshift Survey

Matthew Colless*

Mount Stromlo & Siding Spring Observatories, Australia

1 Survey Design

The 2dF Galaxy Redshift Survey now in progress is using the Two Degree Field 400-fibre spectrograph (2dF) on the Anglo-Australian Telescope to obtain high-quality spectra and redshifts for 250,000 galaxies. These galaxies are a magnitude-limited sample brighter than $b_J=19.5$ (extinction-corrected) drawn from a revised and extended version of the APM galaxy catalogue (Maddox et al. 1990). The bright survey will cover an area of 1700 square degrees selected from both the southern galactic hemisphere APM galaxy survey and its extension into the north galactic hemisphere equatorial region. The approximate arrangement of 2dF survey fields is shown in Figure 1. This bright survey will be supplemented by a faint survey of 10,000 galaxies to R=21 based on APM scans of deep UK Schmidt Telescope films obtained for selected regions. The faint survey is carried out in an over-ride mode making use of the best observing conditions.

2 Goals of the Survey

The survey will address a variety of fundamental problems in galaxy formation and cosmology. The main scientific goals include:

1. Accurate measurements of the power spectrum of galaxy clustering on scales of up to a few hundred Mpc, allowing direct comparisons with microwave background anisotropy measurements of fluctuations on the same spatial scales.

2. Measurements of the distortion of the clustering pattern in redshift space, providing constraints on the cosmological density parameter Ω and the spatial distribution of dark matter.

3. A determination of variations in the spatial and velocity distributions of galaxies as a function of luminosity, type and star-formation history, providing important constraints on models of galaxy formation.

---

* *On behalf of the 2dF Galaxy Redshift Survey Team:* Matthew Colless (MSSSO), Richard Ellis (IoA), Joss Bland-Hawthorn (AAO), Russell Cannon (AAO), Shaun Cole (Durham), Chris Collins (LJMU), Warrick Couch (UNSW), Gavin Dalton (Oxford), Simon Driver (UNSW), George Efstathiou (IoA), Simon Folkes (IoA), Carlos Frenk (Durham), Karl Glazebrook (AAO), Nick Kaiser (IfA), Ofer Lahav (IoA), Ian Lewis (AAO), Stuart Lumsden (AAO), Steve Maddox (IoA), John Peacock (ROE), Bruce Peterson (MSSSO), Ian Price (MSSSO), Will Sutherland (Oxford), Keith Taylor (AAO).

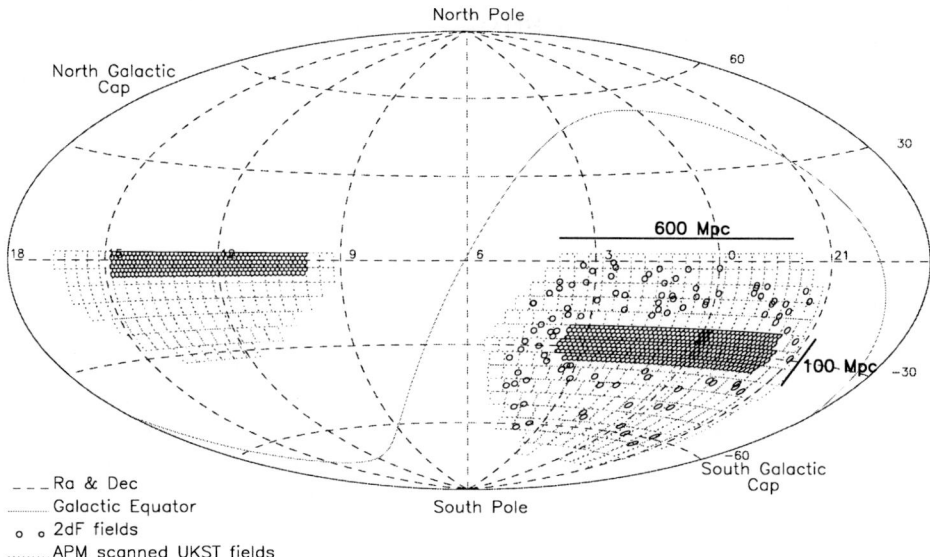

Fig. 1. The 2dF survey fields (small circles) superimposed on the APM survey area (dotted outlines of Sky Survey plates). There are approximately 140,000 galaxies in the 75°×12.5° South Galactic Hemisphere strip centred on the South Galactic Pole, 70,000 galaxies in the 65°×7.5° North Galactic Hemisphere equatorial strip, and 40,000 galaxies in the 100 random 2dF fields scattered over the entire southern region of the APM galaxy survey.

4. Measurements of the evolution of the galaxy luminosity function, clustering amplitude and star formation rates out to a redshift of $z \sim 0.5$. This will be accomplished by combining the bright $b_J < 19.5$ survey with the faint $R < 21$ survey, which is comparable in size to the largest current local surveys.

5. Investigations of the morphology of galaxy clustering and the statistical properties of the fluctuations—for example, whether the initial fluctuations are Gaussian as predicted by inflationary models of the early universe.

6. A study of clusters and groups of galaxies in the redshift survey, in particular the measurement of infall in clusters and dynamical estimates of cluster masses at large radii.

7. Application of novel techniques to classify the uniform sample of 250,000 spectra obtained in the survey, thereby obtaining a comprehensive inventory of galaxy types as a function of spatial position within the survey.

3 Survey Status

As of 30 January 1998 we have obtained ~8000 redshifts for the survey. The rate at which redshifts are gathered will increase rapidly as 2dF is brought up to full operational speed. The survey can be completed within two years of

2dF achieving nominal observing efficiency. We currently expect to complete the survey by the end of 1999.

Two example spectra from the survey are shown in Figure 2. One is a $b_J=19.2$ emission-line galaxy at $z=0.067$ and the other is a $b_J=19.3$ absorption-line galaxy at $z=0.246$. The quality target for the survey spectra is a S/N of at least 10 per 2-pixel resolution element (FWHM$\approx$9Å); most spectra will easily exceed this target.

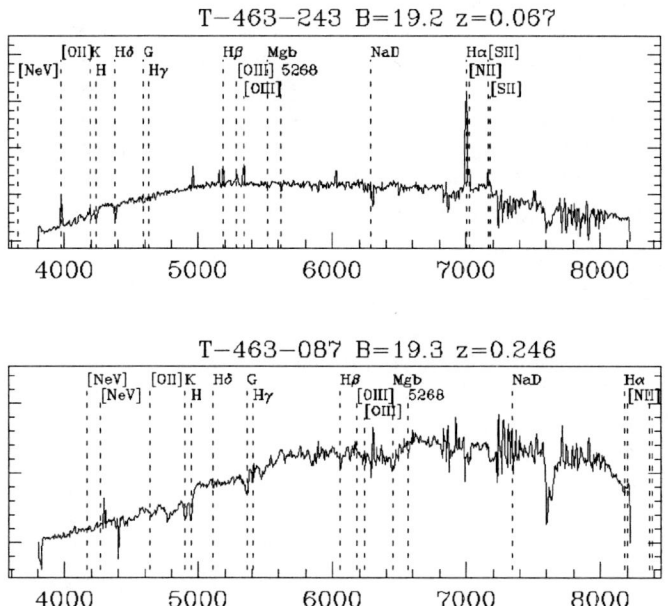

Fig. 2. Example spectra from the survey: a $b_J=19.2$ emission-line galaxy at $z=0.067$ and a $b_J=19.3$ absorption-line galaxy at $z=0.246$.

This minimum S/N permits reliable automatic spectral classification and redshift measurement. Employing both a standard cross-correlation and line-fitting code and a new code which uses Principal Component Analysis and χ^2-fitting to simultaneously classify the spectrum and measure its redshift (Glazebrook et al. 1997), we find we achieve a very high level of reliability. A comparison of the redshifts obtained from these codes with redshifts determined via visual inspection shows a very low level of failures in the automatic algorithms. The success rate in identifying redshifts for survey galaxies is currently 90–95%; the goal is to achieve an overall success rate in excess of 95%.

4 Preliminary Results

The survey is still in its infancy, with only $\sim$8000 redshifts (3% of the total) measured to date. Some preliminary results can, however, provide a hint of the

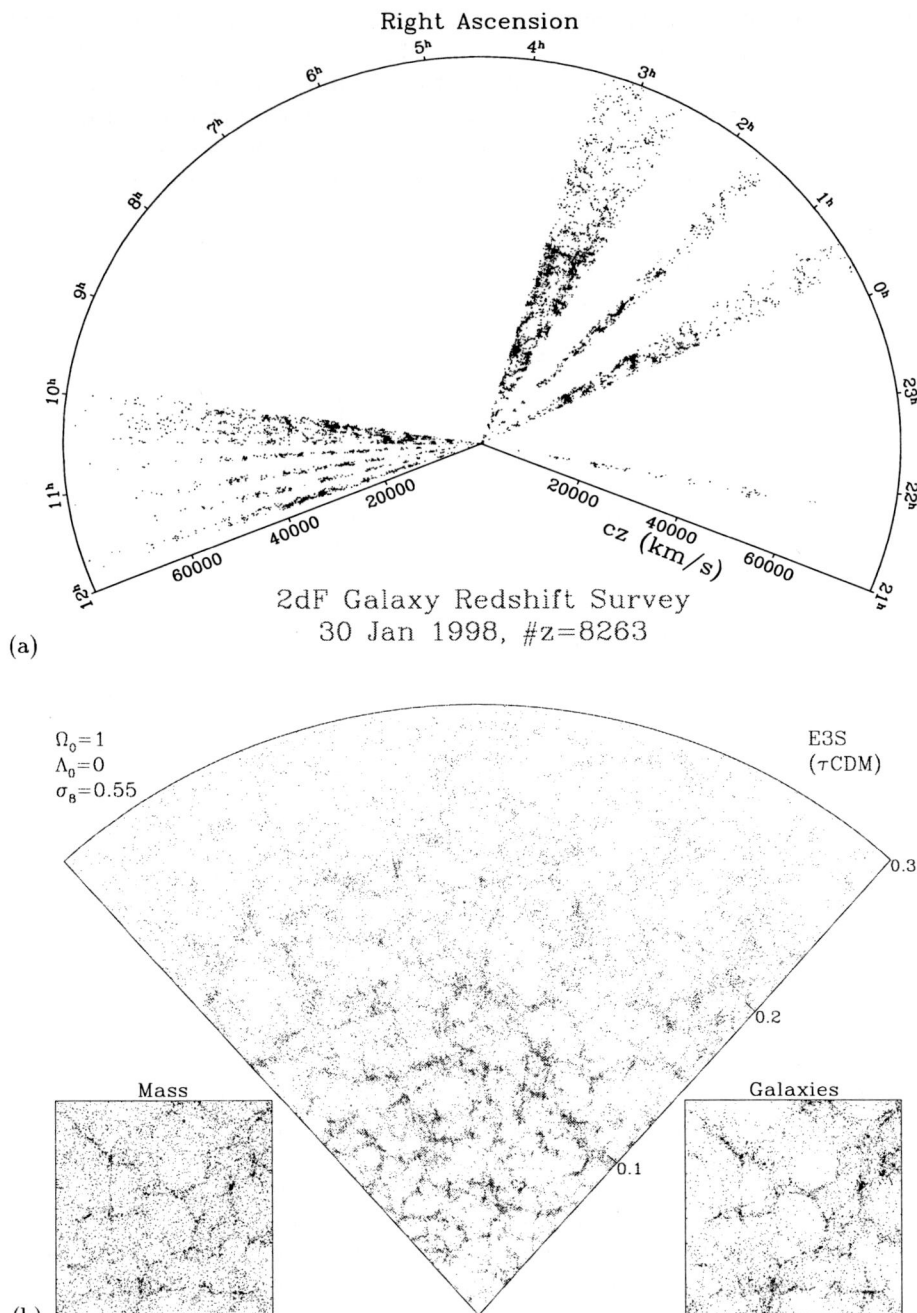

Fig. 3. (a) A redshift slice for the galaxies observed to date, combining northern and southern strips and including ∼8000 galaxies (∼3% of the full sample). (b) A cone plot for a mock 2dF redshift survey from Cole et al. (1998).

power and scope of the full survey.

Figure 3a is a cone plot for these ~8000 galaxies combining both the northern and southern strips. Note the highly incomplete filling of the slice by the 2dF fields so far observed. For comparison, Figure 3b shows a 90°×3° slice through a mock 2dF survey based on a CDM N-body simulation and a recipe for galaxy biasing (see Cole et al. 1998).

Figure 4 shows the combined redshift distribution in comparison to the predicted distribution in a homogeneous universe. There is clearly strong clustering, but the irregular distribution on the sky of the fields we have observed so far makes quantitative analysis difficult.

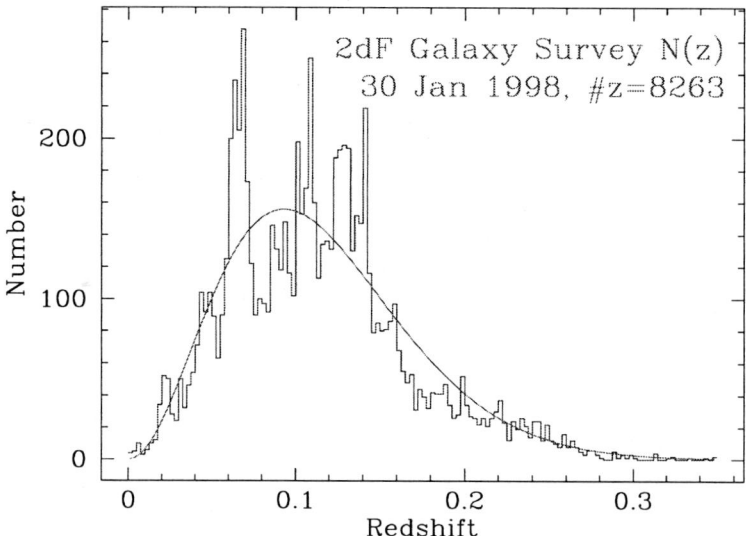

Fig. 4. A preliminary redshift distribution from the 2dF survey. The smooth curve is the predicted redshift distribution neglecting clustering.

Although so far we have observed too few fields to effectively address questions of large-scale structure, we do have a sufficiently large sample of redshifts to begin to look at the properties of local galaxies. Figure 5 shows the galaxy luminosity function at a mean redshift of $z=0.1$. This preliminary determination uses a single global K-correction and is normalized by the number counts from the APM input catalogue, so that the shape and normalization may change in the final analysis. Note however that we are achieving a good determination of the luminosity function 5 magnitudes below L^* even with this small subset of the survey. The LF is generally well represented by a Schechter function with $M^* \approx -19.5$, $\alpha \approx -1.1$ and $\phi^* \approx 0.02$. With the full sample we should be able to tightly constrain the faint end of the local luminosity function for each morphological or spectral type and determine the variation in the luminosity function with local density.

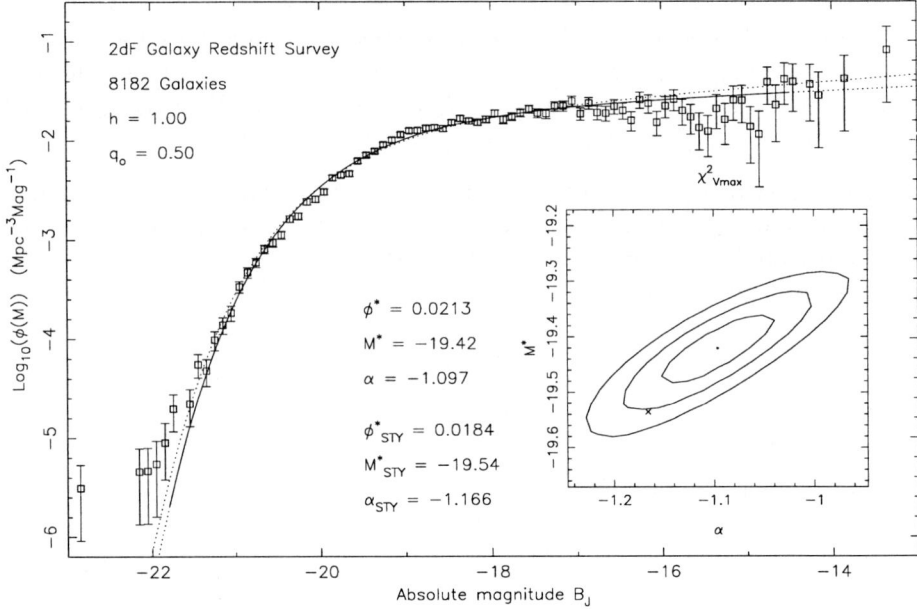

Fig. 5. A preliminary galaxy luminosity function from the 2dF survey using mean K-corrections and normalised to the APM number counts. The points are the $1/V_{max}$ LF, the solid curve is the χ^2 fit of a Schechter function to the points and the dotted curve is the STY fit of a Schechter function. The parameters of these fits are given on the figure. The inset shows the 1, 2 and 3-σ contours of the χ^2 fit in M^* and α; the cross marks the M^* and α obtained from the STY fit.

Further details regarding the 2dF survey can be found in Colless (1997) and Colless & Boyle (1998). Up-to-date information is posted on the WWW at http://msowww.anu.edu.au/~colless/2dF/.

References

Cole S., Hatton S., Weinberg D.H., Frenk C.S., 1998, MNRAS, in press
Colless M.M., 1997, Wide Field Spectroscopy and the Universe, *Wide Field Spectroscopy*, eds Kontizas M., Kontizas E., Kluwer, pp.227–240
Colless M.M., Boyle B.J., 1998, in 'Highlights of Astronomy', Vol.11, in press
Glazebrook K., Offer A.R., Deeley K., 1998, ApJ, 492, 98
Maddox S.J., Efstathiou G., Sutherland W.J., Loveday J., 1990, MNRAS, 242, 43P

Discussion

Tsvetanov: You just mentioned that you need about two years to compile the survey. Does this include weather factors, down-time, etc., disturbing factors?

Colless: Yes, to the extent that is possible. We need about 90 nights over two years to complete the survey, allowing for weather.

Lineweaver: In the redshift histogram of galaxies you said that it was dominated by clustering. Could you show us the curve you expect to see if the distribution of galaxies is homogeneous within the sample?

Colless: The curve is a χ^2-looking distribution. As we begin to sample a larger, more representative volume, the peaks in the redshift distribution will average out and the observed $n(z)$ will approach the smooth curve.

Rocca-Volmerange: Can you comment on the flux-calibration of your spectra observed within the 2dF program?

Colless: We will attempt an approximate flux-calibration, but this is difficult to calibrate. One significant problem is to know just which part of the galaxy the fibre covers.

Tsvetanov: Could you comment on the end-to-end efficiency of the 2dF instrument?

Colless: For point sources the end-to-end efficiency of 2dF is 11 per cent.

Ekers: When will the number of redshifts measured with 2dF exceed all previous redshift observations?

Colless: About one third of the way through the survey, so probably towards the end of 1998. By the end of the survey we will have measured three times as many redshifts as the number measured by all other astronomers since Hubble!

Oliver: The plot comparing survey volumes and number of objects is a bit harsh on the PSC-Z since it shows no increase in volume (over QDOT) despite an improvement in Poisson statistics from increasing the number density by a factor of six. The PSC-Z team would claim a larger volume than illustrated on that plot.

Gioia: What is the rate of correct redshift identification in the pipeline? Do you need manual intervention?

Colless: The current rate of correct redshift identification is about 90–95 per cent. The automatic redshift identification software is very good at identifying the objects for which the redshift estimators are unreliable and flags them for manual intervention.

The 2dF QSO Redshift Survey

B.J. Boyle[1], S.M. Croom[2], R.J. Smith[3], T. Shanks[2], L. Miller[4], N. Loaring[4]

[1] Anglo-Australian Observatory, PO Box 296, Epping, NSW 2121, Australia
[2] University of Durham, South Road, Durham DH1 3LE, UK
[3] Institute of Astronomy, Madingley Road, Cambridge CB3 0HA, UK
[4] University of Oxford, 1 Keble Road, Oxford OX1, UK

Abstract. We are currently carrying out a major redshift survey of 30000 QSOs using the 2dF facility on the Anglo-Australian Telescope. As of January 1998, over 1000 QSO redshifts have been obtained for the survey, making this the largest single homogeneous sample of QSOs yet compiled. When complete the survey will be used to provide important information on the large-scale structure of the Universe.

1 Introduction

With the advent of the 2-degree field (2dF), a 400-fibre multi-object fibre spectrograph at the Anglo-Australian Telescope (Taylor et al. 1994), it is now possible to increase the size of existing QSO surveys by more than an order of magnitude. We have therefore embarked on a large spectroscopic survey of QSO candidates with the 2dF. The observational goal of this survey is to identify and obtain redshifts for ~30000 $B<21$ QSOs in two declination strips at the South Galactic Pole and in an equatorial region at the North Galactic cap. The QSO survey will probe the largest scales in the universe (10Mpc $< r <$ 1000Mpc) over a wide range in redshift space ($0.3 < z < 2.9$). The proposed survey will therefore increase the total number of QSOs known by a factor of ~4, and the number suitable for clustering studies by more than a factor of 10.

The primary scientific aims of the survey are: a) to obtain the primordial fluctuation power spectrum out to COBE scales; b) to determine the rate of QSO clustering evolution in the non-linear and linear regimes, and hence obtain new limits on the value of Ω and b (Croom & Shanks 1996); c) to apply geometric methods to measure Λ (Ballinger et al. 1996); d) to identify large statistical samples of unusual classes of QSOs (e.g. BALs) or absorption line systems (e.g. damped Lyα systems).

2 Survey Strategy

The 2dF QSO survey covers an area of 740 deg^2 comprising two 75°×5° strips on the sky (small areas surrounding bright stars have been excluded from the analysis). QSO candidates were selected from APM measurements of UK Schmidt U, J and R plates/films, on the basis of their anomalous position in the stellar $U-B/B-R$ colour–colour diagram. The vast majority (>85%) of the candidates are selected solely on the basis of their $U-B$ colour (sensitive to QSOs with

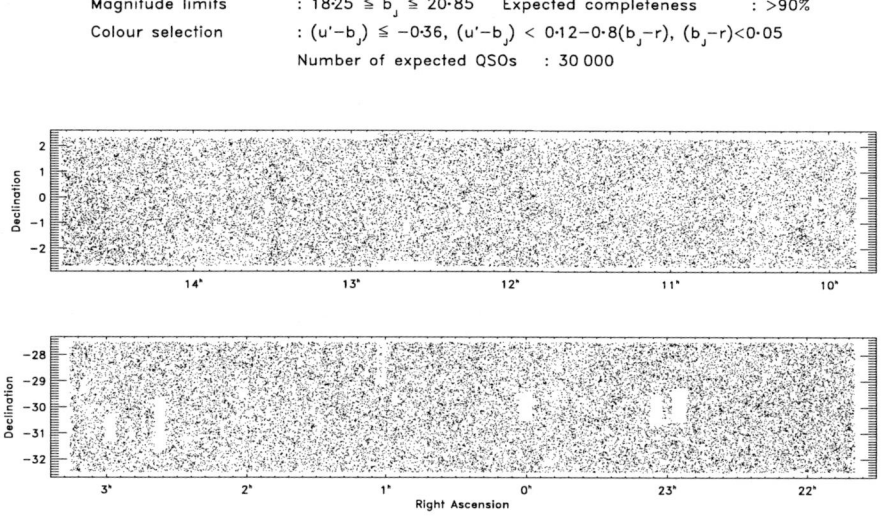

Fig. 1. The UBR-selected objects for the QSO 2dF input catalogue. The RA and Dec of all objects are shown in the two $75° \times 5°$ strips of the catalogue. Increasing contamination by Galactic sub-dwarfs at lower Galactic latitudes gives an obvious gradient (<50%) in number density along the survey strips. Completeness estimates are based on our success at recovering previously identified QSOs.

$z<2.2$), but the additional use of the R magnitude minimises contamination from Galactic stars and extends the redshift range over which QSOs can be selected to $z\sim2.9$. The initial UVX-selected catalogue is shown in Figure 1. In total over 150 U, J and R plates (comprising 30 UKST fields) were used to compile the input catalogue. Great care was taken to keep photometric variations at <0.1 mag level over the entire catalogue (Smith 1998).

The QSO 2dF redshift survey fields form part of the area of sky surveyed by the galaxy 2dF survey, and so the objects in the QSO and galaxy 2dF redshift surveys are being observed simulatenously with 2dF. This leads to significant gains in both efficiency (the combined survey can be carried out in 20 nights less than if the surveys had been separate) and completeness (the QSO survey will pick up compact blue galaxies, the galaxy survey will identify extended low-redshift QSOs).

3 Survey Progress

Over 1000 QSO redshifts have been obtained to date (January 1998) with the 2dF. Coupled with redshifts obtained for brighter QSOs in the survey strips,

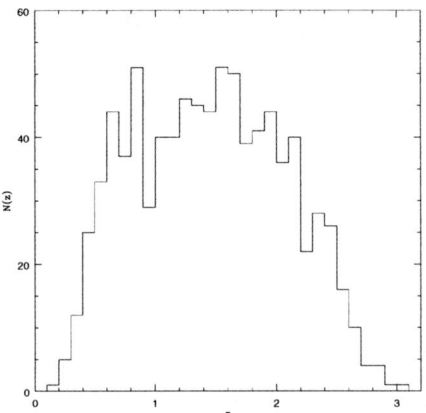

Fig. 2. The number-redshift relation for the QSOs identified to date (January 1998) in the 2dF QSO redshift survey.

radio identifications (from NVSS) and close QSO pairs (< 20 arcsec) almost 1200 QSOs have now been identified from the initial input catalogue.

The number-redshift relation for the survey is shown in Figure 2 and the auto-correlation function for the survey, $\xi(r)$ is shown in Figure 3.

The auto-correlation function is consistent with $\xi(r) = (r/4h^{-1}\text{Mpc})^{-1.4}$, previously derived by Croom & Shanks (1996) from a fit to a composite sample of 1500 QSOs from existing homogenous surveys. There is still no strong evidence for the evolution of the correlation length with redshift (see Figure 4). In linear theory models, this favours a universe with low Ω_0 or high bias.

Survey status is frequently updated on the WWW (URL: http://www.aao.gov.au/aao/rs/qso_surv.html), and the input catalogue will made made available during 1998. Although only 3% complete, the 2dF QSO survey is already the largest single homogeneous QSO survey yet compiled and the survey is due to complete in 2000.

References

Ballinger, W.E., Peacock, J.A., Heavens, A.F., 1996, MNRAS, 282, 877
Boyle, B.J, Mo, H.J., 1993, MNRAS, 260, 925
Croom, S.M., Shanks, T., 1996, MNRAS, 281, 893
Georgantopoulos, I., Shanks, T., 1994, MNRAS, 271, 773
Smith, R.J., 1998, PhD Thesis, University of Cambridge
Taylor, K., 1994, in Wide Field Spectroscopy and the Distant Universe, eds Maddox S.J. et al., (World Scientific), p15

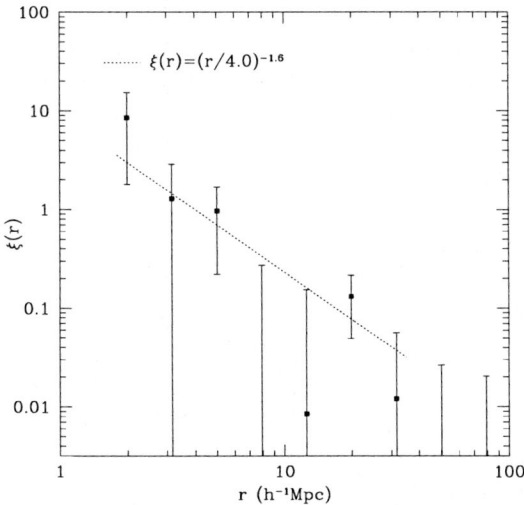

Fig. 3. The spatial auto-correlation function for QSO identified to date in the 2dF redshift survey. The dotted line denotes $\xi(r) = (r/4h^{-1}\,\mathrm{Mpc})^{-1.4}$

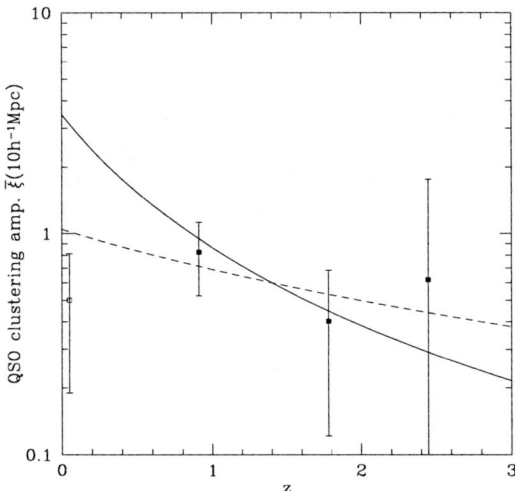

Fig. 4. $\bar{\xi}(10)$, the mean amplitude of the QSO correlation function below $10h^{-1}\,\mathrm{Mpc}$, plotted as a function of redshift. Estimates are based on the 2dF survey data currently available. The point at $z=0$ is derived from the AGN correlation functions of Boyle & Mo (1993) and Georgantopoulos & Shanks (1994). Model predictions for the growth of structure based on linear theory in an $\Omega = 1$ (solid line) and $\Omega = 0.1$ Universe (dashed line) are shown.

Discussion

Tsarevsky: Unfortunately, most high-redshift quasars should be screened through the sieve of your selection criteria. It gives a substantial cut-off for $z > 3 - 4$ and, therefore, restricts all the conclusions of clustering, luminosity function, evolution etc. Any kind of extrapolation would be considered as an extension to the unknown.

Boyle: I agree that we won't be sensitive to QSOs with $z > 3$.

Tsvetanov: Given that QSOs are nearly point sources, what are the prospects of having their spectra also flux calibrated (not just in relative units)?

Boyle: Unfortunately, even relative flux calibration for point sources will be difficult with 2dF. Positioning errors coupled with subtle field-dependent chromatic effects will probably limit relative spectrophotometry to $\pm 20\%$ at best.

Dickinson: Will the 2dF QSO spectra be suitable for low resolution absorption line work, e.g. for damped Lyα system searches?

Boyle: Yes. We expect to identify significant numbers (> 100) of unusual/rare absorption line systems (BAL QSOs, damped Lyα systems).

Lineweaver: You mentioned n, b and Λ as parameters which you can constrain. Why these three?

Boyle: I should also have mentioned Ω. Limits on Ω can, for example, be obtained from the evolution of clustering.

Rocca-Volmerange: Do you derive information on companions from your QSO 2dF survey?

Boyle: Out of the 10 close QSO pairs (< 20 arcsec) identified to date, only one pair have the same redshift, and are a possible gravitational lens system. The relatively low number of lensed QSO pairs can already be used to place quite strong upper limits on Λ.

A Complete 2dF Survey of Fornax

M.J. Drinkwater[1], E.M. Sadler[2], J.I. Davies[3], R.J. Dickens[4], M.D. Gregg[5], Q.A. Parker[6], S. Phillipps[4], R.M. Smith[3]

[1] University of New South Wales, Physics, Sydney 2052, Australia
[2] University of Sydney, Physics, NSW 2006, Australia
[3] University of Wales, Cardiff, Physics, PO Box 913, Cardiff CF2 3YB, UK
[4] University of Bristol, H.H. Wills Physics Lab., Tyndall Av., Bristol BS8 1TL, UK
[5] IGPP, Lawrence Livermore National Laboratory, Livermore, CA 94550, USA
[6] Anglo-Australian Observatory, Coonabarabran, NSW 2357, Australia

Abstract. We are using the 2dF spectrograph on the Anglo-Australian Telescope to obtain spectra for a complete sample of all 14000 objects with $16.5 < B < 19.7$ in a 12 square degree area centred on the Fornax cluster. The aims of this project include the study of dwarf galaxies in the cluster (both known low surface brightness objects and putative normal surface brightness dwarfs) and a comparison sample of background field galaxies. We will also measure quasars, any previously unrecognised compact galaxies and a large sample of Galactic stars. Here we present initial results from the first 680 objects observed, including the discovery of a number of dwarf galaxies in the cluster more compact than any previously known.

1 The Complete Sample

Our primary goal is to obtain a complete sample of galaxies over a large range of magnitude and surface brightness to study the luminosity function and dynamics of both the Fornax cluster and background galaxies. Previous cluster samples were compiled from 2-D images without spectra so it was hard to tell if small galaxies were cluster dwarfs or background giants. We can solve this problem with 2dF which allows us to make a complete spectroscopic survey in the direction of the Fornax cluster. A further limitation of most existing galaxy surveys is that they only considered resolved images, so were biased against compact galaxies. Our survey will measure all images and thus avoids this bias.

One important advantage of our survey is that by including all morphological types it will provide a unique test of the presumed continuity of QSOs and Seyfert-1s as well as measuring the Seyfert luminosity function. Our 2dF spectra have a resolution of 0.85 nm (400 km/s) and can therefore resolve the broad lines of active galaxies. The only other unbiased Seyfert samples have been limited to very bright magnitudes, (e.g. the Hamburg QSO survey to $B < 17$, Kohler et al. 1997). Our sample of 12 square degrees to $B < 19.7$ will contain 200 QSOs and 50 Seyfert 1s, all with spectral classifications.

In this paper we describe the results of our first 2dF observations of a sample of 300 galaxies (1 and 2 hour exposures) and 380 unresolved sources (30 minute exposure). The 380 stellar sources were chosen with a bias to very blue and very red stars.

1.1 Galaxy Sample Results

Our galaxy observations have confirmed many members of the cluster and we have also discovered 7 new dwarf cluster galaxies; these are among the most compact dwarf galaxies known (see Drinkwater & Gregg, 1998). Three of the new cluster members show strong emission lines and are very small blue compact dwarf galaxies and one of these may be the first true dwarf spiral discovered. We are also correlating our optical data with an 843 MHz radio continuum survey of the field with the University of Sydney MOST telescope. Two radio sources we have identified with quasars are shown in Fig. 1.

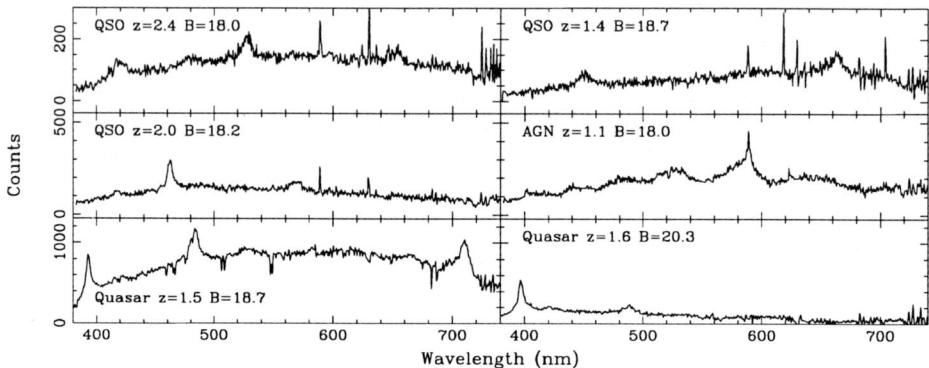

Fig. 1. 2dF spectra of blue stellar objects (upper 4 panels, 30 min exposure) and radio-loud sources (lower 2 panels, 2 h exposure).

1.2 Stellar Sample Results

Many of the bluest stellar images are QSOs; we detected 13 (see Fig. 1). Allowing for the fraction of all stars observed and our magnitude limit, this number is quite consistent with the expected QSO number counts (for $z < 2.2$ and $B < 19.75$) of 55 per 2dF (Boyle et al. 1990). We found one unusual AGN spectrum in the stellar sample, shown in Fig. 1. The broad bands are real, although the peak at 590 nm is next to a poorly removed night sky emission line. This source has previously been identified as an X-ray source at a red shift of $z = 1.1$ from the Einstein Medium Sensitivity Survey (Stoke et al. 1991).

References

Boyle, B.J., Fong, R., Shanks, T., Peterson, B.A. (1990): MNRAS, 243, 1
Drinkwater, M.J., Gregg, M.D. (1998): MNRAS, in press, astro-ph/9801016
Kohler, T., Groote, D., Reimers, D., Wisotski, L. (1997): A&A, 325, 502
Stoke, J.T., Morris, S.L., Gioia, I.M., Maccacaro, T., Schild, R., Wolter, A., Fleming, T.A., Henry, J.P. (1991): ApJ.Sup., 76, 813

Galaxies Behind the Deepest Extinction Layer of the Southern Milky Way

Renée C. Kraan-Korteweg[1], Bärbel Koribalski[2], and Sebastian Juraszek[3,2]

[1] Departamento de Astronomia, Universidad de Guanajuato, Mexico
   kraan@norma.astro.ugto.mx
[2] Australia Telescope National Facility, CSIRO, P.O. Box 76, Epping, Australia
   bkoribal@atnf.csiro.au
[3] School of Physics, Univ. of Sydney, NSW 2006, Australia, sjurasze@atnf.CSIRO.AU

Abstract. About 25% of the optical extragalactic sky is obscured by the dust and stars of our Milky Way. Dynamically important structures might still lie hidden in this zone. Various approaches are presently being employed to uncover the galaxy distribution in this Zone of Avoidance (ZOA). Results as well as the different limitations and selection effects from these multi-wavelengths explorations are being discussed. Galaxies beyond the innermost part of the Milky Way — typically at a foreground obscuration in the blue of $A_B \gtrsim 5^m$ and $|b| \lesssim \pm 5°$ — remain particularly difficult to uncover except for H I-surveys: the Galaxy is fully transparent at the 21cm line and H I-rich galaxies are easy to trace. We will report here on the first results from the systematic blind H I-search ($v \leq 12700$ km s^{-1}) in the southern Zone of Avoidance which is currently being conducted with the Parkes Multibeam (MB) Receiver.

1 Introduction

To understand the dynamics within the local Universe – the mass distribution and the local velocity field with its peculiar and streaming motions – a detailed map of the 3-dimensional galaxy distribution is highly desirable. However, the dust extinction and confusion with stars in the disk of our Galaxy make this very difficult for ∼25% of the sky, and the following questions remain unanswered:

Could a nearby Andromeda-like galaxy have escaped detection to date, hence changing our understanding of the internal dynamics and mass derivations of the Local Group (LG), and the present density of the Universe from timing arguments (Peebles 1994)?

Is the dipole in the Cosmic Microwave Background Radiation (direction and amplitude) entirely explained by the gravity on the LG from the irregular mass/galaxy distribution? As the nearest galaxies ($v < 300$ km s^{-1}) generate 20% of the total dipole moment (Kraan-Korteweg 1989), nearby individual galaxies are equally important as massive groups, clusters and voids.

Is the mass overdensity in the Great Attractor (GA) region – postulated from a large-scale systematic flow of galaxies towards $(\ell, b, v) \sim (320°, 0°, 4500\,\mathrm{km\,s^{-1}})$ (Kolatt *et al.* 1995) – in the form of galaxies, hence does light trace mass?

Does the Supergalactic Plane, other superclusters, walls and voids connect across the Milky Way and might other large-scale structures (LSS) have gone undetected due to this 'zone of avoidance'?

2 Multiwavelength Explorations of the Southern ZOA

Various approaches are presently being employed to uncover the galaxy distribution in the ZOA: deep optical searches, far-infrared (FIR), near-infrared (NIR) surveys and blind H I searches. All methods produce new results, but all suffer from (different) limitations and selection effects.

OPTICAL: Nearly the whole southern ZOA has been systematically surveyed for highly obscured but still visible galaxies using existing sky surveys (*cf.* Woudt 1998, for a detailed overview). These surveys achieve a considerable reduction of the ZOA and have uncovered distinct LSS unrelated with the foreground extinction. Follow-up redshift observations have revealed a number of dynamically important structures such as *e.g.*, the nearby overdensity in Puppis (Lahav *et al.* 1993) and the massive cluster A3627 at the core of the GA (Kraan-Korteweg *et al.* 1996, *cf.* Fig. 1). Deep optical surveys are not biased with respect to any particular morphological type. However, for foreground extinctions above $A_B \gtrsim 5^m$ (H I-column-densities $N_{HI} \gtrsim 6 \cdot 10^{21} \text{cm}^{-2}$), the ZOA remains fully opaque (*cf.* inner contour in Fig. 1). For the southern Milky Way this corresponds roughly to $|b| \lesssim \pm 5°$.

FIR: The IRAS Point-Source Catalog (PSC) has been exploited in the last decade to identify galaxy candidates behind the ZOA. Using different colour selection criteria, galaxy candidates were followed up by HI radio surveys (*e.g.*, Lu *et al.* 1990) or by inspection of plates (*e.g.*, Takata *et al.* 1996). To avoid confusion with Galactic sources, K-band snapshots have proved very efficient (Saunders *et al.* 1994). Confirmed IRAS galaxies can be merged with IRAS galaxy samples outside the ZOA to produce uniform whole-sky samples for LSS studies. But bright spiral and starburst galaxies dominate these samples.

NIR: The recent near infrared (NIR) surveys, 2MASS (Skrutskie *et al.* 1997) and DENIS (Epchtein 1997), provide complementary data. NIR surveys are sensitive to early-type galaxies, are tracers of massive groups and clusters missed in IRAS and H I surveys, have little confusion with Galactic objects and are less affected by absorption than optical surveys.

In a pilot study, we examined the efficiency of uncovering galaxies at high extinctions with DENIS images (*cf.* Schröder *et al.* 1997 & Kraan-Korteweg *et al.* 1998): highly obscured, optically invisible galaxies can indeed be traced to lower latitudes ($|b| \gtrsim 1 - 1°\!.5$) than deep optical surveys. This is not only of interest in charting early-type galaxies but also with respect to the combination of H I data of heavily obscured spiral galaxies detected in blind H I surveys (cf. below) with NIR data, and therewith the possibility to extend the peculiar velocity field into the ZOA via the NIR Tully–Fisher relation.

H I: In the regions of highest obscuration and infrared confusion the Galaxy is fully transparent to the 21-cm line radiation of neutral hydrogen. H I-rich galaxies can readily be found at the lowest latitudes through detection of their redshifted 21-cm emission. Only low-velocity extragalactic sources (blue- and redshifted) within the strong Galactic HI emission will be missed, and – because of baseline ripple – galaxies close to radio continuum sources.

Until recently, radio receivers were not sensitive and efficient enough to attempt large systematic surveys of the ZOA. In a pilot survey with the late 300-ft telescope of Green Bank, Kerr & Henning (1987) surveyed 1.5% of the ZOA and detected 16 new spiral galaxies. Since then a systematic shallow search for nearby, massive galaxies has been completed in the north (Henning et al. 1998), yielding five objects including Dwingeloo 1 (Kraan-Korteweg et al. 1994).

3 The Parkes Multibeam Survey in the Southern ZOA

In March 1997, a systematic blind H I survey began with the Multibeam Receiver (13 beams in the focal plane array) at the 64 m Parkes telescope in the most opaque region of the southern Milky Way ($213° \leq \ell \leq 33°$; $|b| \leq 5°$). The ZOA will be surveyed along constant Galactic latitudes in 23 contiguous fields of length $\Delta\ell = 8°$. The ultimate goal is 25 scans per field where adjacent strips will be offset in latitude by $\Delta b = 1\rlap{.}'5$ for homogeneous sampling. With a total observing time of 1500h, we will obtain an effective integration time of 25 min/beam with a $3\,\sigma$ detection limit of 15 mJy. Roughly 3000 detections are predicted for the covered velocity range of $-1200 \lesssim v \lesssim 12700\,\text{km s}^{-1}$ (Staveley-Smith 1997). This allows the detection of dwarfs with H I-masses as low as $10^6\,M_\odot$ in the local neighbourhood, and will be sensitive to normal Sc galaxies well beyond the Great Attractor region. As a byproduct, the survey will produce a high resolution integrated column density map of the southern Milky Way and a detailed catalog of high velocity clouds (cf. Putman et al. 1998).

3.1 First Results from the Parkes Multibeam Survey

At the time of this meeting, the whole southern ZOA survey had been surveyed twice ($\Delta b = 17'$, rms ~ 20 mJy). The cubes of the Hanning smoothed data (26 km s^{-1} resolution) were inspected visually for 21 of the fields ($220° \leq \ell \leq 4°$) and all galaxy candidates with H I fluxes $\gtrsim 100$ mJy were catalogued.

A total of 87 galaxies were uncovered in this way. Four were seen in more than one cube. Though galaxies up to 6500 km s^{-1} were identified, most of the galaxies (80%) are quite local (v< 3500 km s^{-1}) due to the (yet) low sensitivity. In the low-extinction Puppis region (cf. Fig. 1), a large fraction of the galaxies and their velocities were already known. In the remaining ZOA, about 1/3 have a counterpart in NED or the deep optical surveys.

The distribution of the H I-detected galaxies is shown in the lower panel of Fig. 1. Here we also display the results by Juraszek et al. (in prep.) in the GA region. In this high priority area, defined as $310° \leq \ell \leq 330°$, $|b| \leq 10°$, four scans (rms ~ 15 mJy) were analyzed and 82 galaxies charted. This area hence probes deeper and finds, not unexpectedly, a peak in the velocity distribution between 3000 and 4500 km s^{-1} in the GA direction. The top panel of Fig. 1

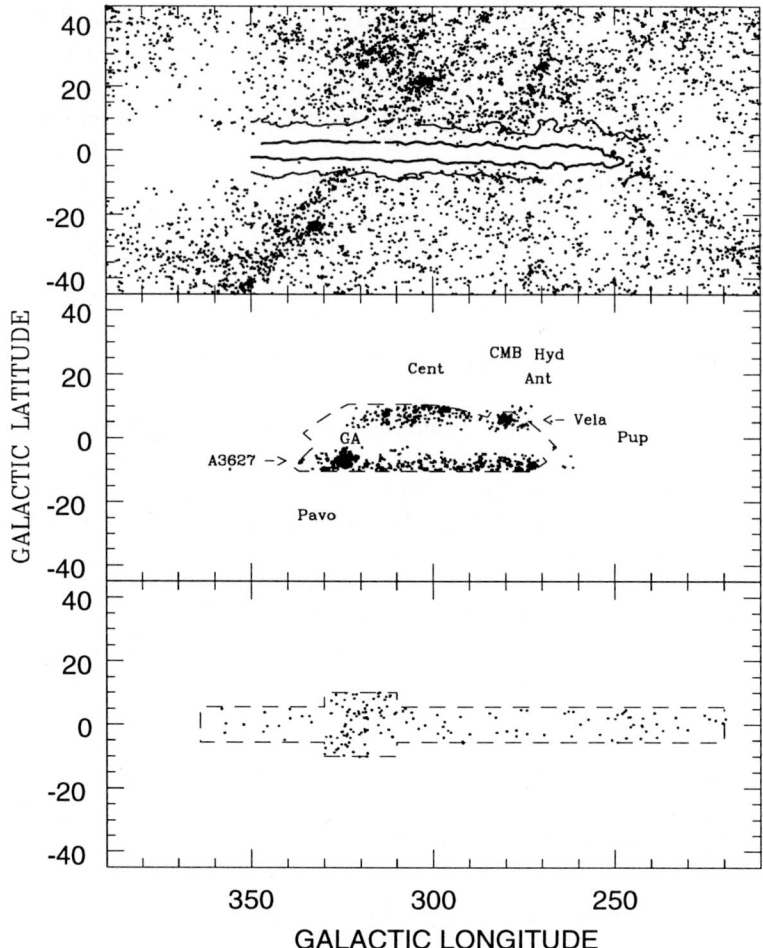

Fig. 1. Galaxies with v<10000 km s^{-1}. Top panel: literature values (LEDA), superimposed are extinction levels $A_B \sim 1\overset{m}{.}5$ and 5^m; middle panel: follow-up redshifts (ESO, SAAO and Parkes) from deep optical ZOA survey with locations of clusters and dynamically important structures; bottom panel: redshifts from shallow MB-ZOA and deeper GA survey in H I with the Parkes radio telescope.

shows the distribution of all galaxies with velocities $v \leq 10000 \, \text{km s}^{-1}$ centered on the southern Milky Way. Note the near full lack of galaxy data for extinction levels $A_B \sim 1\overset{m}{.}5$ (outer contour). The middle panel results from the follow-up observations of the optical galaxy search by Kraan-Korteweg and collaborators. Various new overdensities become apparent at low latitudes. But the innermost part of our Galaxy remains obscured ($A_B \gtrsim 5^m, |b| \lesssim 5°$). Here, the blind H I data (*cf.* lower panel) finally can provide the missing link for LSS studies.

In Fig. 2, the data of Fig. 1 are combined in redshift slices. The achieved

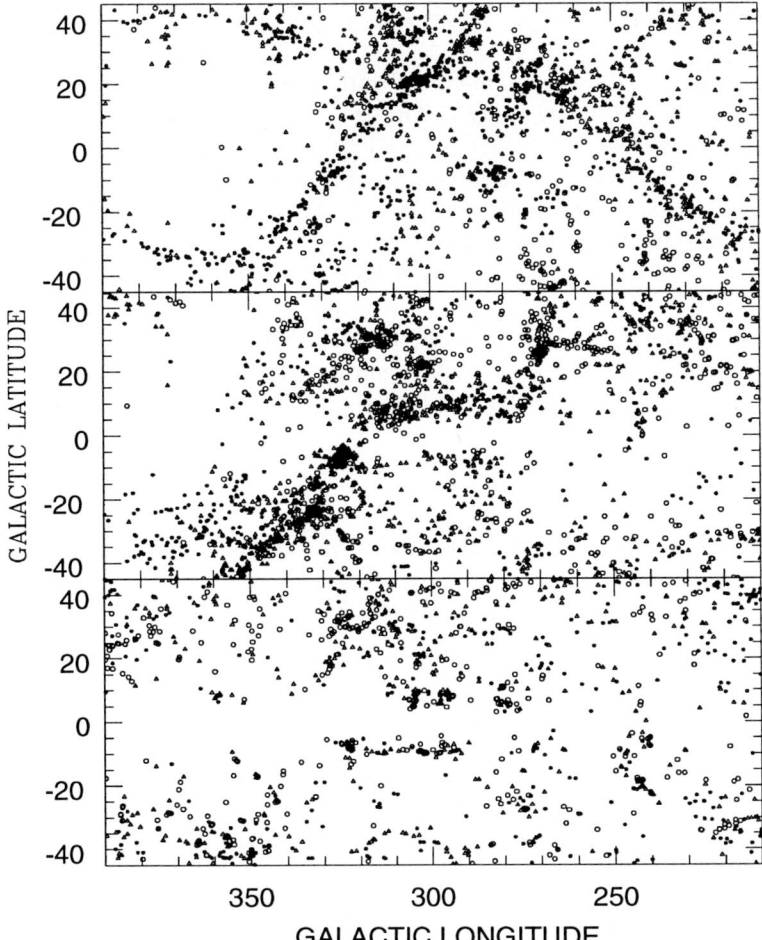

Fig. 2. Redshift slices from data in Fig. 1: $500<v<3500$ (top), $3500<v<6500$ (middle), $6500<v<9500$ km s^{-1} (bottom). The open circles mark the nearest $\Delta v=1000$ km s^{-1} slice in a panel, then triangles, then the filled dots the 2 more distant ones.

sensitivity in the current MB H I-survey fills in structures all the way across the ZOA for the upper panel (v < 3500 km s^{-1}) for the first time. Note the continuity of the thin filamentary sine-wave-like structure that dominates the whole southern sky, and the prominence of the Local Void. This feature is very different from the thick, foamy Great Wall-like structure, the GA, in the middle panel. With the full sensitivity aimed at with the MB-survey, we will be able to fill in the LSS in the more distant panels of Fig. 2 as well.

ATCA follow-up observations of three very extended (20′ to $\gtrsim 1°$), nearby (v < 1500 km s^{-1}) sources revealed them to be interesting galaxies/complexes, with unprecedented low H I column densities (*cf.* Staveley-Smith *et al.* 1998).

4 Conclusions

The combination of the complementary multiwavelength surveys allow a new probing of LSS in the 'former' ZOA. The H I surveys are particularly powerful at the lowest latitudes. But future merging of ZOA data with catalogs outside the ZOA will have to be done with care to obtain 'unbiased' whole-sky surveys.

From the sensitivity attained with the first 2 scans of the ZOA MB-survey it can be maintained that no Andromeda or other H I-rich Circinus galaxy is lurking undetected behind the extinction layer of the southern Milky Way.

Acknowledgements — The help of the HIPASS ZOA team members R.D. Ekers, A.J. Green, R.F. Haynes, P.A. Henning, R.M. Price, E. Sadler, and L. Staveley-Smith is gratefully acknowledged.

References

Epchtein, N. (1997): *The Impact of Large Scale Near-Infrared Surveys* eds. F. Garzon et al. (Kluwer: Dordrecht) p. 15
Henning, P.A., Kraan-Korteweg, R.C., Rivers, A.J., Loan, A.J., Lahav, O., Burton, W.B (1998): AJ **115**, 584
Kerr, F.J., Henning, P.A. (1987): ApJ **320**, L99
Kolatt, T., Dekel, A., Lahav, O. (1995), MNRAS **275**, 797
Kraan-Korteweg, R.C. (1989): *Rev. in Modern Astron.* **2**, ed. G. Klare (Springer: Berlin), p119
Kraan-Korteweg R.C., Loan A.J., Burton W.B., Lahav O., Ferguson H.C., Henning P.A., Lynden-Bell D. (1994): Nat **372**, 77
Kraan-Korteweg, R.C., Woudt, P.A., Cayatte, V., Fairall, A.P., Balkowski, C., Henning, P.A. (1996): Nat **379**, 519
Kraan-Korteweg, R.C., Schröder, A., Mamon, G., Ruphy, S. (1998): *The Impact of Near-Infrared Surveys on Galactic and Extragalactic Astronomy*, ed. N. Epchtein (Kluwer: Dordrecht), in press (astro-ph/9711226)
Lahav, O., Yamada, T., Scharf, C.A. Kraan-Korteweg, R.C. (1993): MNRAS **262**, 711
Lu, N.Y., Dow, M.W., Houck, J.R., Salpeter, E.E., Lewis, B.M. (1990): ApJ **357**, 388
Peebles, P.J.E. (1994): ApJ **429**, 43
Putman, M.E. et al. (1998): Nat, submitted
Saunders, W. et al. (1994): *Unveiling Large-Scale Structures Behind the Milky Way*, eds. C. Balkowski, R.C. Kraan-Korteweg, ASP Conf. Ser. 67, 257
Schröder, A., Kraan-Korteweg, R.C., Mamon, G.A., Ruphy, S. (1997): *'Extragalactic Astronomy in the Infrared*, eds. Trinh Thuân et al. (Frontières: Gif-sur-Yvette), p381
Staveley-Smith, L. (1997): PASA **14**, 111
Staveley-Smith, L., Juraszek, S., Koribalski, B.S., Ekers, R.D., Green, A.J., Haynes, R.F., Henning, P.A., Kesteven, M.J., Kraan-Korteweg, R.C., Price, R.M., Sadler, E.M. (1998): AJ, submitted
Skrutskie, M.F., et al. (1997): *The Impact of Large Scale Near-Infrared Surveys*, eds. F. Garzon et al. (Kluwer: Dordrecht) p. 25
Takata, T., Yamada, T., Saito, M. (1996): ApJ **457**, 693
Woudt, P.A. (1998): Ph.D. thesis, Univ. of Cape Town.

Surveys of Peculiar Velocities of Galaxies

Wolfram Freudling

Space Telescope – European Coordinating Facility
European Southern Observatory
Karl-Schwarzschild-Str. 2
85748 Garching
Germany

Abstract. Substantial progress has recently been made in the understanding of the mass distribution in the local Universe through analysis of the all-sky velocity field. The current status of such peculiar velocity surveys is reviewed in this paper. Recent discrepant conclusions for the density of the Universe are discussed.

1 Introduction

Systematic spectroscopic surveys of galaxies succeeded to estimate redshifts for a large fraction of known galaxies in the local Universe by the mid-80s. Slowly, workers started to supplement this data with photometric CCD imaging and spectroscopic information of higher resolution. The motivation for collecting these data is primarily the estimation of redshift-independent distances, via scaling relations such as the Tully-Fisher relation for spiral galaxies, or similar relationships (Faber-Jackson, $D_n - \sigma$, or Fundamental Plane) for early type galaxies. Early such surveys used either sparse spatial sampling of the sky (*e.g.* Aaronson *et al.* 1986) or concentrated on specific regions in the sky (*e.g.* Freudling *et al.* 1991, Willick 1990). A remarkable effort to combine and homogenize available data was undertaken by the "Mark III" collaboration (Willick *et al.* 1995). Simultaneously, an effort to obtain a new all-sky sample of peculiar velocity measurement using a homogeneous set of selection criteria, observational procedures, standards and data reduction procedures both for field galaxies ("SFI" sample, Wegner *et al.* 1998) and for cluster galaxies (SCI sample, Giovanelli *et al.* 1997) was completed. The availability of these large surveys allows a new suit of cosmological tests to be carried out in the local Universe. In addition, these new data will be used for the calibration of distance relations and investigation of possible differences between the relations for field and cluster galaxies. This will be used as the basis to search for evolution in the scaling relations with 8m class telescopes. Currently, the accuracy of such investigations are still limited by the local calibration.

2 The Local Velocity Field

Redshifts and redshift independent distance measurements from the TF relation can be combined to estimate the radial component of the peculiar velocity $u =$

$cz_{CMB} - d_c$, where d_c is the bias corrected distance estimate (Freudling et al. 1995) and cz is the measured redshift. These radial components of the peculiar velocities can be used to reconstruct a three-dimensional peculiar velocity field. For the reconstruction, the standard assumption is that the flows are irrotational. This assumption allows calculation of the scalar velocity potential field from the integral along radial paths of the radial velocity component (Dekel & Bertchinger 1990). Figure 1 shows the reconstructed velocity fields from both the SFI and the Mark III samples. The most significant differences between the velocity field is in the Pieces-Perseus (PP) region (da Costa et al. 1996) and the backside of the Great Attractor region. These velocity fields can be compared to predictions from galaxy redshift surveys, which can predict peculiar velocity fields under the assumption of linear biasing and quasi-linear evolution of density perturbations (e.g. Freudling et al. 1994, Fisher et al. 1995). It turns out that the SFI sample agrees well with the predictions from the IRAS sample for a suitable choice of parameters (e.g. da Costa et al. 1998), while the Mark III sample measures significant deviations from the IRAS model in the PP region (Willick & Strauss 1998). However, this discrepancy could be due to difference in the zero point in one of the original samples which contribute to the Mark III sample (Willick & Strauss 1998).

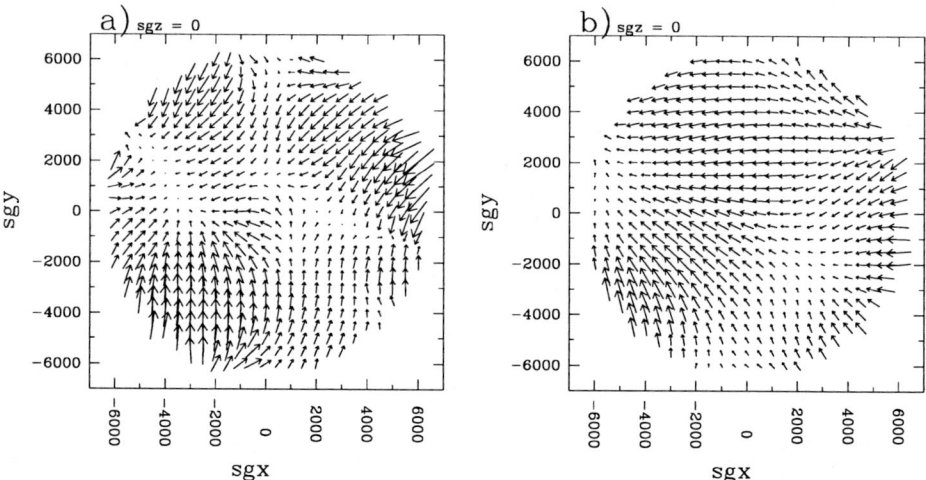

Fig. 1. The POTENT velocity field as derived from the SFI (panel a) and Mark III (panel b) samples.

3 The Density Parameter from Comparison with the IRAS Velocity Field

Predictions of the peculiar velocity field from redshift surveys depend on an assumed value for the density parameter $\beta = \Omega^{0.6}/b$, where b is the bias factor between galaxies in the redshift sample and matter. Comparison of the radial component of the predicted velocity field and the measurements can therefore be used to determine the density parameter β. Most recent such comparisons found values for β between 0.5 and 0.7 (*e.g.* Davis, Nusser & Willick 1996, Willick & Strauss 1998, da Costa *et al.* 1998). The agreement between results from different samples suggests that the detected differences in the velocity field have little impact on the global comparison with the IRAS velocity field.

4 The Power Spectrum

In the standard picture of cosmology, structure originated from small-amplitude density fluctuations that were amplified by gravitational instability. Peculiar velocity measurements can be used to constrain the power spectrum of these fluctuations. The available peculiar velocity catalogs cover only a limited range of length scales. A maximum-likelihood method to measure the power spectrum from peculiar velocity catalogs can be used to determine the cosmological parameters for families of physical CDM models with or without COBE normalization (*e.g.* Zaroubi *et al.* 1997, Freudling *et al.* 1998).

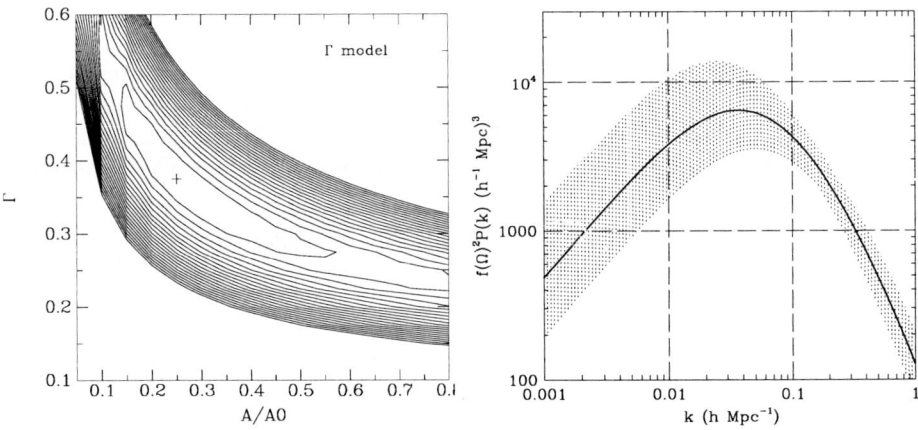

Fig. 2. Left panel: Contour plot of lnL for the Γ model. The best-fit point is marked with a '+'. Right panel: The best-fit PS, with the shaded area marking the uncertainty based on the 90% confidence region of the likelihood contours.

Figure 2 shows the likelihood contours and power spectra for the Γ models derived from the SFI sample (Freudling *et al.* 1998). The robust, model-independent result for both catalogs of peculiar velocities is that the power spectrum at $k = 0.1\,h\,\mathrm{Mpc}^{-1}$ is $P(k)\Omega^{1.2} = (4.4 \pm 1.7) \times 10^3\,(h^{-1}\mathrm{Mpc})^3$, where the random error estimate is based on a 90% confidence level. Integration over the PS leads to $\sigma_8 \Omega^{0.6} = 0.8 \pm 0.25$. This error estimate includes both random and systematic errors. The results are obtained from the peculiar-velocity data independent of the specific shape assumed for the PS, and are consistent with the result of the Γ model (Efstathiou & Bond 1992) independent of the COBE normalization.

The dynamical result of $\tilde{\sigma}_8 \equiv \sigma_8 \Omega^{0.6} \simeq 0.8 \pm 0.25$ can be compared to estimates of the β parameter obtained from the same data in the previous section. With $\sigma_{8g} \simeq 0.7$ for IRAS galaxies, the predicted β from our current constraint on $\tilde{\sigma}_8$ leads to $\beta = \tilde{\sigma}_8/\sigma_{8g} \approx 1.1$, which is marginally inconsistent with the values found from the velocity comparison. This difference in the value for β results from the assumption of larger scatter in the distance relation by the IRAS velocity comparisons as compared to the values used in the PS approach. In other words, the two methods interpret different fractions of the observed deviations from the TF relation as peculiar velocities, which results in different values for β. For the best-fit β value in the velocity comparisons, the assumed scatter has to be significantly higher than the errors originally estimated for the SFI data (da Costa et al. 1998). Alternatively, the results could be explained by peculiar velocities which are not correctly predicted by the IRAS model. One possible reason could be non-linear biasing of the galaxy distribution.

5 Summary

All-sky peculiar velocity surveys have succeeded in mapping the matter distribution in the local universe. Standard assumptions for galaxy biasing and standard cosmological models lead to marginally discrepant results for the density parameter β. This might indicate the impact of non-trivial effects such as non-linear galaxy biasing.

References

Aaronson, M., Bothun, G., Mould, J., Huchra, J., Schommer, R. A., & Cornell, M. E. 1986 Astrophysical Journal, 302, 536
da Costa, L. N., Freudling, W., Wegner, G., Giovanelli, R., Haynes, M. P., & Salzer, J. J. 1996, Astrophysical Journal, 468, L5
da Costa, L. N., Nusser, A., Freudling, W., Giovanelli, R., Haynes, M. P., Salzer, J. J., & Wegner, G. 1998, Monthly Notices of the Royal Astronomical Society, submitted (astro-ph/9707299)
Davis, M., Nusser, A., & Willick, J. A. 1996, Astrophysical Journal, 473, 22
Dekel, A., Bertchinger, E., & Faber, S. M. 1990, Astrophysical Journal, 364, 349
Efstathiou, G., Bond, J. R., & White, S. D. M. 1992, Monthly Notices of the Royal Astronomical Society, 258, 1p
Fisher, K. B., Huchra, J. P., Strauss, M. A., Davis, M., Yahil, A., & Schlegel, D. 1995, Astrophysical Journal Supplement, 100, 69
Giovanelli, R., Haynes, M. P., Herter, T., Vogt, N., Salzer, J. J., Wegner, G., da Costa, L. N., & Freudling, W. 1997a, Astronomical Journal, 113, 22

Freudling, W., Zehavi, I., da Costa, L. N., Dekel, A., Eldara, A., Giovanelli, R., Haynes, M. P., Salzer, J. J., & Wegner, G., in preparation

Freudling, W., da Costa, L. N., Pellegrini, P. S. 1994, Monthly Notices of the Royal Astronomical Society, 268, 943

Freudling, W., da Costa, L. N., Wegner, G., Giovanelli, R., Haynes, M. P., & Salzer, J. J. 1995, Astronomical Journal, 110, 920

Freudling, W., Martel, H., Haynes, M. P. 1991, Astrophysical Journal, 377, 349

Wegner, M. P. et al. 1998, in preparation

Willick, J. A. 1990, Astrophysical Journal 351, L5.

Willick, J. A., Courteau, S., Faber, S. M., Burstein, D., & Dekel, A. 1995, Astrophysical Journal, 446, 12

Willick, J. A., & Strauss, M. A. 1998, Astrophysical Journal, submitted (astro-ph/9801307)

Zaroubi, S., Zehavi, I., Dekel, A., Hoffman, Y., & Kolatt, T. 1997, Astrophysical Journal, 486, 21

ESO Imaging Survey

Luiz da Costa and Alvio Renzini

European Southern Observatory
Karl-Schwarzschild-Str 2, D–85748 Garching bei München, Germany

Abstract. Preparation of statistical samples of astronomical targets suitable for 8-m class telescopes is now of paramount importance. At ESO an effort is being made to address this problem by carrying out a public imaging survey to produce targets for the first year of operation of the VLT. Here the goals, strategy and preliminary results of this effort are presented. Current plans to greatly expand the imaging facilities of ESO in the short and long-term are also briefly described.

1 Introduction

The advent of 8-m class telescopes has pointed out the need to produce suitable target lists to sustain cutting-edge research. This is particularly true for programs that depend on large statistical samples of rare, high-redshift objects, requiring catalogs to be produced in advance. While until recently most programs could rely on samples extracted from photographic plates, this is no longer the case as the new imaging surveys must match the fainter limiting magnitudes reached by the new telescopes. Therefore, modern digital, multicolor, deep imaging surveys have become an indispensable complement to the 8-m telescopes. The new generation of imaging surveys will, without doubt, be the backbone of future research and are likely to be as long-lived as their older counterparts, which have served the astronomical community so well over the past three decades. These surveys are now becoming possible thanks to the new CCD-mosaics mounted on wide-field telescopes.

The strategy of modern surveys in support to 8-m class telescope will have to satisfy different requirements and will complement other digital, all-sky but shallower surveys such as SLOAN. The area and depth of these surveys will be determined primarily by the population of the rarest objects being targeted. Even though covering a limited area, to reach the required faint magnitudes these surveys will still demand generous amounts of telescope time. In order to make them viable it is important to guarantee the maximum use of the data even before the completion of the survey. Since a number of programs may benefit from shallower data, this can be done by making the data public as they become available, thus serving the broadest possible community of users which can use the data to cover a wide range of topics. These programs will generate large amounts of data and the main challenge in the near future is to develop procedures to effectively cope with this flood of data. This requires new tools to process, archive, distribute and translate the results into useful target lists. Equally important, is designing a comprehensive database and data mining tools to maximize the scientific output of the data. The investments are large for most individual institutes and some degree of cooperation is required.

At ESO an experiment has recently been made to tackle this generic problem, whereby a public imaging survey is being carried out at the NTT, to define targets for the first year of operation of the VLT. The ESO Imaging Survey (EIS) has been supervised by a Working Group selected from the European astronomical community, which is responsible for the definition of the survey strategy and for monitoring of the progress of the survey. The main challenges of the project have been: 1) to cover a sufficiently large of area of sky to an appropriate depth using a CCD camera with a relative small field-of-view; 2) to carry out a public survey in a limited amount of time requiring observations, software development and data reduction with the goal of distributing the survey data products before the call for proposals for the VLT. To cope with the ambitious one-year timetable, a novel type of collaboration between ESO and the community has been established which has allowed EIS to combine the scientific and technical expertise of the community with in-house know-how and infrastructure. Such a model could set the example for future surveys. The goal of this contribution is to briefly describe EIS and its preliminary results, and discuss the long-term prospects for imaging surveys at ESO.

2 Goals

The ESO Imaging Survey (EIS) (Renzini and da Costa 1997) is a concerted effort by ESO and the Member State community to provide targets for the first year of operation of the VLT. It consists of two parts: a wide-angle survey (EIS-wide) and a deep, multicolor survey in four optical and two infrared bands (EIS-deep) using the recently installed SUSI2 and SOFI cameras, respectively. Further information is available at "http://www.eso.org/eis".

EIS-wide covers four pre-selected patches of sky, 6 square degrees each, spanning the right ascension range $22^h < \alpha < 9^h$ (see Table 1). The main science goals of EIS-wide are the search for distant clusters and quasars, but important spin-offs include bright high-z galaxies, galactic structure and stellar populations. To achieve these goals the original proposal envisioned the observation of all four patches in V and I, one patch also in B, and 2 square degrees portion of this patch in U. Because of the slow start of the survey due to the unusually bad weather caused by El Niño, some of these goals had to be reassessed by the WG (da Costa et al. 1998). It was decided to limit the observations to the I-band, except for patch B, close to the South Galactic Pole, where observations were conducted in B, V and I.

EIS-deep is a multicolor survey in four optical and two infrared bands covering 75 arcmin^2 of the HST/Hubble Deep Field South (HDFS), including the WFPC2, STIS and NICMOS fields, and a region of 100 arcmin^2 in the direction of the southern hemisphere counterpart of the Lockman Hole, to produce samples with photometric redshifts, to find U- and B-dropout candidates, and galaxies in the redshift range $1 < z < 2$. Observations for EIS-deep will start in August 1998.

3 Survey Strategy

The observations for EIS-wide started in July 1997 and have been conducted using the EMMI camera with a Tektronix 2046 × 2046 chip, mounted on the NTT at La Silla. The pixel size is 0.266 arcsec, providing an effective field-of-view of about $9' \times 8.5'$, because of the strong vignetting at the top and bottom parts of the CCD. In order to cover a large area of the sky the observations have been conducted by a sequence of 150 sec exposures shifted by half the size of an EMMI-field both in right ascension and declination. This leads to an image mosaic whereby each position in the sky is observed twice for a total integration time of 300 sec, except at the edges of the patch. The EIS mosaic thus consists of frames with significant overlaps (a quarter of an EMMI frame). The easiest way of visualizing the geometry of the EIS mosaic is to picture two independent sets of tiles forming a contiguous grid (normally referred to as odd and even) superposed and shifted in right ascension and declination by half the length of an EMMI frame. To ensure continuous coverage, adjacent odd/even frames have a small overlap at the edges ($\sim$ 20 arcsec). Therefore, a small fraction of the surveyed area may be covered by more than two frames. Such a mosaic ensures good internal astrometry ($\sim$ 0.03", Nonino et al. 1998), relative photometry and the satisfactory removal of cosmic hits and other artifacts.

4 EIS Data Reduction Pipeline

An integral part of the EIS project has been the development of an automated data reduction pipeline to handle the large of volume generated by EIS. The pipeline consists of different modules built from preexisting software consisting of: 1) standard IRAF tools for the initial processing of each input image and preparation of superflats; 2) the Leiden Data Analysis Center (LDAC) software, developed for the DENIS (Epchtein et al. 1996) project to perform photometric and astrometric calibrations; 3) the SExtractor object detection and classification code (Bertin & Arnouts 1996); 4) the "drizzle" image co-addition software (Fruchter & Hook 1997, Hook & Fruchter 1997), originally developed for HST, to create coadded output images from the many, overlapping, input frames; 4) the commercially available Sybase, which is used to control the data reduction, the quality of the data, the calibration parameters and as traceback facility on a frame basis.

A major aim of the EIS software has been to handle the generic problem posed by building up a mosaic of overlapping images, with varying characteristics, and extraction of information from the resulting inhomogeneous coadded frames. This has required significant changes in the preexisting software, which have been discussed by da Costa et al. (1998) and Nonino et al. (1998).

5 Current Status

Observations for EIS-wide were completed on March 31, 1998, with the exception of the U-band observations scheduled for this fall. The total coverage is summarized in Table 1, where the J2000 centers of the actually surveyed patches and the area covered (in deg^2) in the different bands are given.

Table 1. Current Sky Coverage

Patch	α	δ	B	V	I
A	22:42:54	-39:57:32	-	1.2	3.2
B	00:49:25	-29:35:34	1.5	1.5	1.6
C	05:38:24	-23:51:00	-	-	6.0
D	09:51:36	-21:00:00	-	-	6.0
	-	-	1.5	2.7	16.8

The observations were carried out in standard visitor mode and data were taken in less than ideal conditions. Therefore, the quality of the data has varied over the period of observations. Regions observed under poor conditions were, whenever possible, re-observed in order to maintain some degree of uniformity in the depth of the survey. It is worth emphasizing that the data for patch A is by far the worst. This can be seen in Figure 1, where the seeing distribution for the different patches is shown.

Data for patch A has already been made publicly available in the form of pixel maps and single frame catalogs. Further information is available at http://www.eso.org/eis/datarel.html. This preliminary release was meant to allow users to have a hands-on experience with the data. All the data for EIS-wide, including the coadded image, faint object catalogs extracted from them and derived catalogs will be available by July 31, 1998.

6 Results

In Figure 2, the galaxy counts derived from the EIS single-frame catalogs of patch A are compared to those of Lidman & Peterson (1996) and Postman *et al.* (1996). There is a remarkable agreement between the EIS galaxy-counts and those of the other authors. The slope of the EIS counts is found to be 0.43±0.01. Also note that the EIS counts extend beyond those of Postman *et al.* (1996) even for the counts derived from single frames (*e.g.*, Nonino *et al.* 1998). These preliminary results, based on single-frame catalogs, indicate that the depth of the survey (Figure 2) is close to that originally expected and is sufficiently deep, especially after co-addition ($I \lesssim 23.5$), to search for distant clusters of galaxies, one of the main science goals of EIS (Olsen *et al.* 1998).

The I-band data for patch A have been used to search for clusters of galaxies over an area of 2.5 square degrees, in the redshift range $0.2 \leq z \leq 1.3$ (Olsen *et al.* 1998). The matched filter algorithm (Postman *et al.* 1996) has been applied

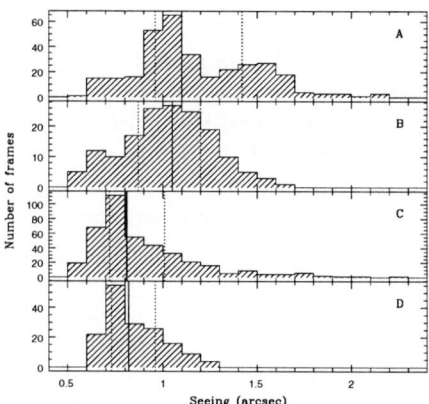

Fig. 1. Seeing distribution for each patch in the survey. Note the great improvement of the seeing distribution, especially for patches C and D.

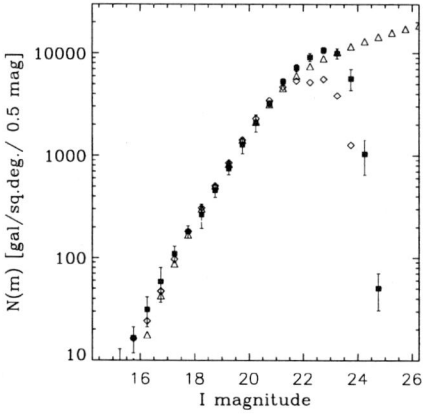

Fig. 2. The EIS galaxy counts (filled squares) based on galaxy catalogs extracted from single-frames (see text), compared to the galaxy counts derived by Lidman & Peterson (1996) (triangles) and Postman et al. (1996) (diamond). The data from the other authors have been converted to the Johnson-Cousins system.

to the even and odd single-frame catalogs to assess the performance of the detection technique, to establish the detection threshold for robust detections and to evaluate the quality of the EIS data for this kind of analysis.

The candidate cluster sample based on 4σ detections consists of 21 objects, yielding a surface density of 8.4 candidates per square degree, with a median redshift of $z = 0.4$. When all 3σ detections are considered, 39 candidates are found, leading to a surface density of 16 per square degree and a median redshift of $z = 0.6$, comparable to the findings of Postman *et al.* (1996). Cutouts for the cluster candidates are available at "http://www.eso.org/eis/datarel.html". These results should be considered preliminary as significantly better data are available for the other EIS patches. More importantly, the use of catalogs extracted from the coadded images will allow a deeper cluster search to be carried out, thereby extending the redshift range for the cluster sample. Clearly, the EIS data more than fulfills the science requirements of the survey, as originally planned.

7 Future Prospects

7.1 The Wide Field Imager at the 2.2m Telescope (WFI@2.2)

After lacking wide field imaging facilities for so many years, the ESO community will soon face the opposite problem posed by the enormous flow of data that is about to start. The new 8k×8k camera (WFI@2.2) at the 2.2m telescope will provide a field of view of $0.54° \times 0.54°$. When factoring in telescope aperture, field of view, throughput of the optics and QE of the CCDs, the WFI@2.2 will be ~ 6 times more efficient than EMMI for wide angle (survey) work. Perhaps most importantly, the 2.2m telescope will be the first, worldwide, of its class to be fully dedicated to imaging with a large format camera. For comparison, with the advent of the WFI@2.2 the survey potential capability of ESO over the period 1999-2001 is equivalent to ~ 200 EIS. During 1999 the WFI@2.2 data flow is estimated at a rate of ~ 8 Gby/night.

Recently, the EIS Working Group has recommended the completion of EIS by carrying out a public **"Pilot Survey"** during the early phases of operation of the new WFI@2.2. The main science drivers for the Pilot Survey remain the same of EIS, and the following strategy is planned:

1. Cover two EIS patches (C and D) in the U, B, V and Gunn-z bands, to provide additional leverage in the identification of EIS candidate clusters and extend to these patches the search for high-redshift quasars, bright galaxies at high redshift, and interesting objects for stellar population studies.
2. Extend the depth of the I-band observations, reaching a limiting magnitude 1 mag fainter. This will allow further confirmation of the EIS clusters, better characterization of low surface brightness features already detected with EIS, and extend somewhat the redshift limit of the candidate clusters. The re-observation of two patches at the beginning of 1999 will also allow interested teams to search for high-proper motion objects by comparison with EIS data taken one year earlier.
3. Cover 0.25 square degree (one WFI@2.2 field) in 5 optical bands U, B, V, I and Gunn-z to reach $I \gtrsim 25$. The data will allow the definition of a sample of high-redshift galaxies ($z > 3$).

The Pilot Survey will also be useful from the operational point of view as it will allow: 1) the creation of a calibration database; 2) the monitoring of the performance of the system at the start of its operation; and 3) the further development of software to process data from CCD-mosaics and the un-supervised production of object catalogs.

7.2 The VLT Survey Telescope (VST) on Paranal

Though already a respectable facility for survey work, the 2.2m telescope is envisioned as a temporary solution. According to current plans, a new 2.5m telescope will be in operation on Paranal by 2001, as part of a joint project of ESO and the Capodimonte Astronomical Observatory. It is estimated that the 2.5m telescope will have an overall efficiency ~ 12 times higher than that of the WFI@2.2. This figure comes from the combination of better site, higher throughput, and larger field of view and format of the camera (one square degree covered by at least a 16k×16k array of CCDs). Therefore, the availability of the new telescope will mark another quantum jump in the ESO capability of conducting imaging survey work, primarily – though not necessarily exclusively – in support of VLT Science. With its very large detector, the data flow from the VST will rival the data flow of the whole VLT/VLTI. Although the VST with its one square degree camera will not be the largest telescope with a very wide-angle imager, it will likely be the first facility of its class to be fully dedicated to wide field imaging.

The advantages offered by the VST over the 2.2m telescope can be appreciated when considering that a factor ~ 10 gain in efficiency means that either a given survey can be completed ten times faster, or that for given telescope time a ten times larger area can be explored (hence ten times rarer objects can be found), or that a survey can be pushed more than one magnitude deeper, or that more numerous pass-bands or narrower ones can be used. Clearly, such a jump opens a whole unexplored parameter space, offering to the community a variety of opportunities and alternatives all very attractive for VLT science, though not only for it. For example, the large proportion ($\sim 77\%$) and even distribution of photometric nights on Paranal makes the VST uniquely suited for the extensive observation of microlensing events in the Galactic bulge, and the search of extrasolar planets using this technique. In essence, a long-term scientific planning for the use of the VST will bring great benefits to the ESO community, and the experience gained with the 2.2m telescope will be critical in this respect.

8 Summary

In anticipation of the VLT first-light, a collaborative effort between ESO and the astronomical community in its member states has been undertaken to carry out an imaging survey with the aim of preparing targets for follow-up observations. The EIS project has charted new ground both on technical and organizational aspects with the ultimate goal of making long-term public surveys in support of VLT science viable.

EIS has also done the groundwork in the development of software for the un-supervised reduction of large volumes of data either in the form of dithered images or large mosaics. This is an essential first step to provide the tools required for the optimal use of the data that will become available from the 2.5-m class telescopes with large CCD-mosaics fully dedicated for imaging observations. Incremental efforts are planned for a gradual transition from the current survey conducted with a small field-of-view camera, to the WFI@2.2 to the one-square degree camera of the VST. A long road is ahead but the basic framework is already in place and this is, perhaps, one of the most relevant contributions of the EIS project.

Many of the candidate targets extracted from imaging surveys will require spectroscopic confirmation using 4-m telescopes, before investing VLT time on them. The first results of the 2dF, presented in this meeting, demonstrate that this facility will be ideal for pruning samples drawn from imaging surveys, such as EIS and those envisioned with the WFI@2.2 and VST. This complementarity offers new opportunities for fruitful collaboration between the ESO and AAT communities to look deeper and deeper in the Southern Hemisphere.

Acknowledgments

We would like to thank the members of the EIS team for their efforts and dedication to the project.

References

Bertin, E. & Arnouts, S., 1996, A&AS 117, 393
da Costa *et al.* 1998, The Messenger, 91, 49
Epchtein, N., et al., 1996, The Messenger 87, 27
Fruchter, A.S. & Hook, R.N., 1997, in Applications of Digital Image Processing XX, ed. A. Tescher, Proc. S.P.I.E. vol. 3164, 120
Hook, R.N. & Fruchter, A.S., 1997, in ASP Conf. Series, Vol. 125, Astronomical Data Analysis Software and Systems VI, ed. G. Hunt and H.E. Payne (San Francisco: ASP), 147
Lidman, C. & Peterson, B., 1996, MNRAS 279, 1357
Nonino *et al.* 1998, astro-ph/9803336
Olsen *et al.* 1998, astro-ph/9803338
Postman, M., Lubin, L.M., Gunn, J.E., Oke, J.B., Hoessel, J.G., Schneider, D.P., Christensen, J.A., 1996, AJ 111, 615
Renzini, A. & da Costa, L.N., 1997, The Messenger 87, 23
Renzini, A. 1998, The Messenger, 91, 54

Cosmological Tests from the New Surveys

Ofer Lahav

Institute of Astronomy, Madingley Road, Cambridge CB3 0HA, UK

Abstract. We review cosmological inference from galaxy surveys at low and high redshifts, with emphasis on new Southern sky surveys. We focus on several issues: (i) the importance of understanding selection effects in catalogues and matching Northern and Southern surveys; (ii) the 2dF galaxy redshift survey of 250,000 galaxies (iii) the proposed 6dF redshift and peculiar velocity survey of near-infrared galaxies (iv) radio sources and the X-Ray Background as useful probes of the density fluctuations on large scales, and (v) how to combine large scale structure and Cosmic Microwave Background measurements to estimate cosmological parameters.

1 Introduction

It is believed by most cosmologists that on the very large scales the universe is an isotropic and homogeneous system. However, on scales much smaller than the horizon the distribution of luminous matter is clumpy. Galaxy surveys in the last decade have provided a major tool for cosmographical and cosmological studies. In particular, surveys such as CfA, SSRS, IRAS, APM and Las Campanas have yielded useful information on local structure and on the density parameter Ω from redshift distortion and from comparison with the peculiar velocity field. Together with measurements of the Cosmic Microwave Background (CMB) radiation and gravitational lensing, the redshift surveys provide major probes of the world's geometry and the dark matter.

In spite of the rapid progress, there are two gaps in our current understanding of the density fluctuations as a function of scale: (i) it is still unclear how to relate the distributions of light and mass, in particular how to match the clustering of galaxies with the CMB fluctuations, (ii) currently little is known about fluctuations on intermediate scales between those of local galaxy surveys ($\sim 100 h^{-1}$ Mpc) and the scales probed by COBE ($\sim 1000 h^{-1}$ Mpc).

Another unresolved issue is the value of the density parameter Ω. Putting together different cosmological observations, the derived values seem to be inconsistent with each other. Taking into account moderate biasing, the redshift and peculiar velocity data on large scales yield $\Omega \approx 0.3 - 1.5$, with a trend towards the popular value ~ 1 (e.g. Dekel 1994; Strauss & Willick 1995 for summary of results). On the other hand, the high fraction of baryons in clusters, combined with the baryon density from Big Bang Nucleosynthesis suggests $\Omega \approx 0.2$ (White et al. 1993). Moreover, an $\Omega = 1$ universe is also in conflict with a high value of the Hubble constant ($H_0 \approx 70 - 80$ km/sec/Mpc), as in this model the universe turns out to be younger than globular clusters. A way out of these problems was suggested by adding a positive cosmological constant, such that $\Omega + \lambda = 1$, to satisfy inflation. Two recent observations constrain λ : the observed frequency of lensed quasars is too small, yielding an upper limit $\lambda < 0.65$ (e.g. Kochanek 1996), and the magnitude-redshift relation for Supernovae type Ia (e.g. Perlmutter et al. 1998). The next decade will see several CMB experiments (e.g. Planck, MAP, VSA) which promise to determine (in a model-dependent way) the cosmological parameters to within a few percent. We shall focus here on several issues related to clustering and cosmological parameters from new surveys.

2 From 'Biased Surveys' to the 'Real Universe'

Figure 1 shows a compilation of the current and future surveys, indicating for each survey its effective volume (in terms of its solid angle and median redshift) and the number of galaxies with measured redshift.

It is important to emphasize that each survey is selected by different criteria (e.g. wavelength and flux), hence any survey is biased ! One should pay attention to the following aspects in analyzing surveys:
- Source Detection (e.g. 3-σ selection, wavelet filtering, combining radio multi-components)
- Source Classification (e.g. star/galaxy separation - by eye or by Artificial Neural Networks) and multi-wavelength identification
- Incomplete Sky Coverage (e.g. the Zone of Avoidance)
- Matching Northern and Southern catalogues (e.g. the optical UGC/ESO, the radio 87GB/PMN)
- Poisson shot-noise, due to the finite number of galaxies
- Redshift distortion
- Biasing of particular tracer relative to the underlying mass distribution

3 New Surveys

Existing optical and IRAS redshift surveys contain 10,000-20,000 galaxies. The Parkes multi-beam survey in 21cm (HIPASS) will detect about 5,000 galaxies in the Southern hemisphere (see Staveley-Smith in this volume). Another major step forward using multifibre technology will allow in the near future to produce redshift surveys of millions of galaxies. In particular, there are two major surveys on the horizon. The American-Japanese Sloan Digital Sky Survey (SDSS) will yield images in 5 colours for 50 million galaxies, and redshifts for about 1 million galaxies over a quarter of the sky (Gunn and Weinberg 1995). It will be carried out using a dedicated 2.5m telescope in New Mexico. The median redshift of the survey is $z \sim 0.1$. A complementary Anglo-Australian survey, the 2 degree Field (2dF), is described below (see also Boyle & Colless on both the galaxy and quasar 2dF surveys in this volume).

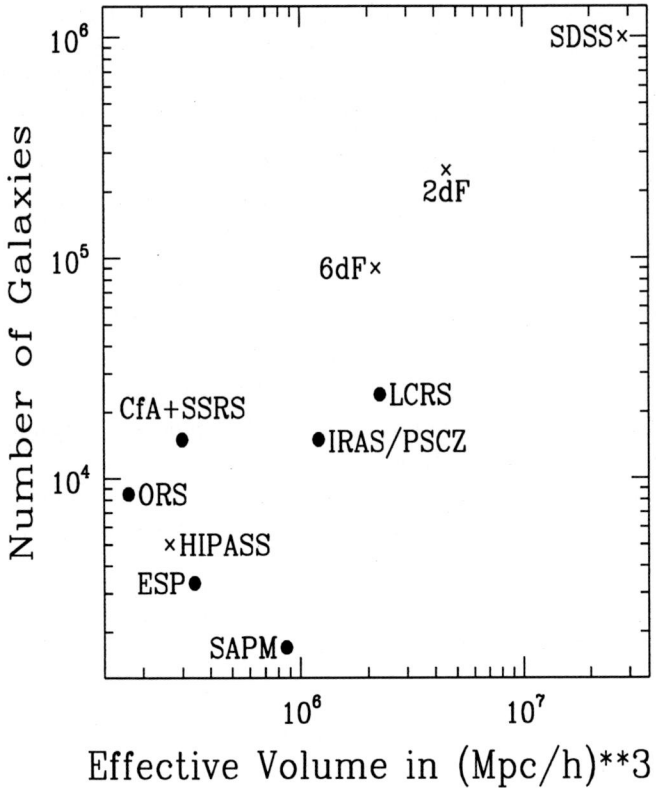

Fig. 1. The effective volume and number of galaxies of completed redshift surveys (solid circles) and surveys in preparation (crosses). The effective volume is defined here as $\omega R_{med}^3/3$, where ω is the solid angle of a survey, and R_{med} its median comoving depth. Optically selected surveys include the Center for Astrophysics (CfA) survey, the Southern Sky Redshift Survey (SSRS), the Optical Redshift Survey(ORS), the ESO Slice Project(ESP), the Stromlo-APM (SAPM) and the Las Campanas Redshift Survey (LCRS), the 2-degree Field (2dF) galaxy survey, and the Sloan Digital Sky Survey (SDSS). The IRAS Point Source Catalogue (PSCZ) is selected in the infrared (60μ), while 6dF/DENIS/2MASS is selected in the near infrared ($\sim 2\mu$). HIPASS is the Parkes multi-beam survey in 21 cm. (A compilation by S. Maddox and O. Lahav).

3.1 2dF

The 2dF galaxy survey will produce redshifts for 250,000 galaxies brighter than $b_J = 19.5^m$ (with median redshift of $z \sim 0.1$), selected from the APM catalogue. The survey will utilize a new 400-fibre system on the 4m AAT, covering $\sim 1,700$ sq deg of the sky. About 8,000 redshifts have been measured so far (March 1998). A deeper extension down to $R = 21$ for 10,000 galaxies is also planned for the 2dF survey. The science goals of the survey are:

- Accurate measurements of the power spectrum of galaxy clustering on scales $> 30h^{-1}$ Mpc allowing a direct comparison with CMB anisotropy measurements such as the recently approved NASA MAP and ESA Planck Surveyor satellites. The power-spectrum derived from the projected APM galaxies (see Figure 2) gives an idea about the scales probed by the 2dF redshift survey.
- Measurement of the distortion of the clustering pattern in redshift space providing constraints on the cosmological density parameter Ω and the spatial distribution of dark matter.
- Determination of variations in the spatial and velocity distributions of galaxies as a function of luminosity, spectral type and star-formation history, providing important constraints on models of galaxy formation.
- Investigations of the morphology of galaxy clustering and the statistical properties of the fluctuations, e.g. whether the initial fluctuations are Gaussian as predicted by inflationary models of the early universe.
- A study of clusters and groups of galaxies in the redshift survey, in particular the measurement of infall in clusters and dynamical estimates of cluster masses at large radii.
- Application of novel techniques (e.g. Principal Component Analysis and Artificial Neural Networks; Folkes, Lahav & Maddox 1996) to classify the uniform sample of $250,000$ spectra obtained in the survey, thereby obtaining a comprehensive inventory of galaxy types as a function of spatial position within the survey.

For more details on the 2dF galaxy survey see
http://msowww.anu.edu.au/colless/2dF/

3.2 6dF

It has recently been proposed by the Anglo-Australian Observatory to automate the FLAIR multi-fibre facility at the 1.2m Schmidt telescope in Siding Spring, Australia. The main purpose of the upgrading is to measure redshifts for $\sim 120,000$ galaxies principally selected in the Near InfraRed from the DENIS survey, and to measure internal motions (hence distance indicators and peculiar velocities) for $\sim 18,000$ galaxies. The unique feature of this survey is the ability to probe *mass* on both galactic and cosmological scales.

The European DENIS project (DEep Near-Infrared Southern Sky Survey) began in 1995 to image the entire Southern sky in three bands in the Near InfraRed (NIR): $I(0.8$ micron), $J(1.25$ micron) and K_s (2.15 micron). NIR light from galaxies is dominated by the old stellar population, hence is more directly related to the underlying mass than surveys in the optical or far-infrared. Moreover, the NIR light is little affected by Galactic extinction, making it ideal to probe galaxies through the Zone of Avoidance. Another large survey at 2 microns (2MASS) is carried out by a US team. In the case of DENIS the main purpose of the survey is to measure redshifts for 90,000 NIR-selected galaxies brighter than $J = 13.7$ (with median redshift $\bar{z} \approx 0.04$) plus perhaps 30,000 additional galaxies over 18000 deg^2 in about 160 spectroscopic nights. Even more exciting than this big redshift survey is the possibility to measure velocity dispersions for elliptical galaxies and rotation velocities for spiral galaxies. This will produce a uniform set of ~ 18000 galaxies with distance indicators (via the Tully-Fisher relation), and hence peculiar velocities. Currently the most recent sample of peculiar velocities (Mark III) includes only ~ 3000 galaxies taken from various subsets. The combination of the redshift and peculiar velocity surveys over an entire hemisphere will allow us to reconstruct maps of NIR galaxy and mass distributions, e.g. by the POTENT (Dekel 1994) and Wiener (e.g. Webster et al. 1997) methods, to estimate the density parameter Ω and to characterize biasing. Assuming that it will take 2 years to build 6dF, the entire redshift survey can be finished by 2001 and the peculiar velocity survey by 2003. Detailed proposals can be found on http://msowww.anu.edu.au/colless/6dF/

4 Probes at High Redshift

The big new surveys (SDSS, 2dF) will only probe a median redshift $\bar{z} \sim 0.1$. It is still crucial to probe the density fluctuations at higher z, and to fill in the gap between scales probed by previous local galaxy surveys and the scales probed by COBE and other CMB experiments. Here we discuss the X-ray Background (XRB) and radio sources as probes of the density fluctuations at median redshift $\bar{z} \sim 1$. Other possible high-redshift tracers are quasars and clusters of galaxies.

4.1 Radio Sources

Radio sources in surveys have a typical median redshift $\bar{z} \sim 1$, and hence are useful probes of clustering at high redshift. Unfortunately, it is difficult to obtain distance information from these surveys: the radio luminosity function is very broad, and it is difficult to measure optical redshifts of distant radio sources. Earlier studies claimed that the distribution of radio sources supports the 'Cosmological Principle'. However, the redshift distribution of radio sources is now better understood, and it is clear that the wide range in intrinsic luminosities of radio sources would dilute any clustering when projected on the sky. In fact, recent analyses of new deep radio surveys (e.g. FIRST) suggest that radio sources are indeed clustered at least as strongly as local optical galaxies (e.g. Cress et al. 1996; Magliocchetti et al. 1998 and in this volume). Nevertheless, on the very large scales the distribution of radio sources seems nearly isotropic. Comparison of the measured quadrupole in a radio sample in the Green Bank and Parkes-MIT-NRAO 4.85 GHz surveys to the theoretically predicted ones (Baleisis et al. 1998) offers a crude estimate of the fluctuations on scales $\lambda \sim 600h^{-1}$ Mpc. The derived amplitudes are shown in Figure 2 for the two assumed Cold Dark Matter (CDM) models. Given the problems of catalogue matching and shot-noise, these points should be interpreted at best as 'upper limits', not as detections. A new Southern radio survey, SUMSS, is described by Sadler in this volume.

4.2 XRB

Although discovered in 1962, the origin of the X-ray Background (XRB) is still unknown, but is likely to be due to sources at high redshift (for review see Boldt 1987; Fabian & Barcons 1992). Here we shall not attempt to speculate on the nature of the XRB sources. Instead, we *utilise* the XRB as a probe of the density fluctuations at high redshift. The XRB sources are probably located at redshift $z < 5$, making them convenient tracers of the mass distribution on scales intermediate between those in the CMB as probed by COBE, and those probed by optical and IRAS redshift surveys (see Figure 2).

The interpretation of the results depends somewhat on the nature of the X-ray sources and their evolution. The rms dipole and higher moments of spherical harmonics can be predicted (Lahav et al. 1997) in the framework of growth of structure by gravitational instability from initial density fluctuations. By comparing the predicted multipoles to those observed by HEAO1 (Treyer et al. 1998) it is possible to estimate the amplitude of fluctuations for an assumed shape of the density fluctuations (e.g. CDM model). Figure 2 shows the amplitude of fluctuations derived at the effective scale $\lambda \sim 600h^{-1}$ Mpc probed by the XRB. One can use this estimate to derive the fractal dimension D_2 of the universe on large scales. Although the fractal dimension of the galaxy distribution on scales $< 20h^{-1}$ Mpc is $D_2 \approx 1.2 - 2.2$, the fluctuations in the X-ray Background and in the Cosmic Microwave Background are consistent with $D_2 = 3$ to within 10^{-4} on the very large scales (Wu et al. 1998).

5 Joint Parameter Estimation LSS/CMB

Observations of anisotropies in the Cosmic Microwave Background (CMB) provide one of the key constraints on cosmological models and a significant quantity of experimental data already exists (e.g. Gawiser & Silk 1998 and Lineweaver in this volume).

Galaxy redshift surveys, mapping large scale structure (LSS), provide another cosmologically important set of observations. The clustering of galaxies in redshift-space is systematically different from that in real-space (Kaiser 1987, Hamilton 1997 for review). The mapping between the two is a function of the underlying mass distribution, in which the galaxies are not only mass tracers, but also velocity test particles. Estimates derived separately from each of the CMB and LSS data sets have problems with parameter degeneracy. Webster et

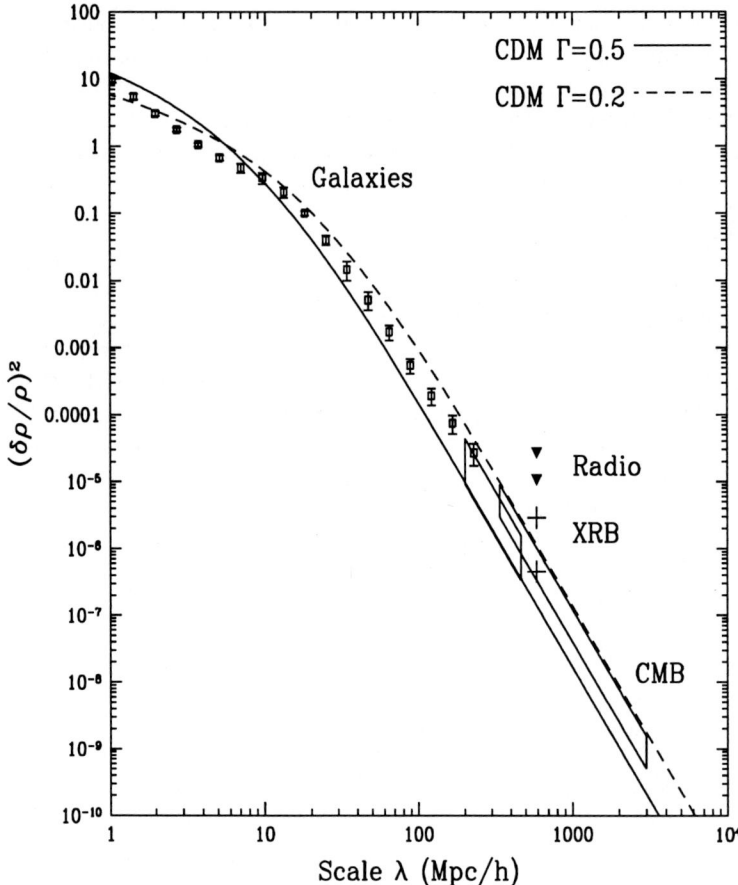

Fig. 2. A compilation of density fluctuations on different scales from galaxy surveys, deep radio surveys, the X-ray Background and Cosmic Microwave Background experiments. The Figure shows mean-square density fluctuations $(\frac{\delta\rho}{\rho})^2 \propto k^3 P(k)$. The solid and dashed lines correspond to the standard Cold Dark Matter power-spectrum (with shape parameter $\Gamma = 0.5$) and a 'low-density' CDM power-spectrum (with $\Gamma = 0.2$), respectively. Both models are normalized with $\sigma_8 = 1$ (at $k \sim 0.15$). The open squares at small scales are estimates of the power-spectrum from 3D inversion of the angular APM galaxy catalogue (Baugh & Efstathiou 1994). The elongated 'boxes' at large scales represent the COBE 4-yr (on the right) and Tenerife (on the left) CMB measurements (Gawiser & Silk 1998). The solid triangles represent constraints from the quadrupole of radio sources (Baleisis et al. 1998). The crosses represent constraints from the XRB HEAO1 quadrupole (Lahav et al. 1997; Treyer et al. 1998). The top and bottom crosses are estimates of the amplitude of the power-spectrum at $\lambda_* \sim 600h^{-1}$ Mpc, assuming CDM power-spectra with shape parameters $\Gamma = 0.2$ and 0.5 respectively, and an Einstein-de Sitter universe (A compilation from Wu, Lahav & Rees 1998).

al. (1998) combined results from a a range of CMB experiments, with a likelihood analysis of the IRAS 1.2Jy survey, performed in spherical harmonics. Their method expresses the effects of the underlying mass distribution on both the CMB potential fluctuations and the IRAS redshift distortion. This breaks the degeneracy inherent in an isolated analysis of either data set, and places tight constraints on several cosmological parameters.

The family of CDM models analysed corresponds to a spatially-flat universe with an initially scale-invariant spectrum and a cosmological constant λ. Free parameters in the joint model are the mass density due to all matter (Ω), Hubble's parameter ($h = H_0/100$ km/sec), IRAS light-to-mass bias (b_{iras}) and the variance in the mass density field measured in an $8h^{-1}$ Mpc radius sphere (σ_8). For fixed baryon density $\Omega_b = 0.0125/h^2$ the joint optimum lies at $\Omega = 1 - \lambda = 0.41 \pm 0.08$, $h = 0.46 \pm 0.06$, $\sigma_8 = 0.65 \pm 0.10$, $b_{iras} = 1.26 \pm 0.06$ (marginalised error bars correspond to 95 percent confidence). For these values of Ω, λ and H_0 the age of the universe is ~ 18.7 Gyr.

6 Discussion

We have shown some recent studies of galaxy surveys, and their cosmological implications. New measurements of galaxy clustering and background radiations can provide improved constraints on the isotropy and homogeneity of the Universe on large scales. In particular, the angular distribution of radio sources and the X-Ray Background probe density fluctuations on scales intermediate between those explored by galaxy surveys and CMB experiments. On scales larger than $300h^{-1}$ Mpc the distribution of both mass and luminous sources satisfies well the 'Cosmological Principle' of isotropy and homogeneity. Cosmological parameters such as Ω therefore have a well defined meaning. With the dramatic increase of data, we should soon be able to map the fluctuations with scale and epoch, and to analyze jointly LSS and CMB data.

Acknowledgments

I thank my collaborators for their contribution to the work presented here. I also thank the conference organizers for the stimulating and enjoyable meeting, and acknowledge the hospitality of Tokyo University, where this paper was written.

References

Baleisis, A., Lahav, O., Loan, A.J., Wall, J.V. (1998): MN, in press, astro-ph/9709205
Baugh C.M., Efstathiou G., (1994): MN, **267**, 323
Boldt, E. A. (1987): Phys. Reports, **146**, 215
Cress C.M., Helfand D.J., Becker R.H., Gregg. M.D., White, R.L. (1996): **473**, 7
Dekel, A., (1994): ARAA, **32**, 371
Fabian, A. C., Barcons, X. (1992): ARAA, **30**, 429

Folkes, S., Lahav, O., Maddox, S.J. (1996): MN, **283**, 651
Gawiser, E., Silk, J. (1998): submitted to Science
Gunn, J.E., Weinberg, D.H. (1995): in *Wide-Field Spectroscopy and the Distant Universe*, eds. S.J. Maddox & A. Aragon-Salamanca, World Scientific,
Hamilton, A.J.S. (1997): review to appear the *Ringberg Workshop on Large-Scale Structure*, in Hamilton, D. (ed.), Kluwer Academic, Dordrecht, astro-ph/9708102
Kaiser N. (1984): ApJ, **284**, L9
Kochanek, C.S. (1996): ApJ, **466**, 638
Lahav O., Piran T., Treyer M.A. (1997): MN, **284**, 499
Magliocchetti, M., Maddox, S.J., Lahav, O., Wall, J.V. (1998): submitted to MN, astro-ph/9802269
Perlmutter, S., et al. (1998): Nature, **391**, 51
Strauss M.A., Willick J.A. (1995): Phys. Rep., **261**, 271
Treyer, M., Scharf, C., Lahav, O., Jahoda, K., Boldt, E., Piran, T. (1998): submitted to ApJ, astro-ph/9801293
Webster, A.M., Lahav, O., Fisher, K.B. (1997): MNRAS, **287**, 425
Webster, M., Hobson, M.P., Lasenby, A.N., Lahav, O., Rocha, G., Bridle, S. (1998): submitted to ApJ Lett, astro-ph/9802109
White, S.D.M., Navarro, J.F., Evrard, A.E., Frenk, C.S. (1993): **366**, 429
Wu, K.K.S., Lahav, O., Rees, M.J. (1998): submitted to Nature, astro-ph/9804062

Discussion

Bland-Hawthorn: Are you concerned about dust in our galaxy for large-scale surveys?

Lahav: Galactic extinction is a problem for quantitative optical studies of large-scale structure at galactic latitude $|b| < 30°$, although galaxies were detected at the zone of avoidance, even at $b = 0°$. Surveys of galaxies in HI or NIR are much less affected by the galaxy.

Ekers: Radio source counts are powerful for measuring large-scale structure because the radio luminosity function is very broad. However the large-scale clustering depends on small differences between very large numbers of objects. Very small differences and non-linearity in the flux scales and detection biases become important. These are non-trivial effects depending on survey resolution and source-detection and parameterising techniques. Future surveys will have to pay much more attention to survey techniques to improve this situation.

Lahav: I agree.

Mapping the Dark Matter with Weak Gravitational Lensing

Peter Schneider

Max-Planck-Institut für Astrophysik, Postfach 1523, D-85740 Garching, Germany

Abstract. The developments summarized with the name "weak gravitational lensing" have led to exciting possibilities to study the (statistical properties of the) dark matter distribution in the Universe. Concentrating on those aspects which require deep wide-field imaging surveys, I will describe the basic principles of weak lensing and discuss its future application (a) to determine the statistical properties of the dark matter halos of individual galaxies, (b) to measure the power spectrum of the matter distribution in the Universe, (c) to measure the bias parameter and investigate its dependence on scale and/or redshift, and (d) to detect mass (dark) concentrations from their mass properties alone.

1 Introduction

The distortion of the images of background galaxies by the tidal gravitational field of the intervening mass distribution can be used to investigate individual mass concentrations, or the statistical properties of the matter between us and those redshifts where the faint galaxies are located. Clusters of galaxies are both sufficiently massive, and sufficiently concentrated, that this weak lensing effect can be used on individual clusters to determine their projected two-dimensional mass distribution (Tyson, Valdes & Wenk 1990, Kaiser & Squires 1993), and about a dozen clusters have been mapped with weak lensing techniques (e.g., Fahlman et al. 1994; Smail et al. 1995; Squires et al. 1996; Seitz et al. 1996; Luppino & Kaiser 1997; Hoekstra et al. 1998). Mass concentrations which are less massive than clusters (like galaxies and groups of galaxies), or which are much less concentrated (like filaments, or more general, the LSS), cannot be investigated individually with weak lensing techniques, but their statistical properties can be inferred by suitably combining the effects of samples of objects.

In the last few years, tremendous progress has been made in the investigation of weak lensing. Observational progress was made possible by the development of optical imaging instruments with extremely good image quality and/or wide-angle capability. These instruments made it possible to obtain images of thousands of faint galaxies (where faint means $I \sim 24.5$, where the number density of galaxies is about 30 per square arcminute). Whereas – by definition – each of these images is only weakly affected by lensing, their shear number can be used to beat down statistical noise to a degree that lensing signals can be detected. Theoretical progress has been made to construct observing strategies and statistical methods to obtain quantitative results from such imaging campaigns. And on the interface of theory and observations, new methods have been developed to measure an 'ellipticity' of a faint galaxy from CCD data, which is affected by (in general anisotropic) PSF smearing (Bonnet & Mellier 1995; Kaiser, Squires & Broadhurst 1995). As an example, the shear caused by the cluster Cl 0024 has been measured to distances larger than 1 Mpc from the cluster center, using 'conventional methods' (Bonnet, Mellier & Fort 1994), as well as the so-called auto-correlation method (van Waerbeke et al. 1997).

I shall briefly describe the current status of several aspects of weak lensing, with particular emphasis on applications to deep and wide-field optical imaging surveys.

2 Tidal Distortion of Galaxy Images

A gravitational lens provides a map from the observer's sky to the undistorted sky, caused by the gravitational field of the deflector. The deflection angle is determined by the mass distribution of the lens, and the mapping from the source to the observable images depends in addition on the redshifts of the sources and the lens. Light bundles are not only deflected as a whole, but due to the tidal component of the gravitational potential, they are distorted. The image of a circular source will, to first order, be an ellipse. Thus, if all background galaxies were circular, the observable ellipticity of galaxy images would provide a direct means to measure the local tidal gravitational field of the lens. Since galaxies are intrinsically not circular, this simple method is impractical. However, since it can be assumed that the orientation of galaxies is random, a tidal field would cause a preferred orientation of the images. The distortions are such that galaxy images are preferentially aligned tangent to the 'mass center', such as is clearly visible for the giant luminous arcs.

Denote by $\epsilon^{(s)}$ the complex intrinsic ellipticity of a galaxy (defined in terms of its second order brightness moments), and let ϵ be the corresponding complex ellipticity of the observed image. The variable ϵ is defined such that for an image with elliptical isophotes of axis ratio r and orientation φ of the major axis relative to a fixed reference direction, $\epsilon = (1-r)/(1+r)e^{2i\varphi}$. The locally linearized gravitational lens equation yields the transformation between $\epsilon^{(s)}$ and ϵ, which for the case of weak distortions reads

$$\epsilon = \epsilon^{(s)} + \gamma , \qquad (1)$$

and γ is the shear or the tidal gravitational field of the lens (for definition, see, e.g., Schneider, Ehlers & Falco 1992) constructed from the trace-free part of the Hessian of the gravitational potential. The random orientation of intrinsic ellipticities then implies that the expectation value of ϵ equals γ, and the noise of this estimator is determined by the width of the intrinsic ellipticity distribution: to measure a shear with precision $\Delta\gamma$, one needs $N \sim \sigma_\epsilon^2/(\Delta\gamma)^2$ galaxy images. Therefore, in order to measure a weak shear signal, N must be large, which can be achieved by either increasing a local 'smoothing scale', or by increasing the number density of galaxy images, that is, by taking very deep images.

3 Galaxy-Galaxy Lensing

The first attempt to detect a weak lensing signal was performed by Tyson et al. (1984) where they searched for a coherent alignment of faint (and thus presumably high-redshift) galaxies relative to the line connecting them with the closest bright (presumably foreground) galaxy. The strength of the expected signal depends mainly on the mean mass of the foreground galaxies within the typical separation of pairs used in the investigation. Tyson et al. failed to obtain a significant lensing signal, and obtained an upper bound on the velocity dispersion σ_* of an L_* galaxy; this bound was revised upwards later by Kovner & Milgrom (1987).

It took more than a decade before another serious attempt in measuring galaxy-galaxy lensing was made (and published): Brainerd, Blandford & Smail (1996) studied this effect on a single deep field of about 9 arcminutes sidelength, and they obtained a statistically significant result. The main differences between these two works are the difference between CCDs and photographic plates, and the considerably better seeing in the latter data set. Brainerd et al. selected 439 galaxies with $20 \leq r \leq 23$ ('foreground galaxies') and 506 ('background') galaxies with $23 \leq r \leq 24$. Although they have detected many more galaxies fainter than $r = 24$, the measurement of their ellipticities is not accurate enough to allow quantitative conclusions. Selecting an annulus with $5'' \leq \theta \leq 34''$, their 3202 foreground-background pairs showed the effect of a preferential tangential alignment of the background galaxy images relative to the direction to the foreground galaxies, at the 99.9% confidence level. Further detections of galaxy-galaxy lensing have been made from the Medium Deep Survey (Griffiths et al. 1996) and the HDF (Dell'Antonio & Tyson 1996; Hudson et al. 1997); owing to the small angular scale of the latter, the detection of galaxy-lensing in the HDF on scales larger than a few arcseconds was possible only by using photometric redshift estimates to distinguish foreground from background galaxies.

The results from these studies have confirmed that galaxies have a dark matter halo, with a typical velocity dispersion of $\sigma_* \approx 220$km/s. More ambitious than the mass scale is an estimate of the size of the dark halo. Whereas the current results do not put strong constraints on the halo size (the characteristic size of an L_*-galaxy is $s_* \gtrsim 20h^{-1}$kpc), it is only a matter of increasing the sample size before more stringent constraints on the halo size can be obtained (Schneider & Rix 1997). Photometric redshift estimates increase the accuracy of the mass and size estimates significantly, and progress in that field will immediately reflect onto galaxy-galaxy lensing. More ambitious still is a comparison of the halo size of field galaxies with that of cluster galaxies (Natarayan & Kneib 1997; Geiger & Schneider 1998a) which would provide constraints on the amount of halo stripping during cluster formation and evolution. Preliminary results (Natarayan et al. 1997; Geiger & Schneider 1998b) suggest that cluster galaxies are 'smaller' than field galaxies, but for significant results, larger samples need to be considered and in principle, the data is all there: e.g., for about 10 lensing clusters WFPC2 imaging data are available.

In the future, the use of galaxy-galaxy lensing will be restricted only by the degree to which systematics of image ellipticity measurements can be controlled. Since even a single wide-field image contains ten thousands of foreground-background pairs with angular separation below, say, 20″, statistical uncertainties can be beaten down to virtually zero, which allows attempting the determination of further parameters of the mass distribution of galaxies, such as the 'Tully-Fischer' index and its evolution with redshift, the scaling of halo size with galaxy luminosity etc.; a first attempt, using the HDF, has been published by Hudson et al. (1997).

4 Cosmic Shear: Lensing by the Large-Scale Structure

Mass distributions on scales larger than clusters are assumed to be too diffuse for them being detectable individually with weak lensing techniques: they are weak lenses indeed. On the other hand, these large-scale structures will impose a distortion field on observable galaxy images. As has been shown by several groups (e.g., Blandford et al. 1991; Miralda-Escudé 1991; Kaiser 1992) the two-point statistics of the galaxy ellipticities (such as ellipticity two-point correlation function, or the rms ellipticity in fields of given size) are related to a projection of the power spectrum of the LSS, weighted by geometrical factors containing the lens efficiency as a function of redshift, and the source redshift distribution. Hence, for each cosmological model one can predict the expected ellipticity two-point statistics which can then be compared with observations. On large angular scales, the linear evolution of the power spectrum suffices to predict the cosmic shear quantitatively, whereas on smaller scales, the non-linear evolution has to be taken into account (e.g., Jain & Seljak 1997).

Cosmic shear is, besides CMB measurements, the cleanest probe of the LSS, since it makes no assumption on the relation between mass and light. Whereas the CMB can probe the LSS on comoving scales above about 10 Mpc, cosmic shear is sensitive also to much smaller scales; in particular, it can probe the non-linear evolution of the power spectrum.

Our ability to measure cosmic shear depends on whether it is possible to measure a shear signal of about one percent or less. Of course, this can only be done at excellent sites and in good conditions. Provided the image distortion of an instrument is known, and the PSF varies smoothly over the field, there is no reason why such a small shear should not be measureable. In three small fields of 2′, a significant shear of about 2.5% rms has been measured (Schneider et al. 1998a) using data from the SUSI imager at the ESO-NTT (Fort et al. 1996). Whereas this result cannot be used to obtain cosmological constraints, due to the large error bars (which are dominated by 'cosmic variance'), it is reassuring that such measurements can be made with existing instrumentation. A more detailed discussion on cosmic shear can be found in Mellier et al., this volume.

Several projects to measure cosmic shear on various angular scales are now underway; amongst them are the MEGACAM project and its precursor with the UH12K at CFHT (see Mellier et al., these proceedings) to measure cosmic shear on a wide-angular scale, an ESO Key Program (PI: Y. Mellier) which will determine cosmic shear on scales of a few arcminutes, and the analysis of HST/STIS parallel program images which we are currently carrying out with colleagues from the ST-ECF. Owing to the small pixel scale of STIS (0.05 arcsecs) and its high sensitivity compared to WFPC2, combined with a superb image quality (the PSF appears to be 'round' to within about 1%), these STIS images will most likely allow a secure measurement of the rms shear on an angular scale of about 50 arcsecs. Since on such small scales non-linear effects in the large-scale structure are dominant, higher-order statistics of the measured shear can be used to discriminate between various cosmological models, by comparing the observed distribution with the simulations performed using ray-tracing in very high resolution N-body simulations for the LSS (B. Jain et al. in preparation).

5 Detection of (Dark) Matter Concentrations

The number density and evolution of clusters of galaxies provides a most important cosmological probe; since structure formation occurs later in high-density universes, the abundance of clusters at intermediate and high redshifts compared to the present day abundance has the potential for measuring the cosmological parameters Ω_0 and Λ_0 (see, e.g., Eke, Cole & Frenk 1996, and references therein). An impressive demonstration of how lensing can discriminate between various cosmologies has been given by Bartelmann et al. (1998); using cluster models from high-resolution N-body simulations, they have shown that in a high-Ω_0 universe the expected number of giant luminous arcs is smaller than the ones already observed, thus excluding a cluster-normalized Einstein–de Sitter universe at high significance.

One of the main difficulties in using clusters as cosmic probes is their selection procedure: clusters (assumed to trace dark matter halos fairly) are selected by their optical or X-ray properties. In order to relate these samples to the results of cosmological simulations, one has to provide the missing link between mass and radiation properties, which involves complicated physics (e.g., star formation, galaxy merging processes; gas stripping; shock heating etc.). A much cleaner comparison with theoretical predictions would be possible with a mass-selected sample of halos. This, in fact, is feasible with weak lensing techniques: a massive halo induces an alignment signal on the ellipticity of background galaxies which can be detected. This idea has been formalized and shown that halos with velocity dispersion in excess of $\sim$ 600km/s can be detected with high statistical significance on deep optical images (Schneider 1996). Indeed, the cluster MS1512 has a velocity dispersion of about 600km/s and shows up very clearly in a weak lensing analysis of WFPC2 images (Seitz et al. 1998). As reported by Luppino & Kaiser (1997), at least one cluster has been selected from its induced shear field on a wide-field image. The current bottleneck for a systematic search for such mass concentrations is the availability of high-quality deep wide-field images and the ability to reduce them in due time.

The lens signal of these mass concentrations contains no information about the redshift of the halo. Given that the preferred lens redshift for this method is $z \sim 0.4$, one should be able to identify these clusters from an excess of galaxies – or one may find truly 'dark' mass concentrations, with very little optical light. In the first case, the redshift distribution of the selected halos provides a clean cosmological probe for structure evolution, whereas in the second case one might discover the tail of the mass-to-light ratio distribution of clusters. If there exists indeed a population of dark halos, the current normalization of the power spectrum of the LSS may need revision.

We have investigated the statistics of dark halos which are detectable by weak lensing searches (Kruse & Schneider, in preparation). Using a variety of cosmological models and power spectrum normalizations, and combining the halo volume number density as a function of mass, as predicted by the Press-Schechter theory, with the 'universal' halo density profile obtained from numerical simulations (Navarro, Frenk & White 1996, 1997), we find that the abundance of mass-detected halos and their redshift distribution depends indeed very sensitively on Ω_0 and λ_0. For example, the abundances of mass-selected dark halos in cluster-normalized open and EdS models differ by about a factor 5, and a search over a few square degrees should be able to distinguish between those models.

6 Cosmic Shear and Biasing

Wide-field surveys for cosmic shear can be used to reconstruct the projected density of the LSS, using methods used for cluster mass reconstructions (an example of this is provided in Mellier et al., these proceedings). This map can then be correlated with the distribution of galaxies in the same area. If these galaxies are brighter than those from which the shear is estimated, they are presumably in the foreground, at least on average, and therefore at redshifts similar to those where the bulk of the lensing signal is due to. Provided galaxies trace the underlying dark matter distribution, they should be correlated with the projected mass map. This correlation was calculated by Kaiser (1992), Sanz et al. (1997), and Dolag & Bartelmann (1997), and the same statistics used for the dark halo search was applied to this correlation (Schneider 1998). As it turns out, this correlation, which is proportional to the bias factor, should be easier to observe than the cosmic shear as described above. On the other hand, a positive detection of this shear-foreground galaxy correlation would be a clear detection of cosmic shear; in fact this correlation is a first-order statistic for the cosmic shear, in contrast to the quadratic statistics mentioned before. However, in order to interpret this correlation quantitatively in terms of cosmic shear, the bias factor needs to be known. Since we are quite ignorant about this bias factor, in particular at medium redshifts, this method is less useful for a quantitative measurement of cosmic shear. As discussed in van Waerbeke (1998), the signal-to-noise ratio of this correlation is larger if non-linear evolution of the power spectrum is taken into account. He furthermore demonstrated that it is possible to study the scale-dependence of the bias factor if one can isolate a population of foreground galaxies with a relatively narrow redshift range – which indeed seems very feasible, given the recent advances of photometric redshift estimates (e.g., Connolly at al. 1995).

7 Conclusions

Weak lensing has matured from a theorist's imagination to a useful tool for cosmology. We are at the beginning of an era where this tool will be used extensively, owing to the availability of large format CCD cameras. Although the number of published results from observational weak lensing projects is still quite small – except for cluster mass reconstructions – the 'learning curve' of how to measure and interpret weak shear signals is steep; it should be kept in mind that the requirements weak lensing puts on image quality are unprecedented and have been largely unexplored up until a few years ago.

Accepting that the Universe is isotropic around us, there is nothing special about 'looking deep in the *Southern Sky*', compared to the Northern counterpart, but with the opening of the first UT of the VLT, the installation of the 8K-camera on the ESO 2.2-m, and the planned square degree camera on the 2.5-m VST at Paranal, extensive weak lensing observing programs from Southern observatories will become possible.

This work was supported by the "Sonderforschungsbereich 375-95 für Astro–Teilchenphysik" der Deutschen Forschungsgemeinschaft.

References

Bartelmann, M, Huss, A., Colberg, J.M., Jenkins, A. & Pearce, F.R. 1998, MNRAS, in press (also astro-ph/9709229).
Blandford, R.D., Saust, A.B., Brainerd, T.G. & Villumsen, J.V. 1991, MNRAS 251, 600.
Bonnet, H. & Mellier, Y. 1995, A&A 303, 331.
Bonnet, H., Mellier, Y. & Fort, B. 1994, ApJ 427, L83.
Brainerd, T.G., Blandford, R.D. & Smail, I. 1996 ApJ 466, 623.
Connolly, A.J., Csabai, I., Szalay, A.S., Koo, D.C., Kron, R.G. & Munn, J.A. 1995, AJ 110, 2655.
Dell'Antonio, I. & Tyson, J.A. 1996, ApJ 473, L17.
Dolag, K. & Bartelmann, M. 1997, MNRAS 291, 446.
Eke, V.R., Cole, S. & Frenk,C.S. 1996, MNRAS 282, 263.
Fahlman, G., Kaiser, N., Squires, G. & Woods, D. 1994, ApJ 437, 56.
Fort, B., Mellier, Y., Dantel-Fort, M., Bonnet, H. & Kneib, J.-P. 1996, A&A, 310, 705.
Geiger, B. & Schneider, P. 1998a, MNRAS, in press (also astro-ph/9707044).
Geiger, B. & Schneider, P. 1998b, MNRAS, submitted.
Griffiths, R.E., Casertano, S., Im, M. & Ratnatunga, K.U. 1996, MNRAS 282, 1159.
Hoekstra, H., Franx, M., Kuijken, K. & Squires, G. 1998, ApJ in press, also astro-ph/9711096.
Hudson, M.J., Gwyn, S.D.J., Dahle, H. & Kaiser, N. 1997, astro-ph/9711341.
Jain, B. & Seljak, U. 1997, ApJ 484, 560.
Kaiser, N. 1992, ApJ 388, 272.
Kaiser, N. & Squires, G. 1993, ApJ 404, 441.

Kaiser, N., Squires, G. & Broadhurst, T. 1995, ApJ 449, 460.
Kovner, I. & Milgrom, M. 1987, ApJ 321, L113.
Luppino, G. & Kaiser, N. 1997a, ApJ 475, 20.
Luppino, G. & Kaiser, N. 1997b, talk given at the "Workshop on Weak and cluster gravitational lensing", Ringberg Castle, Jan. 1997.
Miralda-Escudé, J. 1991, ApJ 380, 1.
Natarajan, P. & Kneib, J.-P. 1997, MNRAS 287, 833.
Natarajan, P., Kneib, J.-P., Smail, I. & Ellis, R.S. 1997, astro-ph/9706129.
Navarro, J.F., Frenk, C.S. & White, S.D.M. 1996, ApJ 462, 563.
Navarro, J.F., Frenk, C.S. & White, S.D.M. 1997, ApJ 490, 493.
Sanz, J.L., Martínez-González, E. & Benítez, N. 1997, MNRAS 291, 418.
Schneider, P. 1996, MNRAS, 283, 837.
Schneider, P. 1998, ApJ, in press (also astro-ph/9708269).
Schneider, P., Ehlers, J. & Falco, E.E. 1992, *Gravitational lenses*, Springer: New York.
Schneider, P. & Rix, H.-W. 1997, ApJ 474, 25.
Schneider, P., van Waerbeke, L., Mellier, Y., Jain, B., Seitz, S. & Fort, B. 1998, A&A in press (also astro-ph/9705122).
Seitz, C., Kneib, J.-P., Schneider, P. & Seitz, S. 1996, A&A 314, 707.
Seitz, S., Saglia, R.P., Bender, R., Hopp, U., Belloni, P. & Ziegler, B. 1998, MNRAS, in press (also astro-ph/9706023).
Smail, I., Ellis, R.S., Fitchett, M.J. & Edge, A.C. 1995, MNRAS 273, 277.
Squires, G. et al. 1996, ApJ 461, 572.
Tyson, J.A., Valdes, F., Jarvis, J.F. & Mills Jr., A.P. 1984, ApJ 281, L59.
Tyson, J.A., Valdes, F. & Wenk, R.A. 1990, ApJ 349, L1.
Van Waerbeke, L. 1998, A&A, in press (also astro-ph/9710244).
Van Waerbeke, L., Mellier, Y., Schneider, P., Fort, B. & Mathez, G. 1997, A&A 317, 303.

Weak Lensing with MEGACAM and the VLT

Yannick Mellier[1,2], Ludovic van Waerbeke[3,4], Francis Bernardeau[5], and Olivier Le Fèvre[6]

[1] IAP CNRS, 98 bis Boulevard Arago, 75014 Paris, France
[2] Observatoire de Paris, DEMIRM, 61 Avenue de l'Observatoire, 75014 Paris, France
[3] MPA Garching, Karl-Schwarzschild-Str. 1, 85740 Garching, Germany
[4] CITA, Mc Lennan Labs, 60 St George Street, Toronto, M5S 3H8, Canada
[5] SPhT CE Saclay, 91191 Gif sur Yvette Cedex, France
[6] LAS CNRS, Traverse du Syphon BP 8, 13376 Marseille Cedex 8

Abstract. In 2000, the wide field camera MEGACAM will be installed at CFHT and the VIRMOS spectrograph will start first observations on the VLT. These instruments will be dedicated to surveys. In this proceeding, we show how we can use MEGACAM+VIRMOS to map the mass distribution on large scales. Simulated projected mass density maps show that mass reconstruction from lensing inversion can recover the input power spectrum using MEGACAM images. The mass reconstruction process does not produce systematic error, even if an intrinsic ellipticity distribution is given to the galaxies. Indeed, the reconstructed projected mass map and projected power spectrum are remarkably similar to the original ones. We thus are confident that the superb images obtained with MEGACAM, combined with the redshift distribution of sources inferred from the VIRMOS survey, will be a unique tool for cosmology.

1 Cosmology with Gravitational Lenses

The analysis of gravitational distortion of distant background galaxies is one of the most promising way to probe directly the mass distribution in the Universe. It permits a comparison with the distributions of galaxies and gas, without any assumption on the dynamical stage of the deflectors, and thus permits to infer the total mass density of the Universe and to study how the amount of dark matter varies with the scale.

Blandford et al. (1991), Miralda-Escudé (1991) and Kaiser (1992) have first pointed out that weak lensing induced by large-scale structures depends on the power spectrum of density fluctuations. More recently, Bernardeau et al. (1997) have used the perturbation theory to analyze the variance and the skewness of the convergence for various cosmologies and various scales. They have shown that these two quantities can provide independently the slope of the power spectrum and the density parameter Ω. When extended to non-linear regime, the signal below 20 arc-minute scale is even higher than the linear prediction (Miralda-Escudé 1991, Jain & Seljak 1997).

Both for large and small scales, the expected amplitude of the cosmic shear ranges in a domain which is already reachable by the best ground-based telescopes. However, there are still many issues which require further developments.

First, the measurement of weak distortion induced by large scale structures demands high image quality and very large field of view. This has partly motivated the new design of the MEGACAM field corrector for CFHT. Second, the lensing inversion itself may be critical and could be seriously affected by the various sources of noise. Last, but not least, the variance and the skewness of the convergence strongly depends on the redshift distribution of the background sources. Spectroscopic and photometric redshifts are therefore indispensable. The VIRMOS survey done jointly with the MEGACAM surveys in the same areas will permit to have the calibration of photometric redshifts and the redshift distribution of the faint lensed galaxies with a high accuracy.

2 The MEGACAM Camera for CFHT

MEGACAM is a 16K×18K CCD camera which will be mounted at the CFHT prime focus. The project is a collaboration between HIA, CNRS, CEA, Obs. de Paris, LAT (Toulouse) and CFHT and includes three major components: the CCD camera, a new wide field corrector and TERAPIX, a data center for MEGACAM image processing (see Boulade et al. 1998 for more details).

CEA is responsible for the camera, with L. Vigroux as PI. It will be equipped with about 36 UV-sensitive thinned 2K×4K CCDs which will fully cover the prime focus field of view (1deg.×1deg.). The CCDs are not chosen yet, but the pixel-size will be close to 0.2 arc-sec/pixel in order to match the expected median seeing (0.6 arc-sec.).

The new wide field corrector has been designed by CFHT in order to reduce the optical distortion by a factor of three and to provide subarcsecond images over the 1.4 degree diameter field of view. The image quality will provide 80% encircled energy within 0.3 arc-second for a 1000 Åfilter width. The tip-tilt incorporated at the focal plane will guarantee a median seeing on the image of 0.6 arc-second from 0.38 μm to 1.1μm. With MEGACAM we then expect to have a perfect instrument for measuring weak shear.

Each MEGACAM image will contain 600 Mbytes. Since a significant fraction of CFHT will be devoted to wide field imaging surveys, we expect about 10 Tbytes of data from 2000 to 2005. The MEGACAM consortium has decided to promote the development of a data center, TERAPIX, which will be in charge of the design of the data reduction pipeline, the development of softwares and utilities and the processing and archiving of the images. TERAPIX is a collaboration which involves IAP, CEA, CFHT, DAO, CADC, LAT and CDS.

The design of MEGACAM combined with the TERAPIX data center will produce subarcsecond seeing surveys which should perfectly match the requirements for weak lensing. From the first estimate done by Bernardeau et al (1997), it will be possible to measure with a $\approx 15\%$ accuracy Ω and the slope of the power spectrum from observations covering 25 square-degrees. With MEGACAM, this can be completed in 25 nights.

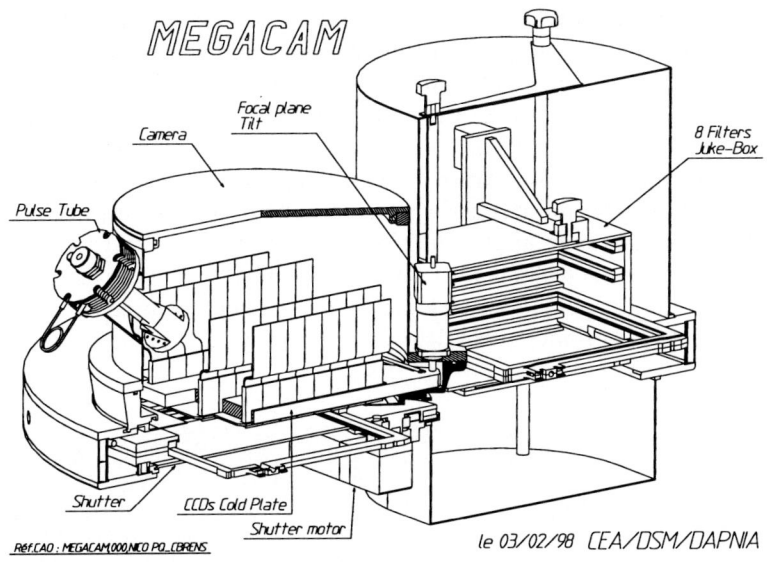

Fig. 1. Layout of the MEGACAM camera.

3 Recovering Mass Maps with MEGACAM

Bonnet & Mellier (1995), Kaiser, Squires & Broadhurst (1995) and van Waerbeke et al. (1997) have discussed in detail how to infer gravitational shear from optimized extractions of galaxy ellipticities. As far as measurement of ellipticities is concerned, it should be possible to measure shear as low as 1% from these techniques with MEGACAM, though there are still some concerns about systematics and PSF correction. However, none of these works have considered what could be the effect of the lensing inversion algorithm and of the sources of noises on the accuracy of the reconstructed mass density.

Van Waerbeke et al (1998) have analyzed in details whether the lensing inversion algorithm could propagate errors and bias the mass reconstruction. Starting from a given power-spectrum for density fluctuations and cosmological parameters, they used simple 2-Dimension Zeldovich approximation to produce projected mass-maps. The size of the simulation is 5×5 degrees, with super-pixel-size of $2' \times 2'$ in order to reproduce the area and the typical galaxy number-density of the MEGACAM deep survey. Once the mass map is generated, the convergence can be computed in oder to create shear-maps. The shear map can then be used in a reverse way, to produce projected mass density maps, obtained from lensing inversion, and to compute the observed power-spectrum. Van Waerbeke et al. have shown that when noise-free maps are used, the reconstructed mass map and the power spectrum are identical to the initial model. It demonstrates that the lensing inversion works remarkably well.

In a second step, they added noise to the mass maps by including the intrinsic ellipticity distribution of background sources as a noise component to the shear predicted for each super-pixel. The comparison between the noise-free mass maps model and the noisy reconstruction from lensing inversion is show in Fig. 2. One clearly see that both the projected mass density and the power spectrum can be recovered with a high accuracy. More details about the skewness and the variance for various cosmological models are discussed in Van Waerbeke et al (1998), in particular the weak lensing approximation, as well as the noise model given by Kaiser (1996) and Seljak (1997), works well. These results definitely demonstrate that, thanks to the image quality of MEGACAM, we can confidently argue that it will be possible to produce wide-field mass-maps with this instrument.

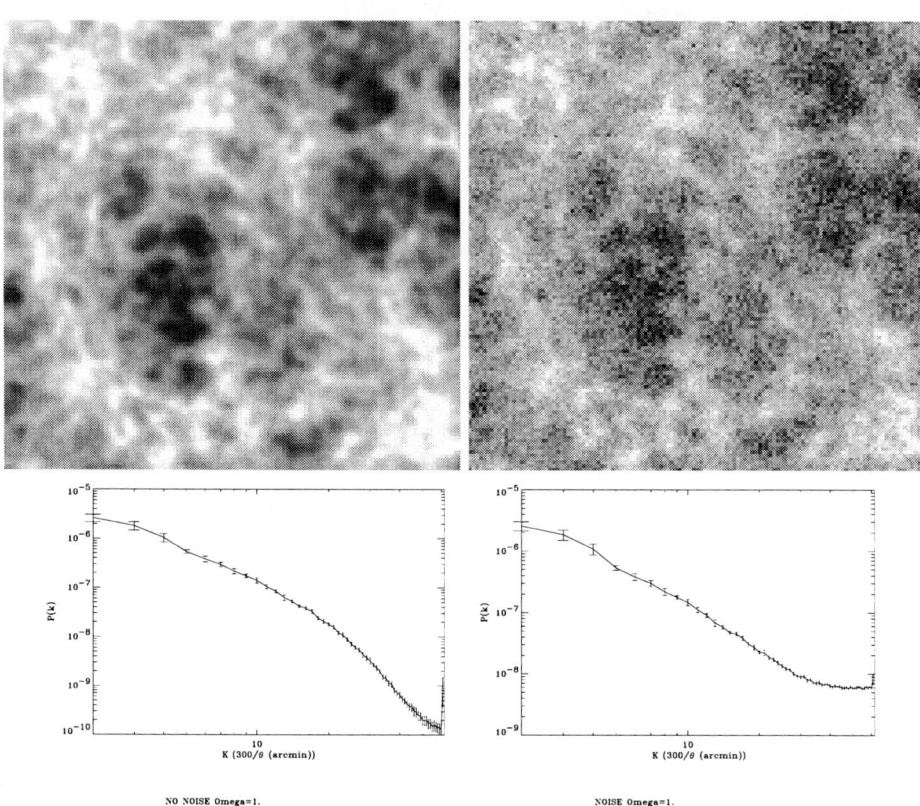

Fig. 2. Simulated mass maps obtained from a flat-Universe evolving using a 2-dimension Zeldovich approximation. The left-hand panels show noise-free simulations and the right-hand panels the reconstructed image and power spectrum once noise is included. The total area covers 25 square-degrees, with a pixel-size of 2 arc-minutes. Within each pixel the ellipticity of 200 galaxies have been averaged.

Fig. 3. View of the VMOS instrument mounted on the VLT platform.

4 The Redshift Distribution of the Sources with VIRMOS

Bernardeau et al. (1997) and Jain & Seljak (1997) emphasized that the variance and the skewness of the convergence does contain cosmological informations, but which strongly depend on the redshift distribution of the sources. It will be provided by the multi-object spectrographs VMOS and NIRMOS which are under construction for the VLT by a French-Italian consortium led by O. Le Fèvre (PI) and P. Vettolani (Co-PI) (Le Fèvre et al 1996).

VIRMOS will be composed of three units: VMOS, a visible (0.37 - 1 micron) MOS covering a total field of view of 4×7'×8', NIRMOS, a near-infrared (1 - 2 microns) MOS covering a total field of view of 4×7'×8' and an integral field spectrograph covering 1'×1' with 6000 micro-lenses. The multiplex gain of VMOS and NIRMOS will range between 800 to 170, depending of the spectral resolution. The VIRMOS consortium has decided to dedicate its guarantee time to a large survey which includes wide field imaging with CFH12K, MEGA-CAM, SOFI and NIRMOS, a deep spectroscopic survey of 10^5 galaxies down to $I_{AB} = 24$, and an ultra-deep small survey with the integral field spectrograph down to $I_{AB} = 26$. This VIRMOS survey will be done in coordination with the MEGACAM weak lensing survey and the XMM survey.

The main scientific goal of the VIRMOS survey is a comprehensive study of formation and evolution of galaxies and large-scale structures. As far as the weak lensing is concerned, the most important issue is the redshift distribution of galaxies. The photometric and spectroscopic data down to $I_{AB} = 26$ will permit to calibrate photometric redshift with UBVRIJK photometry, even for the faintest galaxies. We then are confident that the redshifts distribution of the background sources will not be an issue any longer.

5 Conclusion

The MEGACAM/VIRMOS/XMM survey is a perfect concept for wide field surveys. It combines the study of the mass, the galaxy and the hot gas distributions and their evolution with look-back time. This unique project should have a major impact in cosmology.

Acknowledgments

We thank P. Schneider for stimulating discussions and O. Boulade for detailed technical informations on MEGACAM. YM thanks the local organizing committee for financial support.

References

Bernardeau, F., Van Waerbeke, L., Mellier, Y. 1997, A&A 322, 1.
Blandford, R. D., Saust, A., Brainerd, T. G., Villumsen, J. V. 1991, MNRAS 251, 600.
Boulade, O., Vigroux, L., Charlot, X., de Kat, J., Borgeaud, P., Roussé, J.Y., Mellier, Y., Gigan, P., Crampton, D., SPIE Vol. 3355, "Astronomical Telescopes and Instrumentation", Kona Hawaii, March 1998.
Bonnet, H., Mellier, Y. 1995, A&A 303, 331.
Jain, B., Seljak, U. 1997, ApJ 484, 560.
Kaiser, N., 1992, ApJ 338, 272.
Kaiser, N., 1996, astr-ph/9609043.
Kaiser, N., Squires, G., Broadhurst, T. 1995, ApJ 449, 460.
Le Fèvre, O., Vettolani, P., Cuby, J.G., Maccagni, D., Mancini, D., Mazure, A., Picat, J. P., 1996, proc. Conference "Wide Field Spectroscopy", Kontizas, Kontizas, Morgan, Vettolani, Eds., Kluwer Academic Publishers, p.55.
Miralda-Escudé, J. 1991, ApJ 380, 1.
Seljak, U. 1997, astro-ph/9711124.
Van Waerbeke, L., Mellier, Y., Schneider, P., Fort, B., Mathez, G. 1997, A&A 317, 303.
Van Waerbeke, L., Bernardeau, F., Mellier, Y. 1998, preprint.

Discussion

Rocca-Volmerange: How many redshifts will be measured with VIRMOS on a MEGACAM field to deduce a complete structure of weak lensing?

Mellier: VIRMOS can obtain the redshift of all the galaxies up to $I_{AB} = 24$. VIRMOS can thus measure about 10^5 redshifts plus photometric redshifts of 10 times more using UBVRIJK photometric data obtained with MEGACAM, SOFI at the NTT and NIRMOS.

The Cosmological Uncertainty Principle

Christopher Fluke, Rachel Webster

School of Physics, The University of Melbourne, Parkville, Victoria 3052, Australia

1 Introduction

We are investigating light propagation through evolving cosmological dark matter distributions, in order to determine the magnification properties of the universe, particularly in the weak lensing limit.

On the largest scales, our Universe is homogeneous and isotropic, but on smaller scales the Universe is quite inhomogeneous. A bundle of light rays from a distant source will be gravitationally deflected by these inhomogeneities. Except for the rare occurrence of strong lensing events (multiple images), most light bundles will be weakly lensed, undergoing changes to the shape and area of the bundle's cross-section. The observed flux from a source will be magnified or demagnified *with respect to* a smooth Universe.

We refer to this lensing–induced dispersion as the **Cosmological Uncertainty Principle**, as it causes an intrinsic uncertainty in measurements of the flux from a standard candle source, which in principle can be used to place constraints on the cosmological parameters and the dark matter model.

2 Method

Our method for obtaining the probability distribution function, $P(\mu)$, of magnification μ, for a specific cosmological model (Ω_0, Λ_0, H_0) and initial dark matter spectrum ($P(k)$, σ_8) will be fully described in Fluke, Webster and Mortlock (1998). This is a challenging problem which has been investigated recently with complementary approaches by Wamsbganss, Cen and Ostriker (1996) and Premadi, Martel and Matzner (1997).

We create dark matter–only universes using the **Hydra** SPH N-body code (Couchman, Thomas and Pearce 1995), with output in simulation boxes of size $\sim 100 h^{-1}$Mpc. These boxes are placed sequentially out to the source redshift, representing the matter distribution on a line-of-sight through the universe. Particle positions in each box are projected onto a plane, and the multiple plane lens equation is used to calculate deflections by each particle (treated as point mass lens, but will be replaced by a more appropriate analytic solution).

Photon number is conserved by weak gravitational lensing, so that the magnification of a source is

$$\mu = \frac{\text{Angular image area}}{\text{Angular source area}}. \tag{1}$$

Our images are made from polygons of light rays with N_{vert} vertices, for which the image and source areas may be calculated directly. The accuracy of the calculated μ depends strongly on N_{vert} and the angular size of the images. Using $N_{\text{vert}} = 8$ provides a quadrupole component, which will allow calculations of source shapes.

We include the effect of all point mass lenses within a co-moving radius R_{co} around the location of a ray. The calculate magnification depends strongly on R_{co} and our initial investigation suggests that $R_{\text{co}} > 20 h^{-1} \text{Mpc}$ is necessary (Fluke et al. 1998).

3 Conclusions

We are developing a method with which to study in detail the weak lensing effects of cosmologically distributed dark matter. Our method is computationally less efficient than using a tree-code (Wambsganss, Cen and Ostriker 1996) or solving for a smooth background potential (Premadi, Martel and Matzner 1997), but:

- allows more flexibility in adding analytic lens models to represent strong lensing due to sub-cluster sized objects;
- gives more freedom in defining the geometry of the region from which we choose lenses, as we are not restricted to smoothing our particle distribution onto a grid; and
- enables us to investigate the effects of image/source geometry.

With this method, we can place limits on the cosmological parameters and the dark matter model, by performing a detailed statistical analysis on many numerically simulated model universes.

References

Couchman, H.M.P., Thomas, P.A., Pearce, F.R. (1995): Hydra: an Adaptive-Mesh Implementation of P^3M-SPH. Ap. J. **452**, 797–813
Fluke, C.J., Barber, A.J., Couchman, H.M.P., Mortlock, D.J., Thomas, P.A., Webster, R.L. (1998): *in preparation*
Fluke, C.J., Webster, R.L., Mortlock, D.J. (1998): *in preparation*
Premadi, P., Martel, H., Matzner, R. (1997): Light Propagation in Inhomogeneous Universes I: Methodology and Preliminary Results. astro-ph/9708129
Schneider, P., Ehlers, J., Falco, E.E (1992): *Gravitational Lenses* (Springer-Verlag, New York)
Wambsganss, J., Cen, R., Ostriker, J.P. (1996): Testing Cosmological Models by Gravitational Lensing: I. Methods and First Applications. astro-ph/9610096

Gravitational Lensing in the 2dF Galaxy Redshift Survey

Daniel J. Mortlock and Rachel L. Webster

School of Physics, University of Melbourne, Parkville, Victoria 3052, Australia

The 2 degree Field (2dF) galaxy redshift survey (GRS) will involve obtaining the spectra of $N_g \simeq 2.5 \times 10^5$ galaxies to a limit of $B_J = 19.5$ (Colless & Boyle 1997). As such it will provide a great deal of information about the galaxy population (*e.g.*, luminosity function, clustering properties) in the local ($\langle z_g \rangle \simeq 0.1$) universe. However, the spectral information obtained will not provide any direct information about the mass of the survey galaxies. By searching the galaxy spectra for the presence of background objects, in particular lensed quasars and lensed galaxies, some constraints can be placed on the mass of the survey population, as well as more accurate measurements of the individual lensing galaxies. This possibility first became apparent when the "Einstein Cross", Q2237+0305, was discovered; it was first detected as an anomaly in the lensing galaxy's spectrum (Huchra et al. 1985). The presence of the lensed quasar has allowed unprecedented determinations of the mass distribution of this system (*e.g.*, Schmidt, Webster & Lewis 1998). This paper is concerned with predicting the number of lensed quasars and galaxies that could be detected using only the 2dF GRS spectra as a primary source of data.

Kochanek (1992) calculated that $N_l \simeq 3$ lensed quasars should be discovered in a survey of comparable size to the 2dF GRS. This figure is strongly dependent on both the nature of the 2dF instrument – in particular the angular size ($\sim 2''$ diameter) of the spectrographs' fibres – and the conditions at the site – the seeing at the Anglo-Australian Observatory is typically $> 1''$. Including these in the calculation introduces some curious effects that could result in $N_l \simeq 10$ (Mortlock & Webster 1998). Uncertainties in the mass distribution of the survey galaxies and the nature of the spectral analysis to be used restricts the accuracy of this prediction.

Similarly, it is possible to predict the number of lensed (background) galaxies that will be detected. N_l will be greater than that for quasars as the number density of galaxies is so much greater, but they have less generic spectra. This implies that N_l will depend strongly on the actual algorithms used to search for superimposed (or "secondary") objects in the survey galaxies' spectra. Furthermore, it will be a function of the spectral "prominence" of these sources, or how easily they are detected when superimposed on the spectrum of a brighter survey galaxy. This can be quantified by a maximum magnitude difference, Δm. For quasars, $\Delta m \simeq 2$ (Kochanek 1992), and it is almost certainly less for galaxies, although it must be positive (Mortlock & Natarajan 1998). Nevertheless, a reasonable estimate for the frequency of galaxy-galaxy lensing can be obtained by assuming a standard, no-evolution model for the local galaxy population,

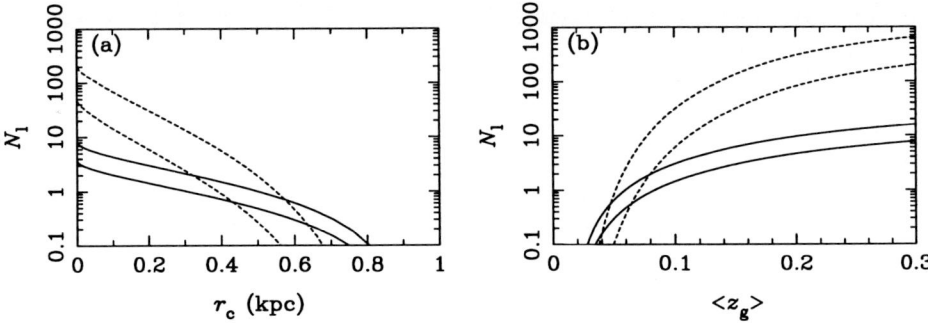

Fig. 1. The number of lensed objects, N_l is plotted as a function of: the typical core radius of the survey galaxies, r_c (a); and the average redshift of the survey galaxies, $\langle z_g \rangle$ (b). The solid lines gives the number of lensed quasars and the dashed lines the number of lensed (background) galaxies. The upper lines in each plot are for spectral "prominence" $\Delta m = 2$ and the lower line is for $\Delta m = 1$.

and modeling the deflectors as smoothed isothermal spheres (Hinshaw & Krauss 1987). This is enough to give the number of background galaxies within the radial caustics of the survey galaxies, but a number of further model assumptions are required.

Fig. 1 shows N_l for both quasars and galaxies as a function of both the typical core radius of the survey galaxies (a) and $\langle z_g \rangle$ (b). From this it can be seen that a potential source of both gravitational lenses and a method to constrain the mass (distribution) of the 2dF GRS galaxies exists. Similar techniques will be relevant to other large redshift surveys, and these statistics can also be used to constrain the distribution of the background galaxies, thus placing limits on their luminosity evolution (Mortlock & Natarajan 1998).

References

Colless, M.M., Boyle, B. (1997): Redshift Surveys with 2dF. Highlights in Astronomy, **11**, in press

Hartwick, F.D.A., Schade, D. (1990): The Space Distribution of Quasars. ARA&A, **28**, 437–489

Hinshaw, G., Krauss, L.M. (1987): Gravitational Lensing by Isothermal Spheres with Finite Core Radii – Galaxies and Dark Matter. ApJ, **320**, 468–476

Huchra, J.P., Gorenstien, M., Kent, S., Shapiro, I., Smith, G., 2237+0305 – A New and Unusual Gravitational Lens. AJ, **90**, 691–696

Kochanek, C.S. (1992): Gravitational Lenses in Redshift Surveys. ApJ, **397**, 381–389

Mortlock, D.J., Natarajan, P. (1998): Gravitational Lensing in Redshift Surveys. MNRAS, in preparation

Mortlock, D.J., Webster, R.L. (1998): Using the 2dF Galaxy Redshift Survey to Detect Gravitationally-Lensed Quasars. MNRAS, in press

Schmidt, R.W., Webster, R.L., Lewis, G.F. (1998): Weighing a Galaxy Bar in the Lens Q2237+0305. MNRAS, in press

A Weak Gravitational Lensing Cluster Survey

G. Squires, P. Rosati, J. Silk, T. Broadhurst

[1] University of California, Berkeley, CA 94720, USA
[2] European Southern Observatory, Karl-Schwarzschild-Str 2,
D-85748 Garching bei München, Germany

Abstract. We are assembling deep, homogeneous optical observations of a statistically complete subset of clusters identified in the ROSAT Deep Cluster Survey. The mass of the clusters is determined via weak gravitational lensing, and correlated with the X-ray and optical properties. This direct probe of the cluster mass distribution will constrain the dynamical and thermodynamical state and evolution of the cluster population.

1 Motivation

Much of the weak lensing cluster studies performed to date have been done in an inhomogeneous fashion; the clusters studied were chosen for their exceptional properties (e.g., very bright in the X-ray, exhibit dramatic strong lensing features, etc), and the observations were done to varying depth and covered a range of physical scales at the cluster redshift. Extracting universal properties from the current observations is, therefore, difficult.

Significant progress has recently been made in X-ray selected surveys of galaxy clusters over a large redshift range. In particular, Rosati *et al.* (1997) have constructed a large X-ray selected cluster sample from a serendipitous search in deep ROSAT-PSPC pointed observations (RDCS). One intriguing result arising from the RDCS is that no evolution is noticed in the X-ray luminosity function for luminosities $\lesssim L^*$ out to $z \simeq 0.8$. However, there is considerable ambiguity in the interpretation of this result, mainly since structure formation theories make clear predictions for the *mass function*, $n(m)$, while additional physics and assumptions are required to relate the luminosity to the mass. At any given luminosity, the inferred mass is uncertain by factors of $\simeq 2-4$. The key to resolving this issue is to directly determine the relation between the cluster mass and X-ray luminosity as a function of redshift.

2 The Survey and Science Goals

We are currently imaging clusters drawn from the RDCS, adopting the following strategy: we are observing a subset ($\simeq 10$ clusters) of the RDCS in the low-z range ($0.2 \leq z \leq 0.4$) at 4m class telescopes (KPNO, CTIO, CFHT, WIYN). To create a homogeneous set of observations, we perform $\simeq 2$ hr integrations in V/I to be complete to $I \simeq 24$. A complimentary cluster sample at high-z ($0.6 \leq z$) is being observed at the Keck 10m telescope. We do $\simeq 2$ hr integrations in V/I to be complete to $I \simeq 25.5$.

The science goals of this survey include: **1)** map the dark matter in each cluster through an analysis of the weak gravitational shape distortion of the background galaxies. Calculate generic quantities such as M/L, and f_b, and compare the total mass determined from the X-ray and lensing observations, **2)** correlate the mass determined from lensing with the cluster X-ray luminosity/temperature in low/high redshift shells and thereby break the degeneracy among theoretical models explaining the lack of evolution in the RDCS, and **3)** constrain the faint galaxy redshift/magnitude distribution, $n(m,z)$, from the strength of the lensing signal in the high-z observations.

3   Current Status

In Figure 1 we display the L_x vs M relation for all of the clusters for which we have published lensing observations. The drawback with this sample is the inhomogeneous selection of clusters. To properly probe the thermodynamical state we need to more carefully select targets with a well known selection function – the RDCS provides such a target list. We currently have data for $\simeq 7$ RDCS clusters, with the analysis ongoing.

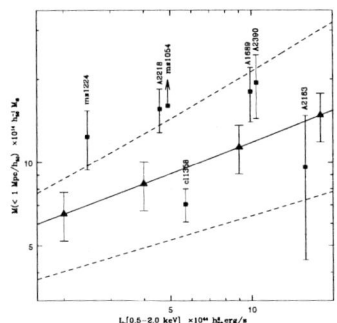

Fig. 1. The L_x vs M relation for clusters with existing lensing observations. The solid line shows the prediction from the local Edge & Stewart (1991) relation, assuming a SIS model for the mass, while the dashed lines show the theoretical uncertainty in the L_x vs M relation (Kitayama & Suto 1997). The squares show existing observations, while the triangles are a simulation for clusters following the nearby L_x vs M relation, with the errorbars appropriate for the high S/N observations we obtain in this survey.

References

Edge, A., & Stewart, G. 1991, MNRAS, 252, 414
Kitayama & Suto 1997, ApJ, 490, 556
Rosati, P. et al. 1997, ApJL, 492, L21

Surveys with the BTC Mosaic CCD Camera at the Blanco 4m Telescope

Malcolm G. Smith

Cerro Tololo Interamerican Observatory, Casilla 603,
La Serena, Chile

Abstract. A series of surveys with the Big Throughput Camera (Wittman et al., 1998) is now under way. Most are extragalactic, including studies of clusters of galaxies, gravitational lensing, distant supernovae, quasars and large-scale structure at $z \sim 1$; the studies nearer to home include investigations into faint dwarf stars and moving objects such as asteroids. Some of these surveys are being specifically co-ordinated with parallel surveys in infrared (JHK) passbands.

The various projects are described, along with their current status. The most immediate result has been the discovery of significant numbers of high-redshift supernovae by two of the teams using this wide-field camera-telescope combination.

1 Introduction — the BTC Mosaic CCD Camera

Wittman et al. (1998) have recently summarized the highly successful first year of operation of the Big Throughput Camera (BTC) as a user instrument on the Blanco 4m telescope at Cerro Tololo Interamerican Observatory (CTIO).

The camera format is a 2x2 array of thinned, broadband, antireflection-coated, 2048x2048 SITe CCDs with 760-pixel gaps between CCDs. At the f/2.8 prime focus of the Blanco telescope, the 24-micron pixels subtend 0.43 arcsec on the sky. The array thus provides a 34.8 arcmin field of view with 5.5 arcminute gaps. The pixels are well matched to the seeing at CTIO.

A single exposure requires roughly 36 MB of disk space, and a good 3-4 night observing run, with flats and biases, totals 15-20 GB. Read noise is about 7e-rms, so that broadband observations longer than a few seconds are background limited. More detailed hardware descriptions are given in Tyson et al. (1992a,b). Information for interested observers may be found at:

http://www.astro.lsa.umich.edu/btc/btc.html.

Illustrations of the BTC geometry and images from combined survey frames are provided in the article by Wittman et al. A combined equivalent total exposure of about 100 minutes over a field 47 arcminutes across yields about 150,000 objects at $\geq 3\sigma$ at high galactic latitude.

2 Description of Individual Surveys in Progress with the BTC at CTIO

In this section, we will begin with a short description of a survey for objects in the solar system, then move on out via surveys of the Milky Way, the Local Group and so on to the most distant reaches of the universe.

2.1 Census of the Kuiper Belt (Beyond Neptune)

Jones, Bernstein and Malhotra are surveying a square degree of the Ecliptic with sensitivity expected to be sufficient to reveal ~ 50 new Kuiper-Belt Objects (KBO's) to $R \sim 27$ in any orbit out to $> 70 AU$. In this way they expect to be able to test for the origins of short-period comets and for the effects of the giant planets on the outer solar system. Their survey will allow them to probe beyond Pluto to where KBOs may still lie unperturbed by the gas-giant planets.

2.2 Brown Dwarfs in Two Young, Open Clusters

Stauffer, Barrado and Navascues are studying the initial mass function for stars in NGC2516 and IC2391, with special emphasis on isolating brown dwarfs. Their survey is expected to reach magnitude limits of: B=25, R=24.5 - 25 and I=23.5 - 24. Young brown dwarfs are expected to be hotter and more luminous than their counterparts in older clusters, and therefore more likely to be accessible to study at optical wavelengths. The above survey limits correspond to brown-dwarf masses between 0.02-$0.03 M_\odot$ in IC2391 and 0.04-$0.05 M_\odot$ in NGC2516.

2.3 Very-Low-Mass Halo-Population Stars as Dark Matter

Boeshaar, Tyson, Dell'Antonio and Kirkpatrick are carrying out deep B_j R and I' multicolor imaging to locate low-metallicity halo subdwarfs. If the halo baryonic fraction exceeds 10%, they expect to find more than 1300 M-type halo subdwarfs in their survey. Such objects can barely be detected by WFPC2 on the HST.

2.4 Outskirts of the Sagittarius Dwarf Spheroidal Galaxy

We now move out from the Milky Way. Mateo, Morrison and Olszewski propose to follow up on recent studies which have identified main-sequence stars located 10-19 degrees from the center of this recently discovered galaxy in Sagittarius. Stars located so far from the previously-known boundaries of this galaxy are thought to correspond to tidal debris from recent close passages of the dwarf spheroidal past the Milky Way. The survey with the BTC is expected to reach magnitude limits of V=I=24 and cover about 60 sq. deg. plus control fields. Among the many goals of this survey is that of constraining the total mass and extent of the Sagittarius dwarf - via measurements of the density of stars in its outer parts.

2.5 Masses of Galaxy Halos via Gravitational Lensing

As discussed in Peter Schneider's talk at this conference and elsewhere, massive halos of dark matter around bright foreground galaxies ($z \sim 0.1$) should produce a coherent shear of about 3% in the shape of distant background galaxies. D. Smith and G. Bernstein propose to measure this weak gravitational lensing effect of bright galaxies at known redshifts to test for the presence of such massive halos. They expect to be able to measure the halo mass within $50h^{-1}$ to a precision of $3.0 \times 10^{11} M_\odot$. They will do this by determining the average shape of about 2100 galaxies with $R < 23.5$ behind 275 foreground galaxies.

2.6 The Faint End of the Luminosity Function in Poor Galaxy Clusters

Marzke and Hudson wish to study the link between galaxy formation and the local environment. In their survey, they are using photometric redshifts from 4-color photometry to identify members of poor galaxy clusters; their approach is expected to ease extraction of faint member galaxies from background noise. The are testing their procedures first on the poor clusters MKW3 and MKW4, which already have redshifts determined from the brightest cluster galaxies. With their survey they expect to reach down to $M_R = -10$, just brighter than the brightest globular clusters.

2.7 Weak Lensing Distortions as a Constraint on Dark Matter Properties (Especially Mass) in Three Nearby, Merging Clusters of Galaxies

Mohr and Fischer propose to test structure formation models via observations of the distribution of gas, galaxies and mass in three, nearby, merging clusters of galaxies. They propose to use X-ray images (as discussed elsewhere at this conference) to constrain the gas, photometry and redshifts to locate the cluster galaxies and weak lensing distortions to constrain the mass. The collisional nature of the cluster gas leads it to behave differently than collisionless components (such as dark matter or galaxies). Numerical simulations predict that after mergers, clusters can exhibit multiple peaks in the collisionless component and a distorted, single-peaked gas distribution. They expect to use only background galaxies with $22 < B_j < 25.5$ in order to avoid cluster contamination at brighter magnitudes.

2.8 Cosmology from Weak Lensing Data on Normal Clusters of Galaxies

Tyson, Fischer, Dell'Antonio and Pildis are seeking to use weak lensing of background galaxies to measure mass distributions in a sample of 10 galaxy clusters of similar x-ray luminosity and redshift and compare these with the corresponding predictions for model clusters produced in N-body simulations of various cosmologies. Specific comparisons will be made of M/L, the radial density gradient and the mass anisotropy of each cluster.

2.9 Cosmological Parameters from SNe Ia at $z > 0.2$

The most active field of research with the BTC at CTIO is in the discovery of distant supernovae. As explained in the talk by Bruno Leibundgut at this conference, two major groups have been using high-redshift type Ia supernovae as standard candles to measure relative distances. Many research papers, reviews, IAU circulars and press releases have been written recently on the results from these groups concerning estimates of the Hubble constant and possible acceleration of the universal expansion. Key factors in this research have been the discovery of the SNe at CTIO coupled with the use of a method developed (also at CTIO) by Mark Phillips to cut the scatter in estimates of the absolute magnitudes of SNe from > 0.3mag to ~ 0.15mag; this method depends on measuring the decline rate of SN Ia light curves shortly after maximum.

The two groups using the BTC at CTIO for this work are the High-z Supernova Search Team led by Brian Schmidt (see, e.g., Riess *et al.* 1998) and the Supernova Cosmology Project headed by Saul Perlmutter (Perlmutter *et al.* 1998).

2.10 Probing Galaxy Evolution from $z \sim 4$

We now turn to observations of the most distant known galaxies.

Giavalisco, Dickinson, Madau and Steidel have been using the Lyman-break with great success as a discriminator to isolate high-redshift galaxies from multicolor surveys. They are carrying out a BTC survey at CTIO through u,B_j,R_T to separate out galaxies in the redshift range $2.8 < z < 3.5$; B_j,V and I filters are used for galaxies in the range $3.5 < z < 5$. They are collecting a sample of ~ 2400 galaxies and will measure their angular correlation function to $\sim 15\%$. This will enable them to constrain the large-scale structure of the first star-forming regions in the universe. They will also be able to study the luminosity function of Lyman-break galaxies at $z \sim 4$ and constrain the onset of galaxy formation and the early phases of evolution.

2.11 The Galaxy Population and Clustering at Redshifts $4 < z < 5$

The BTC team is using "B dropouts" from a BTC survey of faint galaxies to locate those with a Lyman-break feature between the B and V bands, corresponding to redshifts near 4. Extension of this survey to include R and I filters will allow estimates to be made of the space density of galaxies with $z > 4.5$ This work will be followed up with accurate redshift determinations at the Keck observatory and narrow-band imaging of groups of galaxies (early clusters?) with the same redshift.

2.12 Very Deep U-Band Imaging of Deep Survey Fields

Simon Lilly is using the ultraviolet sensitivity of the thinned CCDs in the BTC to complement deep multicolor work being carried out at CFHT and KPNO.

It is expected that the combined survey will yield a catalog of 150,000 galaxies with photometrically estimated redshifts to I_{AB}~25, including 2000 Lyman-break galaxies at $z > 3$ and $R_{AB} \leq 25$. The aims of his imaging survey are to study the clustering and luminosity function of galaxies at $z > 3$, the search for clusters and protoclusters at very high redshifts, the determination of the faint end of the luminosity function and the clustering of galaxies at $z < 1$, the study of weak-lensing effects and the isolation of a very faint quasar sample. Though the faintest objects in the HDF are more than ten times fainter than in Lilly's survey, the areal coverage on the sky is approximately 1400 times greater.

2.13 The NOAO Deep, Wide-Field, IR and Optical Survey

A large team of (primarily) NOAO investigators, led by Buell Jannuzi and Arjun Dey has begun a very deep IR and optical (BRIJHK) survey of 18 square degrees of the sky. The primary goal is to study the evolution of large-scale structure from z ~ 1-4. The detection of large-scale structures at $z > 1$ will provide constraints on theories of hierarchical structure formation and cosmological parameters. In addition, the survey is geared to provide information on the formation and evolution of the red-envelope galaxy population and detect luminous, very distant ($z > 4$) star-forming galaxies and quasars. This large survey will reach 26th magnitude in B, R and I, 21st in J and H and 21.5 in K, for 5σ detection limits in a 2-arcsecond diameter aperture. The data obtained in this survey will be calibrated and distributed rapidly to the general astronomical community.

2.14 A Large, Faint, $z = 3.3 - 5.0$ Quasar Survey

Patrick Hall and collaborators are conducting a more restricted survey - of 8 sq. deg., specifically to search for faint quasars at very high redshifts. Survey limits are expected to be B=26.8 (3σ), I=24.7 (5σ). These data will be combined with data obtained by Bernstein whose survey reached to $R \sim 25.4$ (5σ). U and V data are currently being taken to extend the redshift coverage. The team expects about 130-250 QSOs with redshifts between 3.3 and 4, and 15-45 QSOs with $4 < z < 5$. This will enable them to quantify the shape of the quasar luminosity function and its evolution from z=3.3-5. This in turn will put strong constraints on galaxy formation models and on the quasar contributions to the UV and x-ray backgrounds. Once again, this data will be distributed quickly to the general community.

2.15 A 30 Sq. Deg. Survey for $z > 5$ Quasars, Gravitational Lenses and $z > 0.6$ Clusters

Falco *et al.* have recently completed the BTC observations for this survey and reached 5σ limits of I=24.5, V=24 and B=23. Data reduction work is proceeding to produce photometry, astrometry, catalogs and colors. Well over a million objects have been detected so far. Follow-up spectroscopy of selected objects is being planned. This survey can be seen as a shallower, wider-field complement to some of the surveys listed above.

3 Conclusions

A wide range of survey work is already underway with the BTC camera at Cerro Tololo. This activity will certainly have significant impact on studies of the solar system, the Milky Way, the Local Group and on many aspects of extragalactic studies. The results on supernovae are already having a major impact on observational cosmology. Ever-larger mosaics of CCDs are planned for many large telescopes in the near future, promising even more rapid advances.

References

Perlmutter, S. et al. (1998): Nature, **391**, 51.
Riess, A.G. et al. (1998): A.J., submitted.
Tyson, J.A., Bernstein, G.M., Blouke, M.M., Lee, R.W. (1992a): Proc. SPIE **1656**, 400.
Tyson, J.A., Bernstein, G.M., Blouke, M., Lee, R.W. (1992b): Photonics Spectra, May, 153.
Wittman, D., Tyson, J.A., Bernstein, G.M., Lee, R.W., Dell'Antonio, I.P., Fischer, P., Smith, D.R., Blouke, M.M.(1998): preprint.

Discussion

Boyle: The percentage of 'stellar' objects, 40%, in the 'BTC50' appears to be very high. What is the reason for this?

Smith: Note, firstly, that this is the shallowest of the surveys I have been describing. Secondly, as mentioned by an earlier speaker from ESO, the 'El Niño' effect on Chile's weather this year has resulted in relatively poor seeing, which has in turn affected the depth of the survey. The best data was taken with 0.7 arcsec seeing, the worst with 1.4 arcsec seeing.

Renzini: How long will it take to complete the NOAO optical survey?

Smith: ∼ 2 years *without* 'El Niño' to bother us.

Deep Images of Bright Galaxies

David Malin[1], Brian Hadley[2]

[1] Anglo-Australian Observatory, P.O. Box 296, Epping, NSW 2121, Australia
[2] Royal Observatory, Blackford Hill, Edinburgh, EH9 3HJ, UK

Abstract. The dark skies at Siding Spring and exceptional quality and uniformity of plates from UK Schmidt Telescope result in exposures that reach faint limiting magnitudes, especially for extended objects. This limit can be extended even further by applying special photographic techniques to the plates and combining the enhanced information from several deep plates. This allows us to reach a limiting (B) surface brightness of about 28 B mag arcsec^{-2}, which is deep enough to reveal a number of unexpected peculiarities in nearby, apparently normal galaxies.

1 Background

The UK Schmidt Telescope will have been taking plates for over 25 years when these proceedings appear. For almost all of that time they have exploited the IIIa series of photographic emulsions introduced by Eastman Kodak the 1960s (Sandage and Miller 1966). These products were the first photographic emulsions designed to maximise the detection of faint objects against the sky background, though they were not usable in their as-received condition. The staff of the UK Schmidt Telescope (UKST) were among the first to make them work successfully (Corben et al. 1974, Sim et al. 1976).

This was a major challenge. Although the new products were extremely sensitive to processing variations, both before and after exposure, they were remarkably insensitive to faint light. New hypersensitising processes had to be evolved at a time when the mechanisms involved in low light level sensitivity were just being uncovered in the research laboratories of the manufacturers (Babcock et al. 1974). This effort has been supremely successful, and the uniformity, depth and consistency of plates from the UKST is unrivalled. These UK Schmidt plates were the first to look deep into the southern sky and produced a stream of exciting discoveries, an achievement aided in no small measure by the natural darkness of the Siding Spring site.

The high standards the UKST staff set for themselves ensured that their prime objective, the Southern Sky Survey, was of unparalleled quality. To achieve this, it was often necessary to take several plates of any given field to obtain an 'A-grade' plate acceptable for the survey. In addition, the UKST undertook many non-survey projects, often demanding survey-quality plates on a variety of targets. All these deep plates plates have been archived in the Plate Library of the Royal Observatory, Edinburgh, and are available to the astronomical community. They represent an invaluable and unique resource for looking deep in the southern sky.

2 Other Developments

Shortly after the UK Schmidt began taking deep plates, Malin (1978) developed new ways of revealing ever fainter object on the newly-effective fine grain, high contrast IIIa emulsions. These photographic enhancement processes, especially when applied to the exceptionally uniform UKST plates, quickly led to several important discoveries (e.g. Malin and Carter 1980, Malin et al. 1983, Bothun et al. 1987). In a search for ever fainter objects, Malin (1981) also perfected a way of combining several deep, photographically enhanced exposures of the same field to improve the signal-to-noise of the final image. This in turn has led to several new discoveries.

It is important to emphasise that the archive of the UK Schmidt telescope at the Royal Observatory, Edinburgh maintains a unique collection of plates that cover well over half the sky to unprecedented depths. The images recorded on these plates can be combined digitally or photographically to reveal extended objects that have a surface brightness approaching 28 B mag arcsec^{-2}. This level of detection is still a major challenge for CCDs on large telescopes and achieving it requires generous allocations of telescope time. The purpose of this contribution is to point out that much useful work can be done by exploiting this archival material, without the uncertainty of telescope time allocation panels.

3 Examples

Examples of new discoveries made using this process were shown as slides at the meeting reported in these proceedings. The essential point was made that the large field of Schmidt telescopes, combined with photographic enhancement of plates made at a dark site, remains a powerful way of exploring the low surface brightness universe. This work has been described in detail elsewhere and good reproductions of several new, faint features of bright galaxies have appeared recently (Malin and Hadley, 1995, 1997a, 1997b).

In this paper we choose to show a galaxy group Hickson 90 (Hickson 1982), which contains the galaxies NGC 7172–74 and NGC 7176. Though this group has long been known to be known to be an X-ray source (Marshall et al. 1979) and has been studied previously (Longo et al. 1994), the interaction of the outermost bright member of the group (NGC 7172) was not suspected until this image was made. This galaxy is also a Seyfert 2. The extensive filamentary structure visible between the innermost members were also unknown.

References

Babcock, T.A., Sewell, M.H., Lewis, W.C. and James, T.H. (1974): AJ **79** 1479–1487
Bothun, G.D., Impey, C.D., Malin, D.F., Mould, J.R. (1987): AJ **94** 23–29
Corben, P.M., Reddish, V.C. and Sim, M.E. (1974): Nat **249** 22–24
Longo, G., Busarello, G., Lorenz, H. Richter, G., Zaggia, S. (1994): A&A **282** 418–424
Malin, D.F. (1978): Nat **276** 591–593.

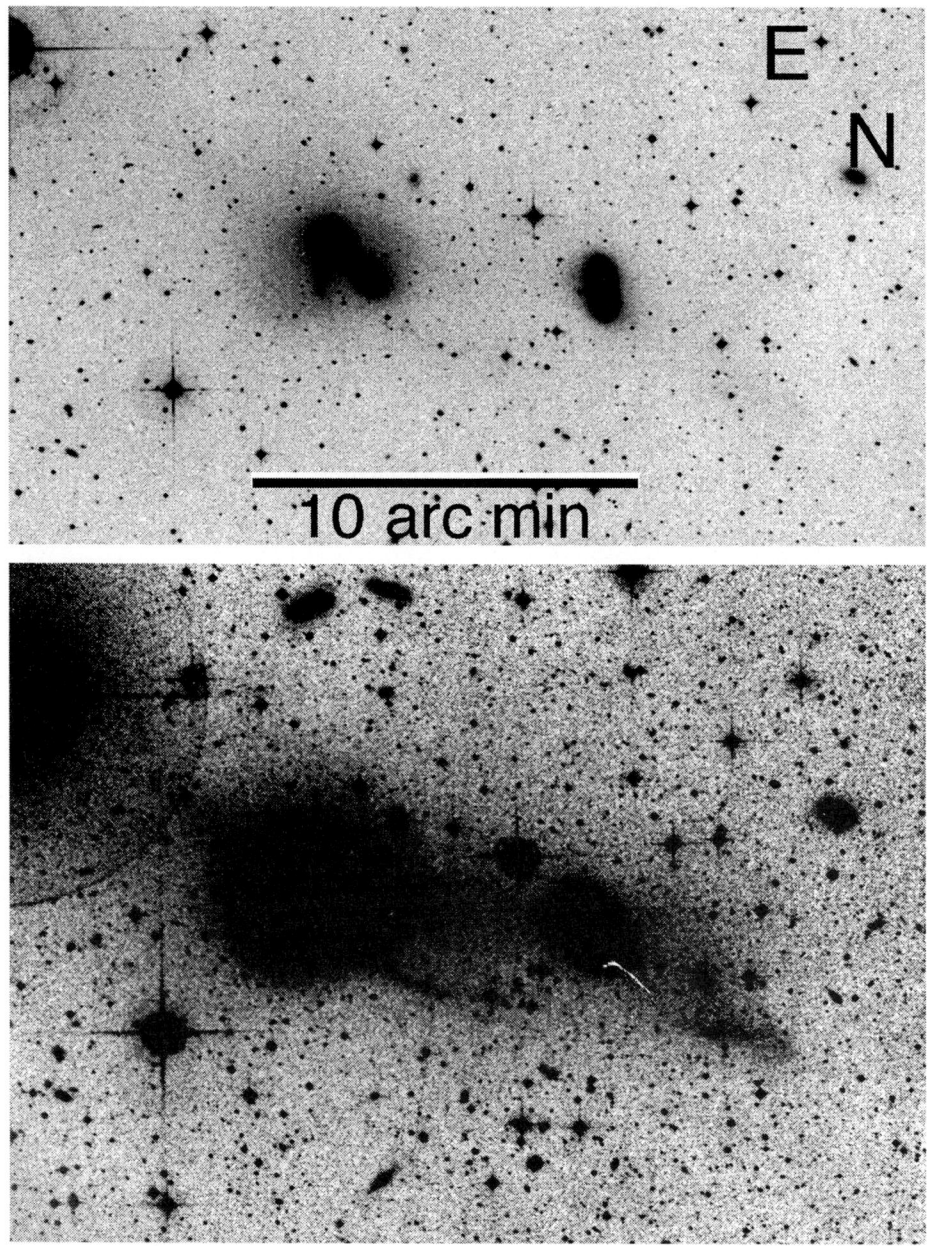

Fig. 1. The Hickson 90 group of galaxies. Upper image from a single UK Schmidt IIIa-J plate. Lower image, enhanced and combined images from four such plates

Hickson, P. (1982) ApJ **255** 382–391
Malin, D.F. (1981): J. Phot. Sci. **29** No.5, 199–205
Malin, D.F., Carter, D. (1980): Nat **285** 643–645
Malin, D,F., Hadley, B.W. (1995): IAU Colloquium 148, ASP Conference Series Vol. 84, Eds: Chapman, J. Cannon, R.D., Harrison, S. J., Hidayat, B. p 436. Astron Soc of the Pacific, San Francisco.
Malin, D,F., Hadley, B.W. (1997a): Proc. Astron. Soc. Aust. **14** 52–58.
Malin, D,F., Hadley, B.W. (1997b): ASP Conference Series, Vol. 116, Eds Arnaboldi, M., Da Costa, G.S., Saha, P. pp460–469. Astron Soc of the Pacific, San Francisco.
Malin, D.F., Quinn, P.J., Graham, J.A. (1983): ApJ **272** L5–L7
Marshall, F., Boldt, E., Holt, S., Mushotzky, R., Rothschild, R., Serlemitsos, P., Pravdo, S. (1979): ApJS **40** 657–665
Sandage, A., Miller, W.C. (1966): ApJ **144** 1238–1240
Sim, M.E., Hawarden, T.G., Cannon, R.D. (1976): AAS Photo-Bulletin, No.11, 3–5

Discussion

Oliver: Do you see any examples of "tidal-like" features which are not connected with interaction/merging phenomena?

Malin: Yes, there are galaxies that have extensive external tidal disturbances but no other obvious signs of interaction. The three last slides I showed (deep images of M83, NGC 253 and M100) fall into that category.

Tsvetanov: Have you tried combining images of NGC 1068, a well known Seyfert galaxy?

Malin: NGC 1068 has signs of faint, diffuse light up to 10 arc min from the nucleus in both the NE and SE quadrants and a distinct diffuse anomalous arm or curved tail about 7 arc min out on the western side of the galaxy.

Rocca-Volmerange: NGC 5102 is a peculiar galaxy: an S0/Elliptical full of gas with evidence of star-formation in it. It is in the same group as NGC 5128 (Cen A). Did you find any peculiar features between the two galaxies?

Malin: I have been obliged to avoid NGC 5102 because the bright halo of iota Cen masks the image on UK Schmidt plates.

Quinn: Have you undertaken "blind" surveys for either faint or very extended galaxies?

Malin: When I compiled the list of $\sim$ 200 galaxies to be included in the Deep Atlas I am preparing I looked carefully at all the film copies of the UKST sky survey fields from equator to pole and noted anything extragalactic I thought would yield something interesting if it was photographically amplified. Of course, this enhancement process reveals faint extended structures that cannot be seen, so no doubt some interesting objects were missed. Unfortunately, it is currently not practical to use this process routinely for all the plates in the archive, however, one soon develops an eye for interesting objects. I also noted 'field' low surface brightness galaxies, of which there are just a few. In the interim others have noted them by surveying the same material and shown some of them to be members of the Local Group.

Hopkins: You mentioned several times combining six plates in this fashion. How many plates can you continue to combine to gain in sensitivity to low surface brightness?

Malin: The gain in detectivity goes up as the square root of the number of plates. Six is typically the number of plates I had access to while preparing the catalogue. Others (e.g. Hawkins, M.R.S. 1995, in ASP Conf. Ser. 84, p.192, "The Future Utilisation of Schmidt telescopes") have combined up to 100 plates of the same field digitally with very great effect, but more than a dozen starts to be difficult using the photographic route.

A Deep Tech Pan Survey of Dwarf Spheroidal Galaxies in Virgo

Quentin A. Parker[1] and Steven Phillipps[2]

[1] Anglo-Australian Observatory
[2] University of Bristol, U.K.

Abstract. We have made a very deep photographic R band survey of a region of the Virgo Cluster by utilising digital stacking of scans of Tech Pan films taken on the U.K. Schmidt Telescope. The objects we detect have the same physical sizes and surface brightnesses as Local Group dwarf spheroidal galaxies. The luminosity function of these extremely low luminosity galaxies (down to $M_R \simeq -11$ or about $5 \times 10^{-5} L_*$) is very steep, with a power law slope $\alpha \simeq -2$, supporting previous observational evidence at somewhat higher luminosities in other clusters.

1 Introduction

A number of current projects at the U.K. Schmidt Telescope are using Tech Pan films for deep imaging. In this paper we describe one of these, a deep photographic survey of significant areas within the Virgo Cluster which span a range of (giant) galaxy densities. The survey, based on digitally stacking UKST films (cf. Bland-Hawthorn, Shopbell & Malin 1993), extends the earlier, seminal, survey of Binggeli, Sandage & Tammann (1985), reaching roughly 3 magnitudes beyond their completeness limits. We are able to detect large numbers of faint galaxies, presumably dwarf spheroidals, at magnitudes down to $M_R \simeq -11$ for an assumed Virgo distance of 18 Mpc (e.g., Jacoby et al. 1992; for consistency we also adopt their value of $H_0 = 75$ km s^{-1} Mpc^{-1} where required).

2 The Data and Image Detection

The data used here are part of a larger photographic survey of the Virgo Cluster (Schwartzenberg, Phillipps & Parker 1995a) using the extremely fine grained, highly efficient Tech Pan films on the 1.2m UK Schmidt Telescope (Phillipps & Parker 1993). Six individual long (1 to 1.5 hour) exposures of the same area, the South East quadrant of the Virgo Cluster, were scanned with the SuperCOSMOS automatic measuring machine at the Royal Observatory Edinburgh (Miller et al. 1992). For convenience, nine separate scan regions 6840 pixels square were created from each film. The pixel scale is 10 microns or $0.67''$ giving a total area for each scan $\simeq 1.3° \times 1.3°$. Two of these scan regions are considered in the present paper, one centred close to M87 and one $3.°1$ to its south south east. Note that, while large in itself, the $\simeq 3.2$ square degrees covered here is only about 1/40 of the full cluster survey area.

The scans from the six separate films were sky subtracted by using a 256×256 pixel spatial median filtered version of the data themselves, then matched in intensity by comparing images of a number of calibrating galaxies and median stacked (see Schwartzenberg, Phillipps & Parker 1996 for details). Median stacking has equivalent noise reduction to simple co-addition and is highly effective in removing artefacts (e.g., due to satellite trails, dust particles adhering to the emulsion and so forth) which affect only one film in the stack. Absolute calibration was via comparison of the images of some brighter galaxies with published CCD photometry, as described by Phillipps & Parker (1993). The final stacked data have an equivalent exposure time of about 7 hours, and the high efficiency of the films, approaching 10% (Parker et al. 1998), results in a pixel-to-pixel sky noise σ_{sky} equivalent to 26.2 R magnitudes per square arc second (henceforth $R\mu$).

Galaxy (and star) images were automatically recovered from the stacked data via a connected pixel algorithm, using a detection threshold $2\sigma_{sky}$ above the sky background, or 25.45$R\mu$, and a minimum area of 25 pixels (11 square arc seconds). Each image thus has a minimum S/N of 10 and has a magnitude $R \leq 22$. (In principle, 3σ detection of images of size around 2″ would reach a magnitude limit of about $R = 24$). Around 28,000 images are detected in each field and the large minimum area ensures that few are spurious.

3 Results

Once the raw catalogues were produced, we 'refined' them by requiring that our images met certain criteria aimed at isolating low surface brightness cluster dwarfs. In particular we kept only those images whose isophotal sizes and isophotal magnitudes were consistent with them having exponential profiles (characteristic of virtually all dwarfs; see Binggeli & Cameron 1991) of scale size $a \geq 2″$ and central surface brightness $\mu_0 \geq 22R\mu$ (cf. Figure 1 of Schwartzenberg et al. 1995b). This reduced the number of potential Virgo Cluster medium to low surface brightness dwarf galaxy candidates to approximately 17,000 (from 56,000 images of all types). Note that we do not separately remove stars as these should disappear along with the higher surface brightness galaxies. The data set includes galaxies with μ_0 down to about 25.2 $R\mu$, but is complete (in the sense that even a 2″ scale size gives images exceeding our area limit) only to 24.5$R\mu$.

In principle it is possible for an LSBG sample to contain cosmologically dimmed normal surface brightness giants at large redshifts or large non-cluster LSBGs in the background. The former would generally appear much smaller than our detection limit (with $a \leq 1″$, cf. Windhorst et al. 1994), while the latter are relatively rare (see Schwartzenberg et al. 1995b) and can be subtracted statistically (Turner et al. 1993). Nevertheless, in order to reduce such contamination problems to a minimum, we have again refined our sample to include only the $\simeq$ 4000 objects with $a \geq 3″$. These images will also be less affected by seeing; even if the scale lengths are slightly increased by the blurring (and the relatively moderate resolution), the central surface brightness will be decreased to compensate, leading to little error in the derived total magnitudes. In effect we will have merely the LF of galaxies limited at a marginally smaller physical size than would have been the case in the absence of seeing.

Since we have pre-selected our dwarf LSBG sample in terms of scale size and central surface brightness, we cannot simply subtract standard number counts for the entire population of field galaxies (e.g., Metcalfe et al. 1995) from the magnitude distribution we obtain, in order to arrive at the cluster LF. We have therefore made a subtraction based on the corresponding distribution of *field* LSBGs parameters found by Schwartzenberg et al. (1995b). This correction turns out to be quite small compared to our total LSBG numbers (a few percent), so is not critical to our final LF, for the simple reason that most background LSBGs appear much smaller than our cluster LSBG candidates (cf. Karachentsev et al. 1995).

Figure 1 illustrates the LFs for the inner cluster region, containing 675 galaxies. We show here only those galaxies with $a \geq 3''$ and $22.0 \leq \mu_0 \leq 24.5 R\mu$, the region of parameter space for which we have a complete and minimally contaminated sample. Note that this is *not* a magnitude limited sample. For instance, we already begin to lose any galaxies with smaller scale sizes at $M_R \simeq -13$, whereas our faintest objects have $M_R \simeq -11$. Of course, even the loss of small objects at faint M_R may not be the whole story as far as the LF goes, since there may exist higher surface brightness dwarfs than we are allowing for (perhaps preferentially at bright M_R), and we will be missing any even lower surface brightness objects at *all* M_R. Indeed, recall that we have many candidates for smaller or lower surface brightness galaxies in our original overall sample (see also Schwartzenberg 1996).

The LFs for the two regions turn out to be very similar so an overall LF of the samples can be used. It is clear that the LF is steep, confirming earlier suggestions for Virgo itself by Impey, Bothun & Malin (1988) and Tyson & Scalo (1988), and for a number of other richer, more distant clusters (e.g. Driver et al. 1994, Smith et al. 1997, Wilson et al. 1997, Trentham 1997). A least squares fit to the combined data gives a power law slope for the range $15.5 < R < 20.0$ (roughly -16 to -11.5 in M_R) of $\alpha = -2.26 \pm 0.13$. The amplitudes of the LF for the two separate areas are also similar (in galaxies per magnitude bin per square degree), with the core sample actually having the lower projected density by a factor $\simeq 0.8$. LSBGs may be adversely affected by the presence of the giant galaxy M87 in the cluster centre region; Thompson & Gregory (1993) have previously found a similar effect in the core of the Coma Cluster. Note, though, that with the very steep LF slope, a relatively small zero point offset in the calibration between fields can have a significant effect on the numbers. For instance an error of $\Delta m = 0.^m 1$ generates a difference of a factor $10^{0.4(\alpha+1)\Delta m} \simeq 1.12$. Such errors would make little difference to the shape of the derived LF, since the background contamination is so small. At the bright end, the number of detected LSBGs is in good agreement with that expected from the Binggeli, Sandage & Tammann (1985, 1988) Virgo Cluster LF, given the small numbers of these objects in our samples (see Figure 1). It is clear that our LF departs from theirs at the expected point where incompleteness and lack of very low surface brightness objects starts to affect their sample, beyond $R \simeq 17$. (We assume here typical early type galaxy colours, $B - R = 1.5$, for the dwarf spheroidals; if they actually have bluer colours, as often seen in low surface brightness galaxies, this would slightly improve the match).

4 Conclusions

By co-adding very deep UKST photographic films we have been able to probe the dwarf population of the Virgo Cluster down to $M_R \simeq -11$ ($\simeq 5 \times 10^{-5} L_\star$). The central surface brightness limit for our sample is $25 R\mu$, corresponding roughly to $26.5 B\mu$ for early type galaxy colours. In both luminosity and surface brightness this is thus one of the deepest surveys yet performed. In particular, our limits allow us to survey well into the regime of the dwarf spheroidal galaxies (Irwin & Hatzidimitriou 1995); the luminosity limit is 25 times fainter than the Fornax dwarf, for instance. We have therefore been able to gather by far the largest sample of dwarf ellipticals/dwarf spheroidals currently known. Of course, in the absence of redshifts, these 'detections' are on a statistical basis only. However, given the paucity of LSBs of moderate to large angular size in the general field, we are confident that the large majority of our candidates are genuine cluster dwarfs.

The luminosity function of the dwarfs is very steep, with $\alpha \simeq -2$, confirming values found over much more limited magnitude ranges in other clusters (e.g., Smith et al. 1997; Trentham 1997). Bernstein et al. (1995) reached similarly faint levels to those discussed here with very deep CCD imaging of a very small area at the core of the Coma cluster (to $M_R = -11.4$). They found a less steep LF than most other deep surveys, $\alpha \simeq -1.3$, but the centre of Coma may be a rather special environment. Though their surveys are less deep, Biviano et al. (1995) and Thompson & Gregory (1993) find steeper slopes for parts of Coma further from the centre.

We might, finally, note that much smaller and fainter galaxies can be *detected* in our data than are present in our photometric samples, Indeed, we can reach down to about $M_R \simeq -8.5$ at the distance of Virgo, the same as the Carina dwarf, the lowest luminosity system currently known. Unfortunately, though, these images are indistinguishable from those of the (very numerous) general background galaxies. However, we should still be able to estimate their numbers through a comparison with an identically observed non-cluster field. This work will be reported in a subsequent paper. At some point we should certainly expect to see a turn over in the LF, since $\alpha = -2$ is the critical value at which the integrated galaxy light formally diverges. If the currently found slopes in the range -2 to -2.5 were to continue down to, say, $M_R = -8$, then the dwarf galaxies fainter than $M_R = 16$ would contain approximately 0.1 to 1.0 times as much light as the brighter galaxies. For a constant M/L (or perhaps more reasonably, a fixed *baryonic* M/L) this would obviously increase the total mass in cluster galaxies by a factor between 1.1 and 2.

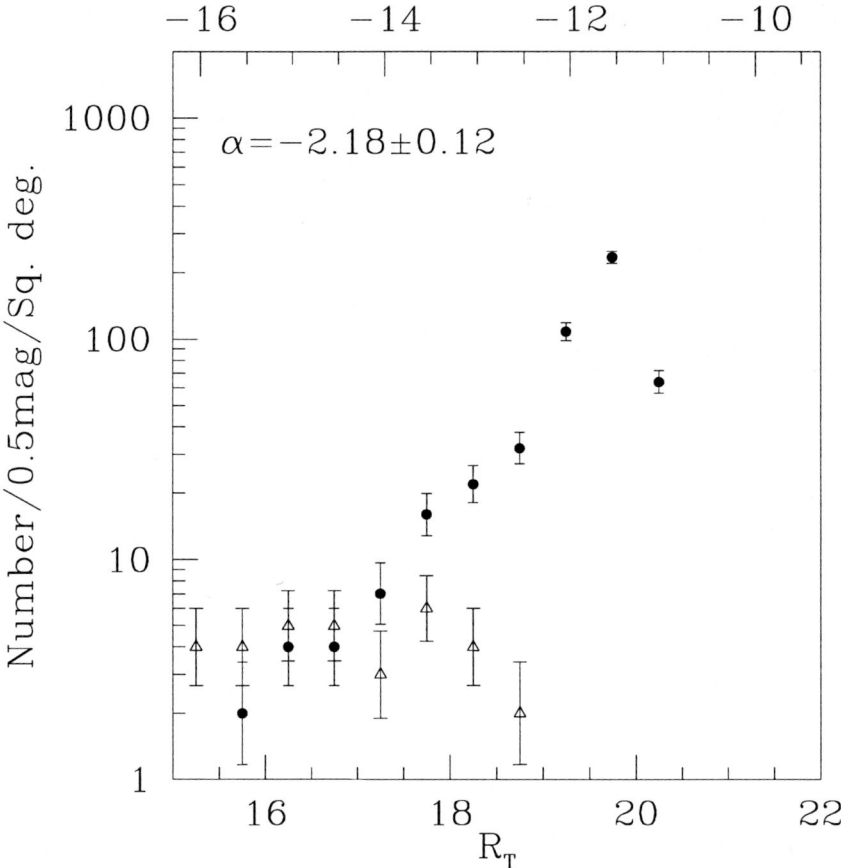

Fig. 1. The absolute magnitude distribution of Virgo LSBGs (as defined in the text), for the inner survey area. Error bars shown are based on Poissonian statistics. The luminosity function of Sandage et al. (1985) is also shown (open triangles) for the Virgo Cluster Catalog galaxies which overlap with the inner field. (We assume for this comparison $B - R = 1.5$).

References

Bernstein G.M. et al., 1995, AJ, 110, 1507
Biviano A. et al., 1995, A&A, 297, 610
Binggeli B., Cameron L.M., 1991, A&A, 252, 27
Binggeli B., Sandage A., Tammann G.A., 1985, AJ, 90, 1681
Binggeli B., Sandage A., Tammann G.A., 1988, ARA&A, 26, 509
Bland-Hawthorn J., Shopbell P.L., Malin D.F., 1993, AJ, 106, 2154
Driver S.P. et al., 1994, MNRAS, 268, 393
Impey C., Bothun G., Malin D., 1988, ApJ, 330, 634
Irwin M.J., Hatzidimitriou D., 1995, MNRAS, 277, 1354

Jacoby G.H. et al., 1992, PASP, 104, 599
Karachentsev I.D. et al., 1995, A&A, 296, 643
Metcalfe N., Shanks T., Fong R., Roche N., 1995, MNRAS, 273, 257
Miller L.A. et al., 1992, in Digital Optical Sky Surveys, Kluwer, Dordrecht, p.133
Parker Q.A. et al. 1998, MNRAS, to be submitted
Phillipps S., Parker Q.A., 1993, MNRAS, 265, 385
Schwartzenberg J.M., 1996, Ph.D. Thesis, University of Bristol
Schwartzenberg J.M., Phillipps S., Parker Q.A., 1995a, A&A, 293, 332
Schwartzenberg J.M., Phillipps S., Parker Q.A., 1996, A&AS, 117, 179
Schwartzenberg J.M. et al., 1995b, MNRAS, 275, 121
Smith R.M., Driver S.P., Phillipps S., 1997, MNRAS, 287, 415
Thompson L.A., Gregory S.A, 1993, AJ, 106, 2197
Trentham N., 1997, MNRAS, 290, 334
Turner J.A., Phillipps S., Davies J.I., Disney M.J., 1993, MNRAS, 261, 39
Tyson N.D., Scalo J.M., 1988, ApJ, 329, 618
Wilson G., Smail I., Ellis R.S., Couch W.J., 1997, MNRAS, 284, 915
Windhorst R. et al., 1994, AJ, 107, 930

Surveys with a 4 m Liquid Mirror Telescope

C. Jean, J.-F. Claeskens and J. Surdej

Institut d'Astrophysique, Université de Liège, Belgium

Abstract. We describe an international project of construction and operation of a 4 m Liquid Mirror Telescope (LMT) led by E. Borra. A LMT, whose main advantage is its very low cost, is particularly well suited for the search and study of gravitational lenses, type Ia supernovae, faint nearby red, brown and white dwarfs, halo stars with high proper motions and, more generally, all variable phenomena like quasars, variable stars, micro-lensing effects, etc.

1 Technical Description of the 4 m LMT

The surface of a reflecting rotating liquid takes the shape of a paraboloid which is the ideal surface for the primary mirror of an astronomical telescope. The focal length F of the mirror is related to the gravity g and to the angular velocity of the turntable ω by

$$F = \frac{g}{2\omega^2} \quad (1)$$

Liquid mirror telescopes cannot be tilted and hence cannot track like conventional telescopes do. In order to track images through narrow- and wide-band filters or slitless spectroscopy, one can use a technique called time delayed integration (TDI), also known as drift scan, in which the CCD detector tracks the charges by electronically stepping its pixels. The information is stored on disk and the night observations can be coadded with a computer to give long integration times (see Borra 1982 and Borra et al. 1992 for more details).

2 Science with a LMT

Thanks to its very low cost ($\sim$ 1–2 millions US$), a 4 m LMT can be entirely dedicated to a scientific project and, in spite of its relatively restricted field of view ($\sim$ 1 °), several scientific drivers could be carried out like statistical determination of the cosmological parameters H_0 and q_0 based upon surveys for gravitational lenses (Surdej and Claeskens 1997) and supernovae, search for low surface brightness and star-forming galaxies, observational studies of quasars and large scale structures, detection of high stellar proper motions, trigonometric parallaxes, a wide range of photometric variability studies (photometry of microlensing effects and of variable AGN over day to year time scales) and also a unique database for follow-up studies with the VLT.

Operation of a LMT from La Silla (latitude of 29 degrees 15 minutes South) would enable to cover approximately 90 square degrees of sky at high galactic latitude ($|b| > 30°$), passing very near to the south galactic pole. At the same time, such a LMT survey would probe regions near the galactic center, offering unique data for studies of the galactic structure, stellar populations, including accurate measurements of stellar proper motions (cf. red, white, brown dwarfs, faint halo stars, etc.), trigonometric parallaxes and detection of stellar microlensing effects caused by bulge stars, dark compact objects, etc.

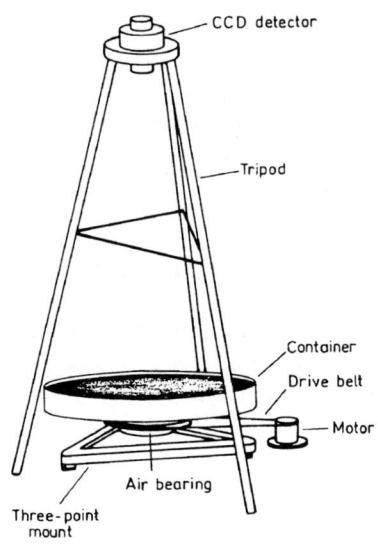

Fig. 1. Entire telescope system.

References

See the WWW bibliography available at the URL:
   http://wood.phy.ulaval.ca/lmt/home.html
   http://vela.astro.ulg.ac.be/grav_lens/grav_lens.html

Visit our LMT web page at the URL: http://vela.astro.ulg.ac.be/lmt/

Borra E. F. (1982): JRASC **76**, 245.

Borra E. F., Content R., Girard L., Szapiel S., Tremblay L. M. & Boily E. (1992): ApJ **393**, 829.

Surdej J. and Claeskens J.-F. (1997): in the proceedings of the Marseille Workshop "Science with Liquid Mirror Telescopes" (April 14-15, 1997) [http://wood.phy.ulaval.ca/lmt/home.html]

The Metagalactic Ionizing Field in the Local Group

J. Bland-Hawthorn

Anglo-Australian Observatory, P.O. Box 296, Epping, NSW 2121, Australia

Abstract. We discuss the sources which are likely to dominate the ionizing field throughout the Local Group. In terms of the limiting flux to produce detectable Hα emission ($\sim 4 - 10 \times 10^3$ phot cm^{-2} s^{-1}), the four dominant galaxies (M31, Galaxy, M33, LMC) have spheres of influence which occupy a small fraction (5 − 10%) of the Local Volume. There are at least two possible sources of ionization whose influence could be far more pervasive: (i) a cosmic background of ionizing photons; (ii) a pervasive warm plasma throughout the Local Group. The Compton y-parameter, measured by *COBE* FIRAS, permits a wide variety of plasmas with strong ionizing fields. It has been suggested (Blitz *et al.* 1996; Spergel *et al.* 1996; Sembach *et al.* 1995, 1998) that a substantial fraction of high velocity clouds are external to the Galaxy but within the Local Group. Deep Hα detections are the crucial test of these claims and, indeed, provide a test bed for the putative Local Group corona.

1 Introduction

In keeping with this Workshop, we concentrate on what one can hope to learn, in the coming decade, from deep spectroscopic studies of diffuse line emission. Such techniques are now used by several groups in both hemispheres. (For progress to date, see Reynolds *et al.* 1997; Bland-Hawthorn 1997.) The spectroscopic methods (e.g. 'staring') achieve extremely deep levels (§2) and can detect the presence of extremely weak ionizing fields.

To encourage a broader campaign of spectroscopic studies, we deduce the expected level of ionizing flux within the Local Group. We provide an inventory of possible sources, and discuss the prospect of a ubiquitous warm plasma. The discovery of such a medium would have fundamental implications:

• A recent review of the cosmic baryon budget finds that a large fraction of 'missing' baryons may well be tied up in warm gas within galaxy groups (Fukugita, Hogan & Peebles 1998). To be consistent with weak x-ray detections (Pildis, Bregman & Evrard 1995), the diffuse gas needs to radiate predominantly at EUV wavelengths.
• Such an ionizing field could render an HI cloud optically visible everywhere within the Local Group. A truly *cosmic* ionizing field could render the same cloud visible over cosmological distances (Circovic *et al.* 1998).
• The existence of warm, tenuous gas in loose galaxy groups has important ramifications for the Lyα forest at low redshift detected along QSO sight lines (Morris *et al.* 1991; Bahcall *et al.* 1991).

- An 'intra-group' medium would indicate the presence of Galactic fountains or winds, or even primordial gas dating back to the formation of the group. If confirmed, this medium is expected to radically influence the star-formation history of the Local Group (e.g. Dressler 1986; van den Bergh 1994).

2 The Deepest Spectroscopic Detections

The deepest spectroscopic limits are achieved by Fabry-Perot 'staring'. Several groups have shown what is possible with modern day optics and detectors (*q.v.*, Bland-Hawthorn 1997). The power of the method comes from dispersing the light of a single emission line onto $\sim 5-10\%$ of a wide area CCD. A hard experimental limit is about 1 mR (1σ) at 1Å resolution which is 8 mag below sky within that narrow band. This is close to the experimental limit of space-borne observations in broad optical bands (Vogeley 1998). Both broad and narrow band limits fall far below the zodiacal light level arising from interplanetary dust and gas.

For our deep limit, what is the corresponding ionizing flux? From the surface of an HI slab optically thick to ionizing photons, the emission measure is $\mathcal{E}_m = \int n_e n_{H+}\, dl = n_e n_{H+} L$ cm^{-6} pc where L is the thickness of the ionised region. The resulting emission measure for an ionizing flux φ_i is then $\mathcal{E}_m = 1.25 \times 10^{-2}\varphi_4$ cm^{-6} pc ($\equiv 4.5\varphi_4$ mR) where $\varphi_i = 10^4 \varphi_4$. The 1 mR spectroscopic limit corresponds to an ionizing flux of less than 10^4 phot cm^{-2} s^{-1}. To put this into perspective, this is the expected level of Hα recombination emission from the interstellar medium *within the Solar System* due to the mean Galactic ionizing field (Reynolds 1984)!

3 Cosmic UV Background

From Hα non-detections towards extragalactic HI, the current best 2σ upper limit (Vogel et al. 1995) for the universal ionizing background is $J^u_{-21} < 0.08$ ($\varphi^u < 2-4 \times 10^4$ phot cm^{-2} s^{-1}). J^u_{-21} is the ionizing flux density of the cosmic background at the Lyman limit in units of 10^{-21} erg cm^{-2} s^{-1} Hz^{-1} sr^{-1}; φ^u ($= \pi J^u_{-21}/h$) is the equivalent photon flux at face of a uniform, optically thick slab. The unexpectedly high number of Lyα absorbers towards 3C273 gives $J^u_{-21} \approx 0.006$ from the proximity effect (Kulkarni & Fall 1995), although this value is uncertain by at least a factor of 5. Sciama (1993; 1998) has suggested a somewhat stronger ionizing field permeates the Universe due to decaying tau neutrinos with masses 24 eV and lifetimes $\sim 10^{23}$ s.

4 An Inventory of the Local Group

The Local Group is a member of the Coma-Sculptor Cloud within the Local Supercluster, a highly flattened structure extending over more than 10 Mpc (Tully 1988). Outside the Local Group, the closest structures are the Sculptor, Maffei/IC342, and M81 groups. While the Maffei group has the biggest dynamical

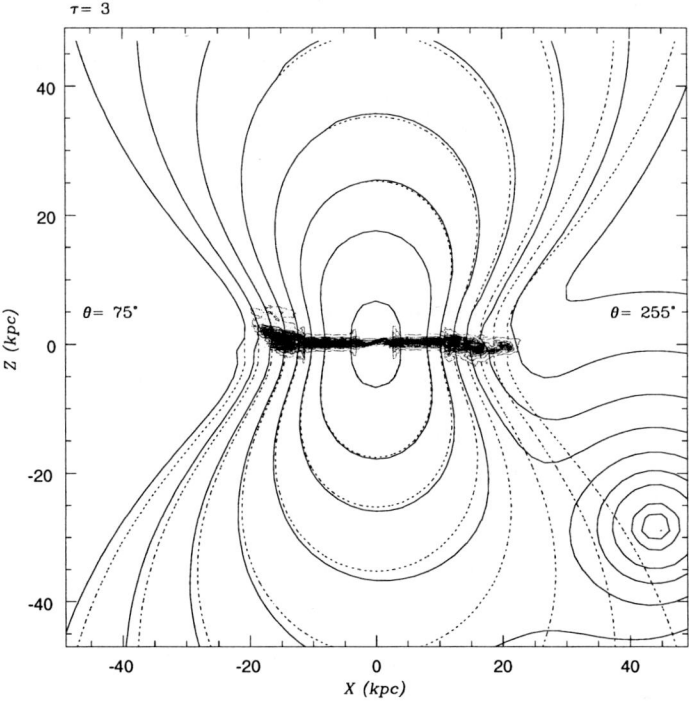

Fig. 1. The Galactic halo ionizing field. The coordinates are with respect to a plane perpendicular to the Galactic disk (with the Galactic Centre at the origin) at a constant galactic azimuth angle (75°, 255°). The dotted lines show the ionizing flux (φ_4) due to the galactic disk; the solid lines include the contribution from the LMC. The opacity of the HI disk (shown in half tone) has been included. The contours, from outside in, are for $\log \varphi_4 = 1, 1.25, 1.5, 1.75, 2, 2.25, 2.5, 3$ phot cm^{-2} s^{-1}. The minor contribution from the Galactic corona is omitted (BM).

influence on us, the Sculptor group is expected to produce the stronger ionizing field at the periphery of our group.

The Local Group has at least 40 members (see appendix; Tully 1988). More than 80% of the visual light emerges from the great rajahs, M31 and the Galaxy. This increases to 95% if we include M33 and the LMC. The remaining subjects are dwarf irregulars, spheroidals or ellipticals which congregate around the royal members, or form small associations near the edge of the group. For convenience, we define the Local Volume as a sphere extending to 1 Mpc radius from the Local Group barycentre (LGB). [1]

---

[1] R.B. Tully (1998, private communication) points out that a criterion for group membership requires total or continuing collapse towards the common potential of the group. However, the distances, space motions, and the underlying mass distribution are not well enough known to be sure of which of the outermost objects are bound.

4.1 Young Stars

Our model of the ionizing background in the Local Group includes the poloidal radiation fields of the Galaxy, M31 and M33. The Galaxy halo includes the contribution from the LMC (see Fig. 1). To a good approximation, the UV radiation field arising from an opaque, stellar disk is given by (Bland-Hawthorn & Maloney 1998; hereafter BM)

$$\varphi^\star = \varphi_o^\star \, e^{-\tau} \, r_{100}^{-2} \cos^{0.6\tau + 0.5} \theta \qquad \text{phot cm}^{-2} \text{ s}^{-1}. \tag{1}$$

The polar angle θ ($0 \leq \theta < \frac{\pi}{2}$) is measured from the spin axis, r_{100} is the radius in units of 100 kpc, and the Lyman limit optical depth is $\tau \leq 10$. In order to explain Magellanic Stream Hα detections and OB star counts, our model favours $\varphi_o^\star = 2.8 \times 10^6$ phot cm^{-2} s^{-1} and $\tau \approx 2.8$ for the Galaxy. For the purposes of this Workshop, the ionizing fields for M31 and M33 are arbitrarily scaled with respect to the Galaxy by their blue luminosities for the same opacity. This crude assumption may well be flawed since the evolutionary histories of these galaxies could be quite different.

While the ionizing fields are expected to be weak, we include the contribution of the dwarfs for completeness. Deep HI surveys may have found gas linked to dwarf galaxies (q.v., Young & Lo 1997) and even a very weak field could render such gas optically visible. Deeper surveys already in progress are expected to turn up more HI associations (Staveley-Smith et al. 1996; Briggs et al. 1997).

For the dwarf irregulars, on-going star formation appears confined to the outermost systems, DDO 210 and Phoenix, which may be evidence enough for a diffuse medium throughout the Local Group. The lack of star formation in the more central dwarfs is a possible manifestation of ram-pressure stripping (van den Bergh 1994; Hirashita, Kamaya & Mineshige 1997).

In dwarf spheroidals, there is evidence for a 'UV upturn' presumably from horizontal branch (HB) stars. In ellipticals (and S0s), the 'UV upturn' population are thought comprise hot HB, post-HB star and post-AGB stars (Dorman 1997; Brown et al. 1997). Our model includes a weak isotropic source for each of the dwarf irregulars and spheroidals.

4.2 Galactic Coronae

The depth of x-ray shadows observed towards high-latitude gas clouds is consistent with a 0.25 keV patchy coronal halo encompassing the Galaxy (Burrows & Mendenhall 1991; Snowden et al. 1991; Snowden, McCammon & Verter 1993). Soft x-ray haloes are also observed in a significant fraction of spirals (Bregman & Pildis 1994; Vogler, Pietsch & Kahabka 1996; cf. Kim et al. 1996).

To explore the effect of a hot galactic corona, we assume that the gas distribution is described by a non-singular isothermal sphere, with a density law

$$n(r) = \frac{n_c}{(1 + r^2/r_c^2)} \text{ cm}^{-3} \tag{2}$$

where r_c is the core radius and n_c is the core gas density. For hot gas temperatures $T_e \geq 2$ keV, the emission is dominated by thermal bremsstrahlung. Only photons in the energy range 13.6 eV to $\sim$250 eV are important to the ionization state of the gas for the column densities of relevance here. The ionizing photon flux on the inner face of a cloud at radius r (for $r/r_c \leq 12$) in kpc is given by (BM)

$$\varphi^{\text{gc}}(r) \approx 18 n_{-3}^2 r_c \mathcal{C}(T_e) T_{\text{keV}}^{-0.225} \left[\frac{\eta + 1.3(r/r_c)^{1.35}}{(1 + r^2/r_c^2)^{1.5}} \right] \quad \text{phot cm}^{-2} \text{ s}^{-1} \quad (3)$$

where $10^{-3} n_{-3}$ cm^{-3} is the core gas density. For an optically thin Galactic disk, $\eta = \pi/4$; when the disk is completely opaque to ionizing photons, $\eta = 0$, which reduces the ionizing photon flux by less than a factor of two.

For the Galaxy, absorption lines towards stars in the Magellanic Clouds have been used to infer coronal temperatures and gas densities; for example, Songaila (1981) estimates $T_e \approx 0.05 - 0.3$ keV and $n_c \sim 3 - 7 \times 10^{-4}$ cm^{-3}. We adopt $r/r_c = 2$, $n_{-3} = 2$, and assume $T_e \approx 0.2$ keV, the virial temperature of the Galactic halo (for an assumed circular velocity of 220 km s^{-1}). This density and temperature distribution matches the emission measure and x-ray luminosity of the diffuse component determined by Wang & McCray (1993).

At such low temperatures ($T_{\text{keV}} < 2$), eq. 3 does not include the strong EUV ionizing lines produced by the gas, and therefore considerably underestimates the true photon flux. We estimate the increase in the effective emissivity using Table 4 in Gaetz & Salpeter (1983). Representative correction factors $\mathcal{C}$ are 28 ($T_e = 0.34$ keV) and 55 ($T_e = 0.22$ keV). At the lower temperature, the expected ionizing flux at 20 kpc radius is $\varphi^{\text{gc}} \sim 2 \times 10^4$ phot cm^{-2} s^{-1}. These values are uncertain: the **MAPPINGS 3** ionization code (Sutherland & Dopita 1993) gives markedly lower correction factors for solar and sub-solar metal abundances.

4.3 Local Group Corona

The extremely accurate blackbody form of the cosmic microwave background (CMB) sets an important constraint on an ionised intergalactic medium. The presence of a hot plasma distorts the microwave background through Compton scattering of the CMB photons (Sunyaev & Zel'dovich 1969).

The departures from blackbody are quantified by the y-parameter such that

$$y = \int_0^t \frac{k(T_e - T_r)}{m_e c^2} \sigma_T n_e \, c \, dt. \quad (4)$$

where σ_T is the Thomson scattering cross section and m_e is the electron mass. The integration is performed over the time taken for the photon to traverse the ionised medium. In most cases, T_e is much greater than the radiation temperature T_r. The expected temperature decrement is then

$$\left(\frac{\delta T}{T}\right)_r = -2 \frac{k T_e}{m_e c^2} \sigma_T N_e(\mu), \quad (5)$$

where N_e is the electron column at an angle Θ_c ($\mu = \cos\Theta_c$) to the LGB sight line. The multipoles, after expanding the sky temperature in spherical harmonics as a function of angular position, are easily derived (Suto et al. 1996). The monopole term reduces to

$$T_0 = \pi\Theta_c\, \sigma_T \frac{kT_e}{m_e c^2} \frac{n_o R_c^2}{x_o} \tag{6}$$

where n_o is the central density of the Local Group corona, x_o is the distance of the Galaxy from the centre, R_c is the core radius of the corona, and $\Theta_c \equiv \tan^{-1}(x_o/R_c)$.

The *COBE* FIRAS data (Mather et al. 1994) imply that the Compton y-parameter is less than 2.5×10^{-5} (95% confidence level). The upper limit translates to (Suto et al. 1996)

$$n_o R_c^2 / x_o \;<\; 1.1 \times 10^{22} T_{\rm keV}^{-1} \left(\frac{1.17}{\Theta_c}\right) \left(\frac{y}{2.5 \times 10^{-5}}\right) \text{ cm}^{-2}. \tag{7}$$

We adopt an identical form to eq. 3 for the Local Group flux, and associate r with x_o such that

$$\varphi^{lg}(\Theta_c) \;<\; 5 \times 10^4 \mathcal{C}(T_e) T_{\rm keV}^{-1.2} n_{-3}(\Theta_c) \mathcal{F}(\Theta_c) \left(\frac{y}{2.5 \times 10^{-5}}\right) \text{ phot cm}^{-2}\text{ s}^{-1} \tag{8}$$

for which

$$\mathcal{F}(\Theta_c) = 1.17 \cos^2\Theta_c \left(0.8 + 1.3 \tan^{1.35}\Theta_c\right)\left(\frac{\sin\Theta_c}{\Theta_c}\right) \tag{9}$$

After Suto, we adopt $n_{-3} = 0.1$, $R_c = 150$ kpc, and $x_o = 350$ kpc. The angular term $\mathcal{F}(\Theta_c)$ is well behaved, with a value of 0.94 towards LGB, peaking at 1.2 near $\Theta_c = 0.55$, falling to 0.75 at Suto's canonical value of $\Theta_c = 1.17$, and to zero beyond here.

The correction factors in §4.2 lead to a high values for the upper limit above, viz. $\varphi^{lg} < 2 \times 10^6$ ($T_e = 0.34$ keV) and $\varphi^{lg} < 5 \times 10^6$ phot cm^{-2} s^{-1} ($T_e = 0.22$ keV). The coronal gas has a long cooling time ($\sim 10^{11}$ yrs) and provides a negligible contribution to the quadrupole anisotropy observed by *COBE* (Bennett et al. 1994).

5 Discussion

If the 3C 273 sight line is representative of the present day Lyα forest, the Galaxy and M31 (φ^*) dominate the cosmic ionizing field φ^u out to at least 700 kpc along the polar axis. This is an order of magnitude smaller than the scale of typical L_* galaxy separations at the present epoch. Although, after orienting the galaxies correctly with respect to the supergalactic plane, we find that while the Galaxy-LMC and M31-M33 pairs may experience significant levels of mutual ionization, this is not expected for the Galaxy-M31 pair.

The galactic ionizing fields are highly elongated for our assumed dust opacity $\tau \approx 3$ (eq. 1). Only a small fraction of the Local Volume (5 − 10%) is influenced by the stellar UV field. At large galactocentric radius, there exists a toroidal shadow region close to the galactic plane (BM). Within this region, the galactic coronal emission, φ^{gc}, could well exceed φ^u.

The COBE y-parameter upper limit permits a wide range of plasma conditions, a subset of which produce significant levels of UV flux, e.g., $\varphi^{lg} \sim 10^5$ phot cm^{-2} s^{-1}. If confirmed, there are some interesting consequences for φ^u: (i) attempts to measure the truly *cosmic* UV background directly (*e.g.*, Henry 1996) cannot be made from our vantage point; (ii) constraints on φ^u from Hα emission (*e.g.*, Vogel *et al.* 1995) require HI clouds which are not associated with galaxy groups.

More than half of all galaxies reside within small groups, and spirals dominate these groups. Could pervasive coronal emission in galaxy groups explain truncated HI disks in spiral galaxies (cf. Maloney 1993)? Could this same medium explain the unusual ionization conditions observed along sight lines towards extragalactic sources (Sembach *et al.* 1998)?

Several authors have suggested that some fraction of HVCs, particularly the more compact clouds, are external to the Galaxy (Blitz *et al.* 1996; Spergel *et al.* 1996; Sembach *et al.* 1995, 1998). Hα non-detections towards these clouds would argue *for* the extragalactic model, and *against* the pervasive 'intra-group' corona, in which case the clouds could be used to set hard limits on the cosmic ionizing field.

Acknowledgments

We are indebted to P.R. Maloney for use of unpublished results arising from collaborations, and to R.B. Tully, G. Da Costa and B. Dorman for their insights. R. Sutherland gave invaluable advice on the use of **MAPPINGS 3**.

References

Bahcall, J.N. *et al.* 1991, ApJ, 377, L5
Bennett, C.A. *et al.* 1994, ApJ, 436, 423
Bland-Hawthorn, J. 1997, PASA, 14, 64
Bland-Hawthorn, J. & Maloney, P.R. 1998, ApJ, submitted (BM)
Blitz, L., Spergel, D.N., Teuben, P.J., Hartmann, D., and Burton, B. 1996, BAAS, 189, #61.01
Bregman, J.N. & Pildis, R.A. 1994, ApJ, 420, 570
Briggs, F.H., Sorar, E., Kraan-Korteweg, R.C. & van Driel, W. 1997, PASA, 14, 37
Brown, T.M. *et al.* 1997, ApJ, 482, 685
Burrows, D.N. & Mendenhall, J.A. 1991, Nature, 351, 629
Circovic, M.M., Bland-Hawthorn, J. & Samurovic, S. 1998, MNRAS, submitted
Dorman, B. 1997, In *Nature of Elliptical Galaxies: Proceedings of the 2nd Stromlo Symposium,* eds M. Arnaboldi, G.S. Da Costa & P. Saha, ASP. Conf. Series, 116, 195

Dressler, A. 1986, ApJ, 301, 35
Fukugita, M., Hogan, C.J. & Peebles, P.J.E. 1998, ApJ, submitted (astro-ph/9712020)
Gaetz, T.J. & Salpeter, E.E. 1983, ApJS, 52, 155
Henry, R.C. 1996, ARAA, 29, 89
Hirashita, H., Kamaya, H. & Mineshige, S. 1997, MNRAS, 290, L33
Kim, D.-W. et al. 1996, ApJ, 468, 175
Kulkarni, V.P. & Fall, M. 1995, ApJ, 453, 65
Maloney, P.R. 1993, ApJ, 414, 41
Mather, J.C. et al. 1994, ApJ, 420, 439
Morris, S.L. et al. 1991, ApJ, 377, L21
Pildis, R.A., Bregman, J.N. & Evrard, A. 1995, ApJ, 443, 514
Reynolds, R.J. 1984, ApJ, 282, 191
Reynolds, R.J. et al. 1997, PASA, 15, 14
Sciama, D.W. 1993, QJRAS, 34, 291
Sciama, D.W. 1998, *Modern Cosmology & The Dark Matter Problem*, Cambridge University Press, 2nd. Ed.
Sembach, K.R., Savage, B.D., Lu, L. & E.M. Murphy, 1995, ApJ, 451, 616
Sembach, K.R., Savage, B.D., Lu, L. & E.M. Murphy, 1998, ApJ, submitted
Snowden, S.L., McCammon, D. & Verter, F. 1993, ApJ, 409, L21
Snowden, S.L. et al. 1991, Science, 252, 1529
Songaila, A. 1981, ApJ, 248, 945
Spergel, D.N., Blitz, L., Teuben, P.J., Hartmann, D., and Burton, B. 1996, BAAS, 189, #61.02
Staveley-Smith, L. et al. 1997, PASA, 13, 243
Sunyaev, R. & Zel'dovich, Ya.B. 1969, Comm. Astr. Spac. Phys., 4, 173
Sutherland, R.S. & Dopita, M.A. 1993, ApJS, 88, 253
Suto, Y., Makishima, K., Ishisaki, Y & Ogasaka, Y. 1996, ApJ, 461, L33
Tully, R.B. 1988, *Nearby Galaxies Catalog*, (Cambridge University Press: New York)
van den Bergh, S. 1994, ApJ, 428, 617
Vogel, S. N., Weymann, R., Rauch, M. & Hamilton, T. 1995, ApJ, 441, 162
Vogeley, M.S. 1998, ApJ, submitted (astro-ph/9711209)
Vogler, A., Pietsch, W. & Kahabka, P. 1996, A&A, 305, 74
Wang, Q.D. & McCray, R. 1993, ApJ, 409, L37
Young, L.M. & Lo, K.Y. 1997, ApJ, 490, 710

A The Local Group

The Local Group has at least 40 members. Some of the outliers may not be bound to the group. The supergalactic coordinates (SGX,SGY,SGZ) are deduced from the NASA/IPAC Extragalactic Database.

name	type	SGX	SGY	SGZ	$\log L_B$
M31	S	0.68	-0.30	0.17	10.48
Galaxy	S	0.00	0.00	0.00	10.30
M33	S	0.71	-0.43	0.00	9.78
LMC	Irr	-0.03	-0.02	-0.03	9.48
SMC	Irr	-0.04	-0.04	-0.01	8.85
IC 10	Irr	0.58	-0.06	0.19	8.70
NGC 3109	Irr	-0.66	0.59	-0.89	8.48
NGC 205	E	0.69	-0.30	0.17	8.48
M32	E	0.68	-0.31	0.17	8.30
NGC 6822	Irr	-0.19	-0.21	0.44	8.30
WLM	Irr	0.13	-0.93	0.13	8.30
NGC 404	E	2.15	-1.15	0.27	8.30
NGC 185	E	0.60	-0.18	0.16	8.30
Leo A	Irr	0.66	1.82	-0.93	8.30
NGC 147	E	0.57	-0.17	0.16	8.00
IC 5152	Irr	-0.77	0.50	-0.01	8.00
IC 1613	Irr	0.30	-0.35	-0.62	8.00
Pegasus	Irr	0.98	-1.36	0.76	7.95
Sextans A	dIrr	-0.30	0.88	-0.80	7.90
Sextans B	dIrr	-0.09	0.94	-0.78	7.70
DDO 210	dIrr	-0.12	-0.37	0.47	7.48
1001-27	dIrr	-0.67	0.56	-0.87	7.00
Fornax	dSph	-0.01	-0.12	-0.07	6.85
DDO 187	dSph	-0.27	1.94	0.89	6.85
DDO 155 (GR8)	dIrr	-0.34	1.49	0.12	6.60
Sculptor	dIrr	-0.01	-0.08	-0.01	6.60
Andromeda I	dSph	0.52	-0.29	0.14	6.30
Andromeda II	dSph	0.52	-0.24	0.13	6.30
Andromeda III	dSph	0.62	-0.31	0.14	6.10
SAGDIG	dIrr	-0.26	-0.23	0.51	6.30
Phoenix	dIrr	-0.11	-0.39	-0.15	6.30
Leo I	dSph	0.08	0.23	-0.12	6.00
Leo II	dSph	0.00	0.19	-0.13	5.90
Tucana	dSph	-0.62	-0.68	-0.01	5.78
Andromeda IV	dSph	0.53	-0.31	0.05	5.78
LGS 3	dIrr	0.46	-0.41	0.04	5.70
Sextans	dSph	-0.02	0.06	-0.05	5.70
Draco	dSph	0.04	0.04	0.05	5.60
Ursa Minor	dSph	0.04	0.04	0.03	5.48
Carina	dSph	-0.04	-0.02	-0.07	5.30

Discussion

van der Hulst: Do your results on the H^+ in the outer path of galaxies imply that the sharp HI edges are still present or does the total gas density distribution smooth out?

Bland-Hawthorn: The total hydrogen distribution is smooth, it simply goes through a rapid phase transition at a critical column.

Staveley-Smith: What is the total mass of the warm ionized gas component in spiral galaxies?

Bland-Hawthorn: 21cm observations have full spatial coverage whereas we only have spot detections in H^+. A simple extrapolation from the Reynold's layer to the outer disk suggests the mass in H^+ can be comparable to HI.

Part 3

COORDINATING MULTI-WAVELENGTH OBSERVATIONS

The New Molonglo Radio Survey

Elaine M. Sadler

School of Physics, University of Sydney, NSW 2006, Australia

Abstract. The Sydney University Molonglo Sky Survey (SUMSS) is a deep imaging survey of the southern radio sky at 843 MHz. Over the next few years, it will cover 8000 square degrees of sky south of declination $-30°$. The survey's resolution (43 arcsec) and sensitivity (3σ detection limit 3–5 mJy) are similar to those of the northern NRAO VLA Sky Survey (NVSS). Some of the scientific problems which can be tackled with such a survey include the clustering of faint radio sources, identification of diffuse and giant sources, and the space distribution and evolution of star–forming galaxies. SUMSS data will be made publicly available as the survey progresses.

1 Introduction

The most powerful radio galaxies act as 'beacons' which can be seen to enormous distances, so radio surveys have played a key role in developing our understanding of the high–redshift universe. Most of what we know about distant radio galaxies comes from studying those detected in 'classic' radio surveys with detection limits of 1 Jy or more. However, such sources are relatively rare (the well–studied 3CR catalogue has only 298 extragalactic sources; Spinrad et al. 1985) and their radio luminosities are often so extreme that observations of the underlying galaxies are strongly affected by the presence of an active nucleus.

Large–area radio–source surveys at mJy levels offer some significant advantages for cosmological studies. Not only do they yield such large numbers of sources that detection of large–scale structure is possible (Cress et al. 1996), they also sample objects which lie at similar redshifts to the Jy–level sources but are 100 to 1000 times less radio–luminous. Thus they provide much more representative probes of galaxy formation and evolution. Furthermore, mJy–level surveys also probe a second cosmologically–significant radio source population, that of star–forming galaxies, which are extremely rare in the strong–source surveys.

A new generation of large–area radio imaging surveys (see Table 1) is now mapping the whole sky with unprecedented sensitivity. Only two of these, the NRAO VLA Sky Survey (NVSS) and the Sydney University Molonglo Sky Survey (SUMSS) cover the southern hemisphere, and only SUMSS covers the region south of declination $-40°$. Here we describe the SUMSS survey, which began in mid–1997, and discuss some of the science which can be done with the survey data.

	FIRST	NVSS	SUMSS	WENSS
Frequency (MHz)	1400	1400	843	325
Area (deg^2)	10,000	33,700	8,000	10,100
Resolution	5″	45″	43″	54″
Detection limit	1 mJy	2.5 mJy	3.5 mJy	15 mJy
Coverage	$\delta > +22°$	$\delta > -40°$	$\delta < -30°$	$\delta > +30°$
Sources/deg^2	90	60	50	21
Reference	(a)	(b)	(c)	(d)

Table 1. Large–area mJy–level radio surveys currently in progress. References: (a) Becker et al. 1995; (b) Condon et al. 1998; (c) This paper, Bock et al. 1998; (d) Rengelink et al. 1997.

2 The Sydney University Molonglo Sky Survey (SUMSS)

The survey uses the Molonglo Observatory Synthesis Telescope (MOST), a 1.6 km east–west cylindrical parabolic reflector which is operated by the University of Sydney and located near Canberra, Australia. In 1996, the telescope was upgraded to provide a 2°.7 diameter field of view (for more details, see Large et al. 1994 and Bock et al. 1998).

Table 2 summarizes some characteristics of synthesis observations with the MOST in its wide–field mode. Each full synthesis observation takes 12 hours and, because it is an east–west array, the MOST performs best at declinations south of −30°. The SUMSS will therefore cover the declination region −30° to −90°.

The rms noise in individual MOST images is a combination of thermal noise and source confusion from both the main beam and sidelobes of the MOST. First results from the survey data show that at southern declinations (near −70°), the rms noise in a single wide–field image is typically 0.7–1.0 mJy/beam near the centre, rising to 1.0–1.2 mJy/beam near the edges of the field. Further north (at declination −35°), the rms noise level is higher (as expected), with a typical value of 1.5 mJy/beam near the centre, rising to 2.5 mJy/beam at the edges. For SUMSS data, the final survey images will be formed into 4° × 4° mosaics to recover uniform sensitivity in the overlap regions.

Frequency	843 MHz (36 cm)
Bandwidth	3 MHz
Polarization	Right–hand circular (IEEE)
Declination range (12 hr synthesis)	$-90° \leq \delta < -30°$
Synthesized beam (FWHM)	$43″ \times 43″\mathrm{cosec}|\delta|$
Image area	Elliptical, $163' \times 163'\mathrm{cosec}|\delta|$
Noise after 12 hr (1σ)	0.7–1.5 mJy/beam
Dynamic range (typical)	250:1

Table 2. Key characteristics of individual SUMSS observations

The New Molonglo Radio Survey   105

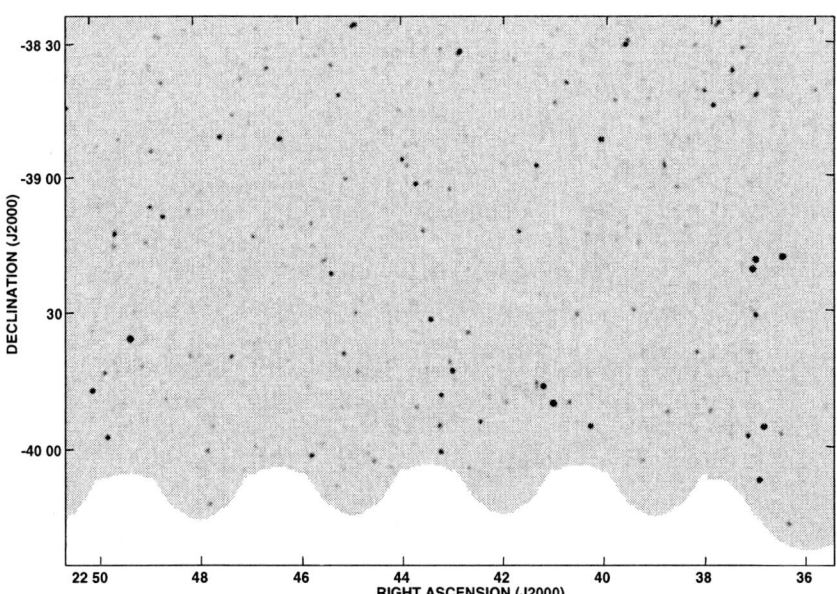

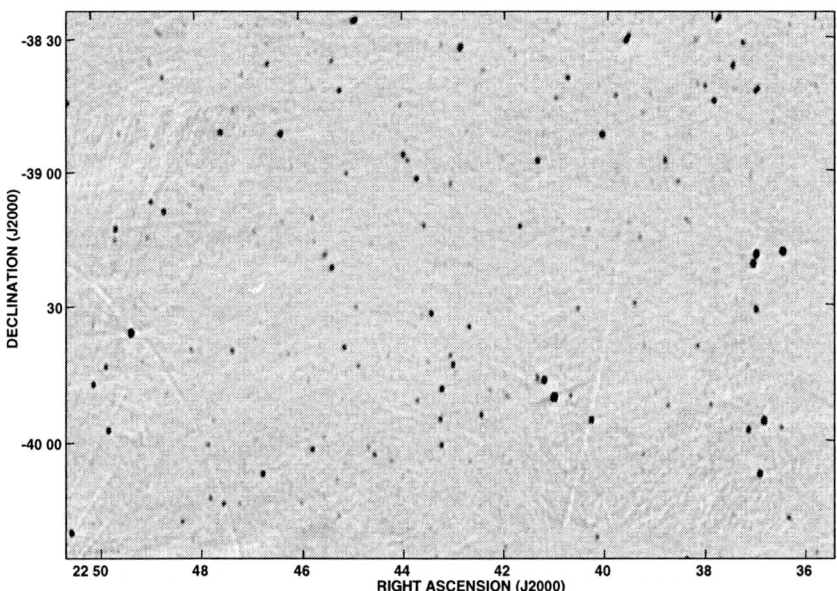

Fig. 1. NVSS and SUMSS radio images of the $6\,\mathrm{deg}^2$ ESO Imaging Survey (EIS) field centred at (J2000) 22 42 54 −39 28 00. The southern limit of the NVSS survey at declination $-40°$ is responsible for the gaps in the top image.

Figure 1 shows radio continuum images from the NVSS (1.4 GHz) and SUMSS (843 MHz) for a 6 deg^2 region overlapping one of the ESO Imaging Survey (EIS) fields. The two surveys are well matched in sensitivity and resolution, and overlap in the declination zone $-30°$ to $-40°$.

There are 2713 SUMSS field centres south of declination $-30°$, each with a fully–synthesised area of $2°.7 \times 2°.7\mathrm{cosec}(\delta)$ ($= 7.3\,\mathrm{cosec}\delta$ deg^2) at declination δ. We use hexagonal close packing for the survey field centres, and the rate of progress (i.e. area added to that already surveyed) is 4.4 square degrees per field. The survey observations are made at night to reduce the effects of solar interference. We can therefore cover about 1000 deg^2/year. The survey began in mid–1997, and our aim is to complete the 1118 fields with Galactic latitude $|b| > 30°$ by mid–2000. The entire southern–sky survey is nevertheless a long-term project which will take up to eight years to complete.

3 The Faint Radio Source Population

Deep radio surveys of a few small areas of sky at 1.4 GHz (Windhorst et al. 1985, Condon 1984; see also Condon 1992) show that there are two astrophysically distinct populations of extragalactic radio sources. Over 95% of the sources above about 50 mJy are classical radio galaxies and quasars (median redshift z$\sim$1) powered by "active galactic nuclei" (AGN), while the remaining sources are identified with star–forming galaxies (median z$\sim$0.1). The fraction of star-forming galaxies increases rapidly below 10 mJy, and below 1 mJy they begin to be the dominant population.

In a poster paper presented at this meeting, Jackson & Wall (1998) predict the radio source population mix at 843 MHz as a function of flux density. Their work suggests that at the lowest flux densities probed by SUMSS (3–5 mJy), about 20% of sources will be associated with star–forming galaxies and the remainder with various kinds of AGN (FR I and FR II radio galaxies, quasars and BL Lac objects).

4 Science with the New Survey

While some important science can be tackled with the SUMSS data alone, the value of the survey will be enormously enhanced when it is combined with data at other wavelengths.

From the SUMSS data alone, we can study the clustering of faint radio sources on scales of a few hundred Mpc. If we can tie the SUMSS and NVSS data together effectively (using the overlap zone at $-30°$ to $-40°$ declination), then we will also be able to produce the deepest–ever 'all–sky' radio image.

In the overlap zone we can also combine the NVSS and SUMSS data to measure a radio spectral index between 843 MHz and 1.4 GHz. Figure 2 shows the results for a small region near the South Galactic Pole. About 2–3% of sources appear to have very steep spectra ($\alpha < -1.6$), and are candidate ultra-steep spectrum (USS) sources. Many of these are likely to be very distant (e.g. Röttgering et al. 1997a).

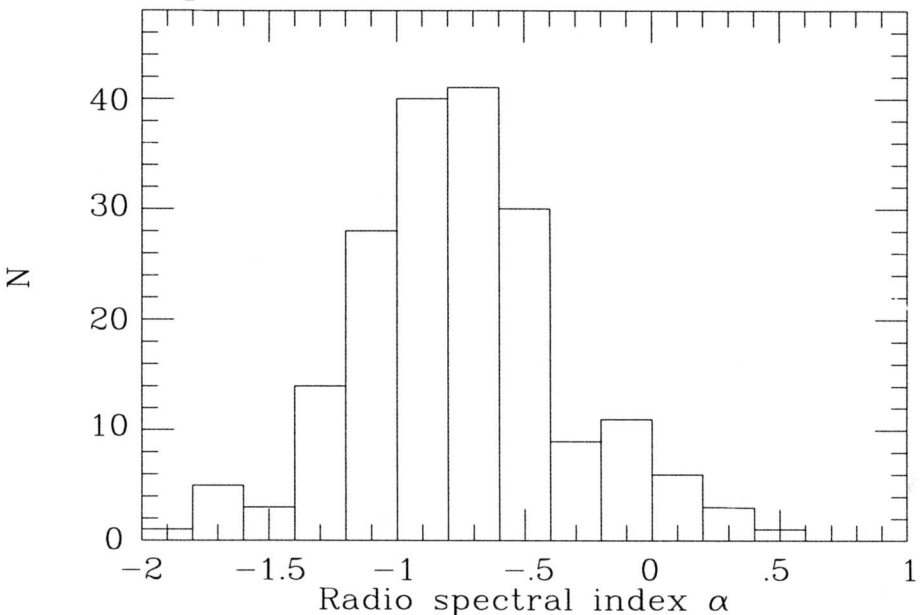

Fig. 2. Radio spectral index distribution for 195 radio sources with $S_{843} > 10$ mJy observed by both SUMSS and NVSS in the South Galactic Pole region. The spectral index $\alpha_{843}^{1.4}$ has a median value of -0.8.

Because of its good coverage of short spacings in the (u, v) plane, the MOST is more sensitive to large-scale, diffuse radio emission than other synthesis telescopes. SUMSS should therefore be efficient at detecting Mpc-scale diffuse radio sources like the giant cluster source A3667 (Röttgering et al. 1997b).

The SUMSS radio positions are accurate enough (typical rms errors of 1–2 arcsec for flux densities above 5 mJy) that unambiguous optical identifications can be made from digitized sky survey (DSS) data. The identification rate is about 30% down to a limiting magnitude of $b_J \sim 22$. Deep CCD data (e.g. from the ESO Imaging Survey) should increase the identification rate to as high as 80–90%.

With follow-up optical spectroscopy we can distinguish AGN from starburst galaxies, measure redshifts, construct luminosity functions and study the evolution of different radio-source populations. We are currently collaborating with the 2dF Galaxy Redshift Survey team (see Colless 1998, this meeting) to obtain spectra of the 2% of their target galaxies which are also NVSS or SUMSS radio sources. This will ultimately yield about 5000 spectra of radio-detected galaxies brighter than b_J ~19.5. We also plan to carry out a deeper 2dF study in a smaller area of sky to probe fainter radio–source counterparts down to b_J ~22.5. Galaxies fainter than this will probably require 8 m class telescopes for follow–up.

The SUMSS data are well–suited to multi–wavelength studies, such as cross–identification with X–ray or infrared catalogues. The SUMSS images and data catalogue will be made publicly available as the survey proceeds, and we welcome collaboration with other groups. Examples of possible collaborative projects could be deeper radio observations of fields which will be studied in detail at other wavelengths, requests for specific areas or objects to be observed early in the survey, or follow–up of targets of opportunity such as X–ray transients.

I would like to acknowledge the efforts of my colleagues in the SUMSS team at the University of Sydney (Duncan Campbell–Wilson, Lawrence Cram, David Crawford, Gene Davidson, Ralph Davison, Anne Green, Richard Hunstead, Carole Jackson, Michael Large, Vincent McIntyre, Bruce McAdam, Barbara Piestrzynski, Gordon Robertson, Tony Turtle, Jeff Webb and Michael White), who have made this survey possible.

References

Bock, D., Large, M.I., Sadler, E.M. (1998). In preparation.
Becker, R.H., White, R.L., Helfand, D.J. (1995). ApJ 450, 559.
Colless, M.M., 1998. This meeting.
Condon, J.J. (1984). ApJ 287, 461.
Condon, J.J. (1992). ARA&A 30, 575.
Condon, J.J., Cotton, W.D., Greisen, E.R., Yin, Q.F., Perley, R.A., Taylor, G.B., Broderick J.J. (1998). AJ, in press.
Cress, C.M., Helfand, D.J., Becker, R.H., Gregg, M.D., White, R.L. (1996). AJ 473, 7.
Jackson, C.A. & Wall, J.V., 1998. This meeting.
Large, M.I., Campbell–Wilson, D., Cram, L.E., Davison, R., Robertson, J.G. (1994). PASA 11, 44.
Rengelink, R.B., Tang, Y., de Bruyn, A.G., Miley, G.K., Bremer, M.N., Röttgering, H.J.A., Bremer, M.N. (1997). A&AS 124, 259.
Röttgering, H.J.A., van Ojik, R., Miley, G.K., Chambers, K.C., van Breugel, W.J.M., de Koff, S. (1997a). A&A 326, 505.
Röttgering, H.J.A., Wieringa, M.H., Hunstead, R.W., Ekers, R.D. (1997b). MNRAS 290, 577.
Spinrad, H., Djorgovski, S., Marr, J., Aguilar, L. (1985). PASP 97, 932.
Windhorst, R.A., Miley, G.K., Owen, F.N., Kron. R.G., Koo, D.C. (1985). ApJ 289, 494.

Discussion

Boyle: Cross-correlating the SUMSS with the HQS would yield a very powerful sample of both 'radio-loud' and 'radio-quiet' QSOs/AGN at low redshift.

Sadler: Yes, we will certainly look at ways to do this.

Tsarevsky: What is the intersection size between the Molonglo Radio Survey and the NRAO-VLA Sky Survey?

Sadler: The most northern declination of MRS is -30 degree, therefore the two surveys have a 10 degree wide overlap zone.

Lahav: What algorithm is used to detect sources in SUMSS?

Sadler: At the moment, we are using the USAD package in AIPS; i.e. the same algorithm used by the NUSS.

Testing Models of Radio Source Space Density Evolution with the SUMSS Survey

C.A. Jackson[1] and J.V. Wall[2]

[1] School of Physics, University of Sydney, NSW 2006 Australia
[2] Royal Greenwich Observatory, Madingley Road, Cambridge CB3 0EZ, UK

Abstract. We present the population data as predicted by our space density analysis at the SUMSS frequency of 843 MHz. This data demonstrates the potential of the SUMSS survey to trace in detail the change-over in radio source populations from 'radio monsters' at high flux densities to a local starbursting population at the survey limit. The exact form of the observed change-over will be used to refine our models of radio source evolution.

Extragalactic Radio Sources at 843 MHz

Our recent analyses (Wall & Jackson 1997 and Jackson 1997) have shown that the powerful radio source populations are well described by a dual-population unification scheme. By adopting an elegantly simple aspect-dependent paradigm we have determined that the two 'parent' populations have undergone quite separate cosmic evolution histories with the result that their space density distributions are very different. Whilst our working hypothesis is straightforward, the results from our analysis are found to be comprehensively supported by cosmological tests embodied in radio source count and identification data over a wide radio frequency range (151 MHz – 8.4 GHz).

The Sydney University Molonglo sky survey ('SUMSS', E. Sadler these proceedings) at 843 MHz can be considered to be an 'intermediate' radio frequency survey, lying between the 'low' frequency surveys ($\nu < 200$ MHz, $e.g.$ 3C, 6C, 7C) and those at 'high' frequency ($\nu > 2$ GHz, $e.g.$ Parkes, 87GB, PMN). The change in radio source types which comprise the source count between the low and high frequencies has been shown to be a natural consequence of the increasing contribution from the preferentially-aligned sources which are Doppler-beamed along or very close to our line-of-sight. Surveys below ~ 200 MHz are almost completely dominated by extended radio galaxies whilst amongst the brightest radio sources at 5 GHz there are almost equal numbers of compact and extended sources.

We have applied our dual-population space-density model to the SUMSS frequency and show the predicted integral population mix in Figure 1. The fraction of bright ($S_{843\,\mathrm{MHz}} > 1$ Jy) sources which will appear flat-spectrum ($\alpha_{843\,\mathrm{MHz}}^{5\,\mathrm{GHz}} > -0.5$) is determined to be ~ 10 % from a fit to the well-defined source count at 1.4 GHz. The changing contribution from each source type is a result of the differential evolution of the underlying 'parent' populations in the radio luminosity function.

Figure 1 shows that at high flux density ($S_{843\,\mathrm{MHz}} > 0.1$ Jy) SUMSS will be dominated by the high-power population: FRII radio galaxies with strong optical emission lines (spectral class 'A', Hine & Longair 1979) and quasars, their beamed counterparts. There is also a significant contribution from the lower-power population: radio sources which have only weak, if any, optical emission lines (*i.e.* FRIs and low-excitation FRIIs, spectral class 'B', Hine & Longair 1979 and Laing *et al.* 1994). In contrast, at the milli-jansky flux density level the dominant populations are expected to be the starburst and Seyfert galaxies, the low-excitation radio galaxies and their beamed counterparts (BL Lac-type sources).

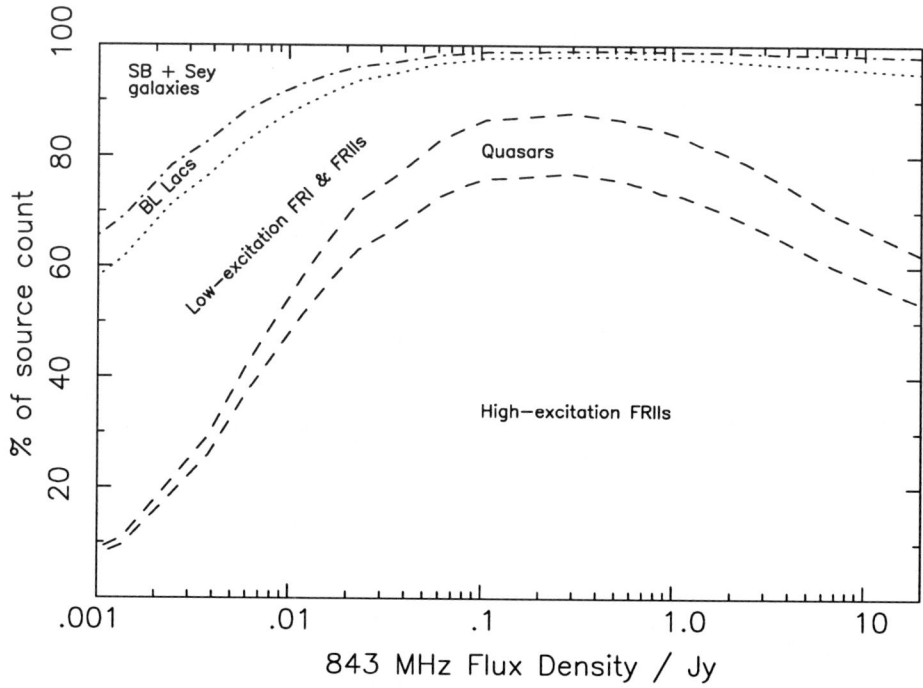

Fig. 1. Predicted population mix at 843 MHz as a function of flux density.

References

Hine, R. G., Longair, M. S. (1979) *MNRAS* **188**, 111–130
Jackson, C. A. (1997) PhD thesis, University of Cambridge
Laing, R. A., Wall, J. V., Jenkins, C. R., Unger, S. W. (1994) *ASP Conf Ser Vol 54, The Physics of Active Galactic Nuclei* (ASP, San Francisco), 201–208
Wall, J. V., Jackson, C. A. (1997) *MNRAS* **290**, L17–L22

Variance and Skewness in the FIRST Survey

M. Magliocchetti[1], S.J. Maddox[1], O. Lahav[1], J.V. Wall[2]

[1] Institute of Astronomy, Madingley Road, Cambridge CB3 0HA
[2] Royal Greenwich Observatory, Madingley Road, Cambridge CB3 0EZ

Abstract. We investigate the large-scale clustering of radio sources by analysing the distribution function of the FIRST 1.4 GHz survey. We select a reliable galaxy sample from the FIRST catalogue, paying particular attention to the definition of single radio sources from the multiple components listed in the FIRST catalogue. We estimate the variance, Ψ_2, and skewness, Ψ_3, of the distribution function for the best galaxy subsample. Ψ_2 shows power-law behaviour as a function of cell size, with an amplitude corresponding a spatial correlation length of $r_0 \sim 10h^{-1}$ Mpc.
We detect significant skewness in the distribution, and find that it is related to the variance through the relation $\Psi_3 = S_3(\Psi_2)^\alpha$ with $\alpha = 1.9 \pm 0.1$ consistent with the non-linear growth of perturbations from primordial Gaussian initial conditions. We show that the amplitude of clustering (corresponding to a spatial correlation length of $r_0 \sim 10h^{-1}$ Mpc) and skewness are consistent with realistic models of galaxy clustering.

Introduction and Results

A simple and highly informative statistical description of galaxy clustering is given by the galaxy distribution function (*Counts in Cells*) i.e. the probability for finding N galaxies in a cell of particular size and shape. Furthermore it can be shown (Peebles, 1980) that the higher order moments of the galaxy distribution can be used as a test of non-linear models for large-scale structure We present here the counts in cells analysis carried out for the FIRST radio survey; we will focus on the second and third moment of the distribution function of the sample and compare to the predictions of cosmological models.

To ensure completeness we have considered only those sources with fluxes greater than 3mJy. The high resolution of the survey complicates our clustering analysis because multicomponent sources are split into several objects which are listed as separate entries in the catalogue. We generated a new sample in which we consider pairs of sources as physically associated if they have a separation less than a well defined flux-separation relationship (Magliocchetti et al., 1998) and fluxes satisfying the relationship $\frac{flux1}{flux2} \leq 4$. We combine each of these associations into a single object which has the sum of the individual fluxes. Isolated sources are left unchanged.

We then divided the sample into square cells and evaluated the galaxy distribution function for cell sizes Θ ranging from $0.3°$ to $3°$. The normalized variance of the distribution, $\Psi_2 = (\mu_2 - \bar{N})/N^2$, is related to the two-point correlation function $w(\theta) = A\theta^{(1-\gamma)}$, through $\sigma^2 = \Omega^2 \Psi_2 = \int A\theta^{(1-\gamma)} d\Omega_1\, d\Omega_2$, with $\Omega = \Theta \times \Theta$. From the analysis of figure 1(a) we get: $\gamma = 2.48 \pm 0.06$; $A = (1.13 \pm 0.07) \cdot 10^{-3}$.

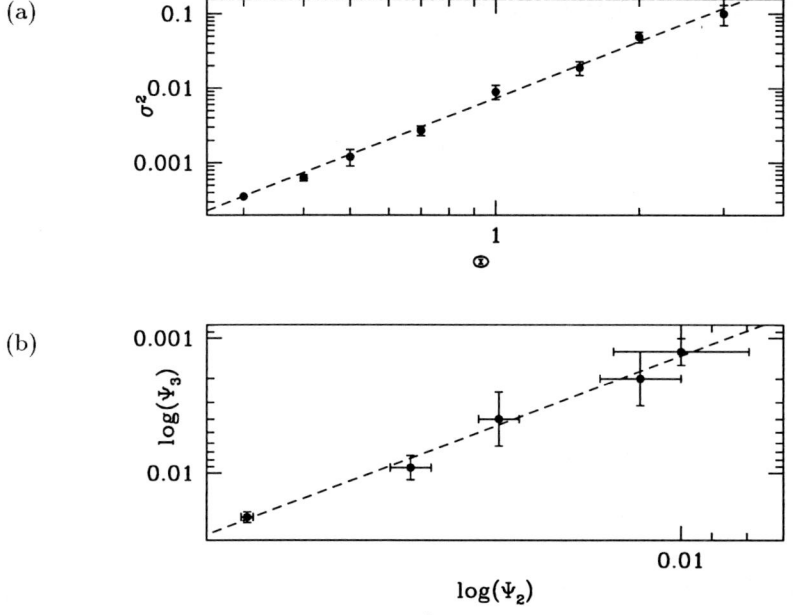

Fig. 1. (a) The normalised variance σ^2 vs. the cell size Θ. (b) Skewness Ψ_3 vs. the variance Ψ_2. Errors in both the cases are estimated from the variance in four random subset.

Other important information on the non-linear clustering of objects can be gained through the analysis of higher moments, such as the skewness of the distribution function. The observed distribution function shows a significant tail towards higher numbers of sources per cell. The tail becomes more noticeable as the size of the cells is reduced. In more detail, plotting the skewness Ψ_3 as a function of the variance Ψ_2 (figure 1(b)) shows that they are related by the expression $\Psi_3 = S_3 \Psi_2^\alpha$, with $\alpha = 1.9 \pm 0.1$, and $S_3(\Theta) = const$ as predicted by the theory of Gaussian hierarchical primordial perturbations. By then assuming a "standard" slope $\gamma = 1.8$ the angular values lead, when deprojected in the case of constant clustering in proper coordinates ($\epsilon = 0$), to a correlation length $r_0 \sim 10h^{-1}$ Mpc and to a spatial skewness $S_3^*(R) \sim 4.2$.

References

Magliocchetti M., Maddox S.J., Lahav O., Wall J.V., 1998; submitted to MNRAS
Peebles P.J.E., 1980; *The Large-Scale Structure of the Universe*, Princeton University Press

The AT–ESP Radio Survey: Goals, Description, First Results

I. Prandoni[1], L. Gregorini[1], P. Parma[1], R.H. de Ruiter[2], G. Vettolani[1], M.H. Wieringa[3] and R.D. Ekers[3]

[1] Istituto di Radioastronomia del CNR, Via Gobetti 101, 40129 Bologna, Italy
[2] Osservatorio Astronomico di Bologna, Via Zamboni 33, 40126 Bologna, Italy
[3] Australia Telescope National Facility, P.O. Box 76, Epping NSW 2121, Australia

Abstract. We used the Australia Telescope Compact Array to deeply image at 20 cm the region covered by the ESO Slice Project (ESP) galaxy red-shift survey (Vettolani et al. 1997). The survey produced 16 radio mosaics with $\sim 8'' \times 14''$ resolution and ~ 70 μJy sensitivity (1σ), over a total area of ~ 26 sq. degrees at $\delta \sim -40°$ in the South Galactic Pole (SGP) region.

First results indicate that we will detect ~ 3000 radio sources above 6σ (~ 0.4 mJy), more than 1000 being sub-mJy sources. The availability of a new homogeneous and relatively deep catalogue containing thousands of radio sources will allow us to undertake reliable statistical studies in the field of low luminosity and/or high red-shift radio sources, and will be especially useful in understanding the nature of the sub–mJy population.

On the other hand, the availability of red-shift information for the ESP sample (3342 galaxies down to $b_J \sim 19.4$) represents a good starting point for the optical identification follow up. More generally, it allows us to study the existing correlations between optical (luminosity, spectral activity, etc.) and radio properties of the ESP galaxies ($z < 0.2$) to be compared with what is known from local samples of elliptical and spiral galaxies. Such analysis represents in fact a second independent goal of our survey.

In the present paper we give a detailed description the AT–ESP radio survey and report some of its preliminary results.

1 Introduction

Normalized radio number counts show a flattening below a few mJy (see solid line in Fig. 1), corresponding to a steepening in the actual observed counts. This flattening is generally interpreted as due to the occurrence of a new population of radio sources, which does not appear at higher flux densities (Windhorst et al. 1990, Kellermann & Wall 1987).

Classical radio galaxies, powered by AGN and typically hosted by giant elliptical galaxies, are known to dominate the counts at high fluxes (99% above 60 mJy), but their contribution is negligible at fainter fluxes. There is a general agreement in interpreting the radio emission from the sub–mJy sources as being due to star-bursts, even though some nuclear activity cannot be excluded (e.g. Hammer et al. 1995). On the other hand, very controversial is the distance distribution of such objects. They could be either local ($z < 0.1$) low luminosity radio sources, (Wall et al. 1986; Subrahmanya & Kapahi 1983), or actively star forming galaxies at intermediate to high redshifts (e.g. Windhorst 1984; Kron et al. 1985; Windhorst et al. 1985, 1987; Oort 1987, Rowan–Robinson et al. 1993). In the latter case, the sub–mJy population would be of fundamental cosmological importance, the flattening in the radio counts being tightly related to the well–known problem of the excess in the faint blue counts.

Due to the long observing times required to reach deep fluxes, the available samples in the sub–mJy region are small (10–100 objects, Windhorst et al. 1995, de Ruiter et al. 1997, Gruppioni et al. 1997) and the nature of the sub–mJy population is still under debate.

The main goal of this project is therefore to produce a large sample of sub–mJy sources to be followed up at other wavelengths, so as to allow a reliable statistical study of this population.

This is now possible thanks to the radio mosaic observing technique, which is a powerful tool for imaging large areas of sky in a reasonable amount of time.

Mapping the region covered by the ESP red-shift survey allows us to get immediate information on the distance distribution of local ($z < 0.2$) radio sources, so as to assess their contribution to the sub–mJy population as well as their radio–optical properties.

2 The Survey

We used the 6 km array of the Australia Telescope Compact Array (ATCA) to image the region covered by the ESP survey at 20 cm.

The mosaic observing mode, recently implemented at the ATCA, allows efficient coverage of large areas of sky by interleaving short observations (snapshots) of a grid of pointings. The main issue to be decided in planning a mosaic experiment is the pointing grid pattern, that depends on the scientific use of the observations.

When dealing with a point source detection experiment aimed to statistical studies, the two main requirements are *deep* and *uniform* sensitivity.

As far as the pointing grid is properly set, the mosaic technique allows to get uniform noise. It can be demonstrated that this requirement is satisfied when the pointing spacings are close to $FWHP/\sqrt{2}$, where $FWHP$ is the beam full width at half power (for ATCA 20 cm observations $FWHP = 33'$). We then carried out simulations to find the best grid configuration in terms of both minimum number of pointings required to complete the survey and uniform sensitivity. The final choice was a rectangular grid with $20'$ spacings.

The $22° \times 1°$ and $5° \times 1°$ areas of the ESP survey would thus be covered with 69×4 and 15×4 pointings. On the other hand, a region of 1.3 sq. degrees, corresponding to 4×4 pointings, has not been observed due to the presence of a strong radio source which prevented us from reaching the deep noise level required. This reduced the total number of fields observed to 320.

Observing times of the order of 1.2 h per field (2×128 MHz bandwidth) were required to reach a sensitivity of $\sim 70\mu$Jy (1σ).

The observing campaign (33 blocks of 12^h) started in November '94 and was completed in January '96. A log of the observations is given in Table 1 (dates, arrays used and observing frequencies). The two 128 MHz observing bands were set in the most interference–free region of the 20 cm band. Some readjustments were required for the second band due to technical problems.

Table 1. Log of the observations

Date	Obs. Time	Array	ν_1 (MHz)	ν_2 (MHz)
18/11/94–21/11/94	3×12^h	6D	1344	1452
23/12/94–04/01/95	13×12^h	6A	1344	1452
15/12/95–01/01/96	17×12^h	6C	1344	1448

Due to the large amount of data to be processed, semi–automated procedures were built for the reduction. The radio survey produced 16 overlapping mosaiced maps with spatial resolution $\sim 8'' \times 14''$. As expected, the noise level ($\sim 70\mu$Jy) is fairly uniform within each map and from map to map ($< 10\%$ variation). Dynamic range problems cause slightly higher levels around strong sources (typically $\sim 80\,\mu$Jy for $S_{peak} > 50$ mJy sources).

3 The Catalogue

We are currently listing all the radio sources above a 6σ-threshold (~ 0.4 mJy), present in the region surveyed, using the automated algorithms for radio sources extraction and components parameterization which recently became available (IMSAD, VSAD, etc.).

To date the list of all the radio sources down to $S_{peak} = 1$ mJy is available (1752 fitted components). This represents a complete and very reliable catalogue ($S/N > 10$), that will be especially useful to study the population of mJy radio sources.

For fainter sources the extraction is more critical and problems such as incompleteness and contamination should be carefully investigated before releasing the final catalogue. Nevertheless, from our preliminary list we expect a total number of $\sim 1000/1500$ sub-mJy sources in the 26 sq. degrees area surveyed.

4 Number Counts

The $\log N - \log S$ relation derived from our preliminary catalogue is in very good agreement with previous determinations at 1.4 GHz (Fig. 1). The AT-ESP counts (filled circles) are compared to the FIRST ones (White et al. 1997, triangles) and to fainter counts obtained by Gruppioni et al. (1997) in the Marano Field (squares). The interpolation determined by Windhorst et al. (1990) from a collection of 10575 radio sources belonging to 24 different surveys at 1.4 GHz, is also shown (solid line).

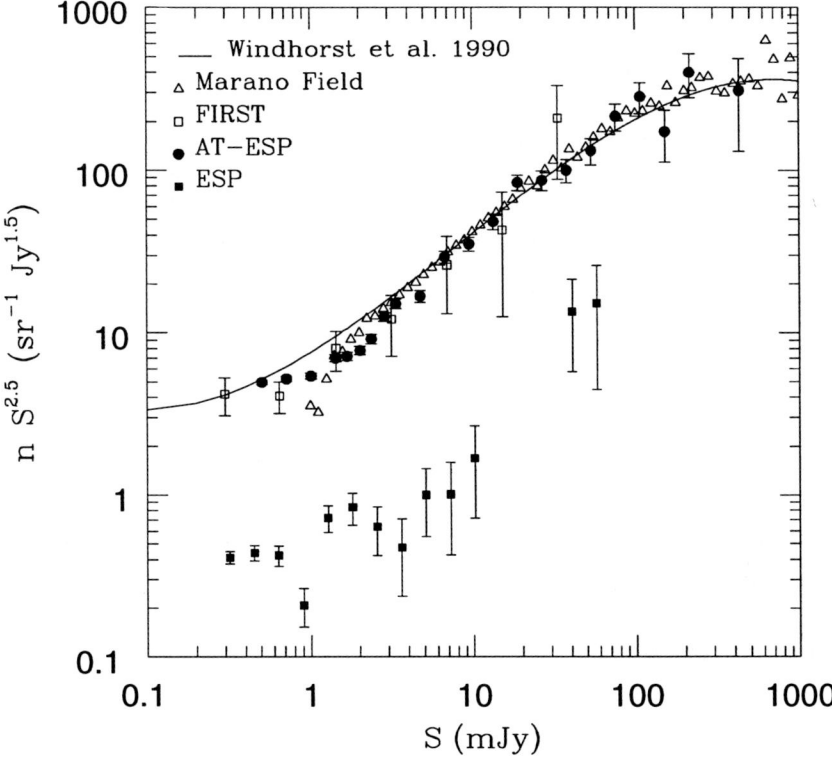

Fig. 1. 1.4 GHz normalized radio number counts as a function of flux for different samples: the FIRST (triangles), the Marano Field (squares), the AT–ESP (filled circles) and the ESP radio counterparts only (filled squares). The fit obtained by Windhorst et al. (1990) is also shown (solid line).

Over the flux range 2–30 mJy, the FIRST counts are the most accurate available today (their statistical errors are very small and have not been plotted to prevent overcrowding). Our counts, on the other hand, will provide, when in their final form, the best determination of the counts shape at fainter fluxes, where the FIRST counts become incomplete. Such shape will represent a strong observational constraint on the evolutionary models for the sub–mJy sources.

Another observational constraint comes from the counts obtained from the radio sources associated with the ESP galaxies (filled squares in Fig. 1; this sub–sample is described in more detail in Sect. 5). Although a direct comparison between a radio– and an optically–selected sample is not possible, we can nevertheless consider the ESP radio counterparts as the local ($z < 0.2$) contribution to the general population of radio sources. Such a contribution is estimated to be $\sim 10\%$ at the sub–mJy level, ruling out the non–evolving scenario for the sub–mJy sources (Wall et al. 1986, Subrahmanya & Kapahi 1983).

5 Radio Properties of ESP Galaxies

As already mentioned, a second goal of the AT–ESP survey is the analysis of the radio–optical properties of the ESP galaxies.

The radio counterparts of ESP galaxies have been searched down to ~ 0.2 mJy (corresponding to a 3σ-threshold, as allowed when sky positions are known). To extract the identifications' sample we used a modified version of the well-known likelihood ratio test (de Ruiter et al. 1977). Such a statistical approach allowed us to define a sample of 491 radio sources associated with the ESP galaxies, corresponding to a detection rate of $\sim 15\%$. The completeness of the sample is estimated to be $\simeq 96\%$, and the reliability $\simeq 92\%$.

Typically, radio detected ESP galaxies are associated with very faint and point-like radio sources ($\sim 80\%$ of them have $S_{peak} < 1\,\mathrm{mJy}$).

As expected, a large fraction ($\sim 60\%$) of the galaxies detected shows one or more emission lines. This suggests that, in normal galaxies, radio emission is mostly induced by star formation, traced by the [OII] line (Kennicutt 1983, Kennicutt 1992).

The same evidence comes from the cumulative distribution of galaxies with and without emission lines as a function of the radio to optical luminosity ratio, R (Condon 1980). For R values below ~ 100 the probability of being a radio source is higher for galaxies which show line activity than for galaxies which do not (see Fig. 2).

References

Condon, J.J., 1980, ApJ, 242, 894
Hammer, F., Crampton, D., Lilly, S.J., Le Fevre, O., Kenet, T., 1995, MNRAS, 276, 1085
Kennicutt, R.C., 1983, A&A, **120**, 219
Kennicutt, R.C., 1992, ApJ, **388**, 310
Kron, R.G., Koo, D.C., Windhorst, R.A., 1985, A&A, 146, 38
Gruppioni, C., Zamorani, G., de Ruiter, H.R., Parma, P., Mignoli, M., Lari, C., 1997, MNRAS, in press
Kellermann, K.I. & Wall, J.V., 1987, in *Observational Cosmology*, IAU Symp. No. 124, eds. Hewitt et al., p. 545
Oort, M.J.A., 1987, Ph.D. Thesis, University of Leiden, Leiden
Rowan–Robinson, M., Benn, C.R., Lawrence, A., McMahon, R.G., Broadhurst, T.J., 1993, MNRAS, 263, 123
de Ruiter, H.R., Willis, A.G., Arp, H.C., 1977, A&AS, 28, 211
de Ruiter, H.R., Zamorani, G., Parma, P., Hasinger, G., Hartner, G., Trumper, J., Burg, R., Giacconi, R., Schmidt, M., 1997, A&A, in press

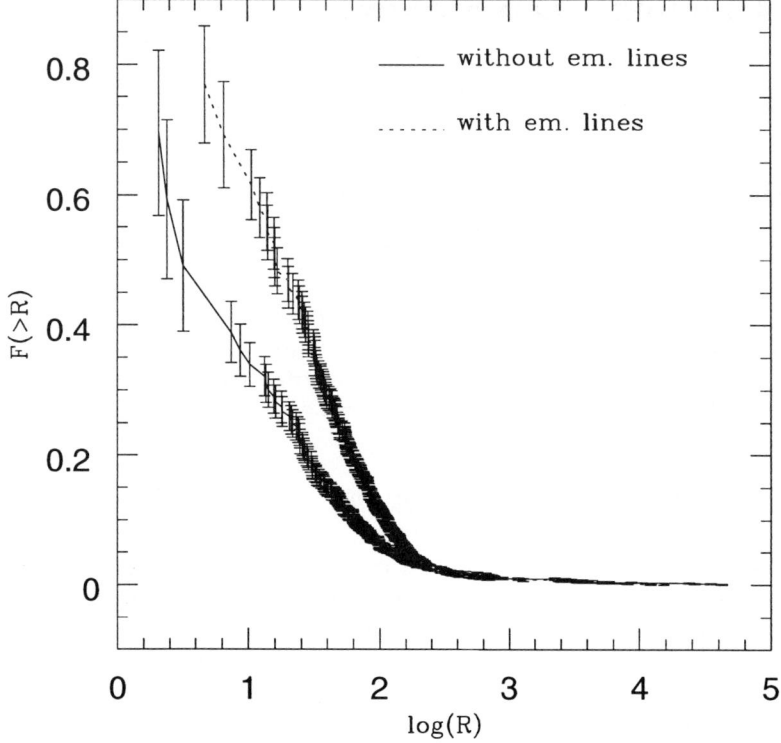

Fig. 2. Bivariate radio luminosity function for ESP galaxies with (dot line) and without (solid line) emission lines as a function of the "radio excess" R.

Subrahmanya, C.R., & Kapahi, V.K., 1983, in *The Early Evolution of the Universe and its Present Structure*, IAU Symp. No. 104, eds. Abell G.O. & Chincarini G., p. 47
Vettolani, G., Zucca, E., Zamorani, G., Cappi, A., Merighi, R., Mignoli, M., Stirpe, G.M., MacGillivray, H.T., Collins, C.A., Balkowski, C., Cayatte, V., Maurogordato, S., Proust, D., Chincarini, G., Guzzo, L., Maccagni, D., Scaramella, R., Blanchard, A., Ramella, M., 1997, A&A, 325, 954
Wall, J.V., Benn, C.R., Grueff, G., Vigotti, M., 1986, in *Highlights of Astronomy*, ed. Swings J.P., Vol. 7, p. 345
White, R.L., Becker, R.H., Helfand, D.J., Gregg, M.D., 1997, ApJ, 475, 479
Windhorst, R.A., 1984, Ph.D. Thesis, University of Leiden, Leiden
Windhorst, R.A., Dressler, A., Koo, D.C., 1987, in *Observational Cosmology*, IAU Symp. No. 124, eds. Hewitt et al., p. 573
Windhorst, R.A., Fomalont, E.B., Kellermann, K.I., Partridge, R.B., Richards, E., Franklin, B.E., Pascarelle, S.M., Griffiths, R.E., 1995, Nature, 375, 471
Windhorst, R.A., Miley, G.K., Owen, F.N., Kron, R.G., Koo, D.C., 1985, ApJ, 289, 494
Windhorst, R.A., Mathis, D., Neuschaefer, L., 1990, in *Evolution of the Universe of Galaxies*, ed. R.G. Kron, p. 389

The Phoenix Deep Survey

Andrew Hopkins[1], Lawrence Cram[1], Bahram Mobasher[2]
and Antonis Georgakakis[2]

[1] School of Physics, University of Sydney, NSW, Australia 2006
[2] Blackett Laboratory, Imperial College, London SW7 2BZ, England

Abstract. A brief description of the *Phoenix Deep Survey* project is presented. A model of the bivariate (radio/optical) luminosity function is described and used to predict the bivariate source counts and the redshift distributions for different combinations of limiting flux densities. Comparisons are drawn with observations and discussed.

1 Observations and Catalogues

The *Phoenix Deep Survey* (PDS) is a multiwavelength survey covering a two degree diameter region of sky. The PDS (named for the constellation in which it lies) includes observations with the Anglo-Australian Telescope (AAT) at R- and V-band, with the Molonglo Observatory Synthesis Telescope (MOST) at 843 MHz, and with the Australia Telescope Compact Array (ATCA) at 1.4 GHz. In addition, spectroscopy of selected sources and observations at H- and K-band for a number of sources have been carried out. The details of the radio observations are described in Hopkins et al. (1998). Catalogues of 1079 radio sources brighter than 0.1 mJy and ~ 40000 optical galaxies brighter than $R = 22$ have been compiled. 544 of the optical galaxies have been identified as counterparts to radio sources, and 133 redshifts have been obtained. Investigations of numerous aspects of this database are progressing, but just one is presented here.

2 The Bivariate Luminosity Function

The bivariate luminosity function (BLF) of galaxies, $\Phi(L_1, L_2, z)$, is defined as the volume density of galaxies per unit of luminosity squared at redshift z with luminosities in the intervals $[L_1, L_1 + \mathrm{d}L_1]$ and $[L_2, L_2 + \mathrm{d}L_2]$. The BLF has the property that it must integrate over each of the luminosity variables to reproduce the luminosity function of the other variable, i.e.,

$$\int_{L_1} \Phi(L_1, L_2, z) \mathrm{d}L_1 = \Phi(L_2, z), \quad (1)$$

and vice versa. It can be used to generate a *bivariate* source count distribution (BSC), analogously to the way source counts are predicted from a univariate luminosity function. The BSC is the number of galaxies per unit area of sky with flux densities in the range $[S_1, S_1 + \mathrm{d}S_1]$ and $[S_2, S_2 + \mathrm{d}S_2]$. In a similar fashion to the BLF, the BSC integrates over each flux density variable to reproduce the source counts of the other flux density variable.

The model BLF was constructed by first considering the function

$$\Psi(L_1, L_2, z) = \frac{\Phi(L_1, L_2, z)}{\Phi(L_2, z)}. \qquad (2)$$

If both $\Phi(L_2, z)$ and $\Psi(L_1, L_2, z)$ are known, the BLF can be constructed simply by rearranging Equation (2). This method for constructing a BLF is not new, and has been used by other authors with $\Psi(L_1, L_2, z)$ given by the observed distribution of $r = L_1/L_2$, (i.e. $\Psi(L_1, L_2, z) \Rightarrow \Psi(r)$), (Corbelli et al. (1991), Toffolatti et al. 1987, Meurs & Wilson 1984, Elvis et al. 1978). This method can be used for multiple galaxy populations, the final BLF being the sum of those from each population. The goal of constructing a model radio-optical BLF, $\Phi(L_{1.4}, L_R, z)$, now becomes one of choosing suitable functions for $\Psi(r = L_{1.4}/L_R)$ (or $\Psi(r, z)$ in general) and $\Phi(L_R, z)$.

2.1 The Model

A two population model consisting of "starbursts" and "AGN" has been adopted. The assumption is made that starbursts occur predominantly in spiral galaxies and AGN in ellipticals. This allows the $\Phi(L_R, z)$ chosen for the starburst and AGN populations to be the optical (Schechter) luminosity functions for the corresponding morphological types. Models for $\Psi(r)$ consistent with observations from Corbelli et al. (1991), (starbursts), and Sadler et al. (1989), (AGN), were developed. It is emphasised here that this is an initial attempt at constructing $\Phi(L_{1.4}, L_R, z)$ and that some of the early assumptions will require refinement in future modelling, particularly for the models of $\Psi(r)$. Also, it is well established that the radio source counts cannot be reproduced unless the radio luminosity function is evolving (Dunlop & Peacock 1990, Rowan-Robinson et al. 1993), so such evolution must be incorporated into the BLF.

2.2 Predictions

The BSC has been predicted from the model BLF, and is shown in Figure 1. The horizontal axes are both in units of mJy, and the vertical axis has units of the log of galaxies steradian^{-1} Jy^{-2}. The observational BSC is also shown in Figure 1. There remain discrepancies between the model and predictions, even after incompleteness at the faintest flux densities is accounted for. Nevertheless, even this initial model still predicts a reasonable distribution for the BSC, indicating that further refinements of the model will produce more realistic predictions. The redshift distribution for these two galaxy populations can also be predicted from the model BLF. Figure 2 shows how the redshift distribution changes for a constant $S_{1.4_{\min}} = 0.1\,\mathrm{mJy}$ as the optical limit is lowered. Figure 3 is the analogous situation for a constant $R_{\min} = 22$. Note the predominance of the AGN population at high radio flux densities and how the starbursts emerge at moderate redshifts as the radio flux density limit is decreased. Figure 4 shows the

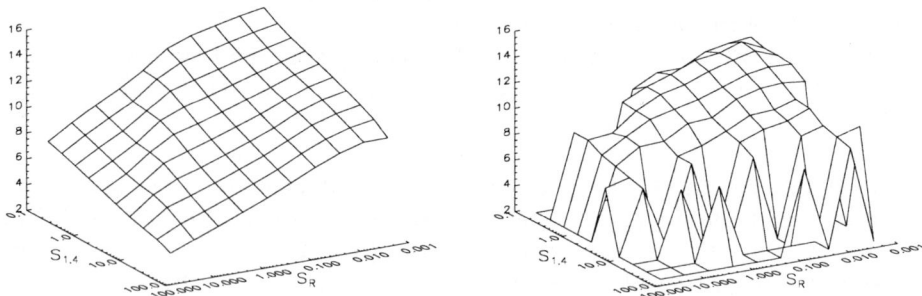

Fig. 1. Left: Model BSC predicted from model BLF. Right: Observational BSC constructed from PDS data. All three axes are logarithmic.

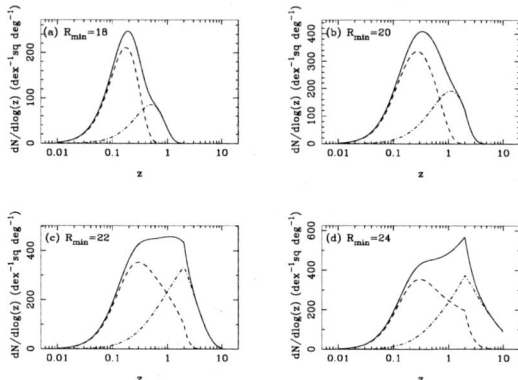

Fig. 2. Redshift distributions, with $S_{1.4\,\mathrm{min}} = 0.1\,\mathrm{mJy}$ in each graph. Dashed line: starbursts; Dot-dashed line: AGN; Solid line: AGN plus starbursts. R_{min} is given for each graph.

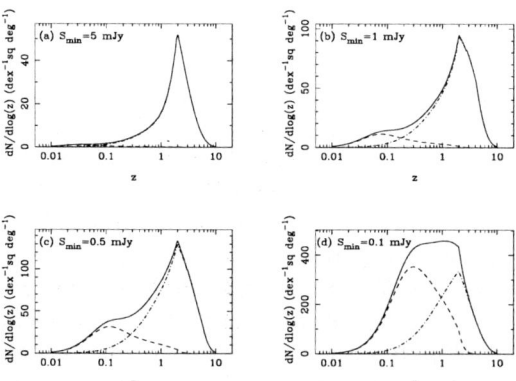

Fig. 3. Redshift distributions, using $R_{\mathrm{min}} = 22$ in each graph. The lines are as for previous figure. $S_{1.4\,\mathrm{min}}$ is given for each graph.

observational redshift distribution compared with the BLF predictions. Not all optically identified radio sources have their redshifts measured at present, so the distributions have been normalised to the total number of radio sources with optical IDs. Those with measured redshifts can be considered a representative sample of sources brighter than about $R = 20$ and $S_{1.4} = 0.3\,\mathrm{mJy}$. Although there are some similarities between the observations and the predictions, there are still quite pronounced differences. There is a large spike in the observed redshift distribution for starbursts above the predictions at $z \sim 0.2$, and the AGN population seems to have a markedly nearer median redshift than the model predicts. These are discrepancies which need to be investigated as the model BLF is refined, and more redshifts for the sample become available.

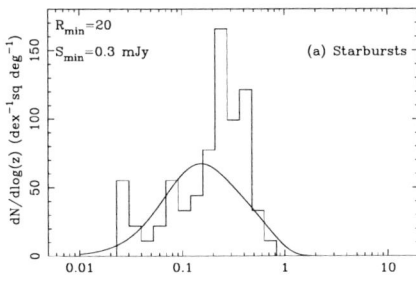 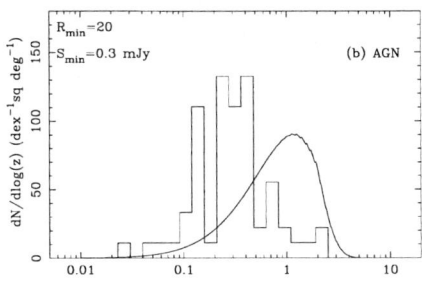

Fig. 4. Observed redshift distribution for (a) starbursts and (b) AGN, shown with predictions from the BLF overlayed. The flux-density limits used are shown.

3 Conclusions

A deep multiwavelength survey, the *Phoenix Deep Survey*, is well underway, and is producing useful catalogues and interesting results. One of these results includes a model radio/optical BLF which requires refinement but still makes predictions which, with caveats, compare moderately well with the PDS observations.

References

Corbelli, E., Salpeter, E., Dickey, J. (1991): ApJ, 370, 49
Dunlop, J. S., Peacock, J. A. (1990): MNRAS, 247, 19
Elvis, M., Maccacaro, T., Wilson, A. S., Ward, M. J. (1978): MNRAS, 183, 129
Hopkins, A., Cram, L., Mobasher, B., Rowan-Robinson, M. (1998): MNRAS (in press)
Meurs, E. J. A., Wilson, A. S. (1984): A&A, 136, 206
Rowan-Robinson, M., Benn, C. R. Lawrence, A., McMahon, R. G., Broadhurst, T. J. (1993): MNRAS, 263, 123
Sadler, E. M., Jenkins, C. R., Kotanyi, C. G. (1989): MNRAS, 240, 591
Toffolatti, L., Franceschini, A., DeZotti, G., Danese, L. (1987): A&A, 184, 7

Discussion

Ekers: You have very clearly separated the two main populations (star forming galaxies + AGN, Radio Galaxies) in your bivariate RLF. These two populations may have completely different evolution – have you included this? Can this explain the remaining discrepancies between your model and the observations?

Hopkins: Yes, I actually forgot to point out that we've been very careful in defining the evolution of the two populations. This won't account for the discrepancies between model predictions and observations, although further refinement may reduce the discrepancies somewhat. There are actually less well constrained aspects of our model which, upon refinement, should have a greater effect.

Prandoni: You have claimed that your redshift distribution prediction for starburst galaxies is in good agreement with the observations. It seems to me that the observed distribution has a larger peak at higher redshifts that the predicted one. Can you comment on that?

Hopkins: This is true – the model predictions certainly have differences from the observations. Rather, I think the point is that, qualitatively at least, there are similarities in the distribution. This leads us to believe that we are on the right track in our modeling, and that further refinements, constrained by observations recently completed and currently in progress, will reduce the obvious discrepancies.

All-Sky Radio Surveys

Richard Wielebinski

Max-Planck-Institut für Radioastronomie, Auf dem Hügel 69, 53121 Bonn, Germany
E-mail: rwielebinski@mpifr-bonn.mpg.de

Abstract. One of the essential data bases in radio astronomy are the all-sky surveys. These have been traditionally made by combining some northern survey with a southern one, often made with different instruments at some nearby frequency. The combinations led to many calibration problems since beams and observing methods were often not identical. For successful all-sky surveys a well organised cooperation between a northern and southern observatory, before the observations start, is desirable. Observing methods, calibration procedures and data reduction should be made jointly for best results. In this talk the development of this field of research and the present status of several all-sky survey ventures will be discussed.

1 The Early All-Sky Radio Surveys

Radio sky surveys provide fundamental information and were made from the beginnings of radio astronomy. Karl Jansky's observations were reanalysed and published as a map of the northern sky. Grote Reber made radio continuum maps which showed that the position of the Galactic centre was off by more than 30°! Sky surveys occupied the early radio astronomers in Cambridge, Sydney and Dwingeloo. The first combination of northern and southern radio continuum data to an all-sky survey was made by Dröge and Priester (1956) who used their own observations made in Kiel and the southern data of Allen and Gum (1950). This map at 200 MHz showed the dominant Galactic plane with a concentration of the continuum emission to the Galactic centre and a pronounced North Polar Spur feature. Maxima in the directions of the Cygnus X and Vela X complexes were interpreted as tangential directions of the local spiral arm. Another milestone of the early north-south combinations was the H I distribution in the plane of the Galaxy presented by Kerr and Westerhout (1964). This data was used finally to pinpoint the position of the New Galactic Centre, a basic result of modern astronomy. An all-sky combination of radio continuum at 150 MHz by Landecker and Wielebinski (1970) has shown numerous additional spurs of emission, interpreted to be nearby supernova remnants, as well as a pronounced north-south anomaly. The radio continuum emission as seen in the northern sky shows a very broad component in the direction of the Perseus (local) arm which was interpreted to be a signature of a strong spherical halo. In the southern sky two distinct minima are seen above and below the Galactic plane as would be expected from a thin halo or even a thick disk only. A combination of a northern 38 MHz survey and a southern 30 MHz map by Cane (1978) gave a low-frequency result which supported the suggestion of the north–south anomaly.

The modern era of all-sky surveys was heralded by the use of computers to store and reduce the large volumes of data. In addition, baseline fitting techniques were developed to give the surveys a much better dynamic range. The Haslam et al. (1982) 408 MHz radio continuum survey was completely computer reduced and done by one team of astronomers in the north and the south (using Jodrell Bank, Effelsberg and Parkes telescopes). A combination of northern and southern HI surveys with telescopes of the 25–30 m class was presented by Colomb et al. (1980). This showed that the H I gas is distributed in the form of numerous shells rather than spherical clouds, more like the continuum emission.

The mapping of the galactic plane with high angular resolution has a long tradition. Here again we must use instruments in the northern sky and combine this with observations made in the southern hemisphere. The 1390 MHz survey of Westerhout (1958) is still a basis of nomenclature of H II regions. The corresponding southern survey was made by Hill (1968) at 1410 MHz and the source list by Manchester (1969). Several low-frequency high resolution surveys were made of the southern Galaxy (e.g. Shain et al. 1961 at 19.7 MHz; Jones and Finlay 1974 at 29.9 MHz) showing low-frequency absorption by H II regions. Multifrequency northern surveys were made by Altenhoff et al. (1970; 1979) with improved angular resolution by using higher frequencies at the largest radio telescopes. The southern complements were the series of surveys by Thomas et al. (1969), Goss and Day (1970) and Haynes at al. (1978). A large map of the southern sky at 2.3 GHz was presented by Jonas et al. (1985). The 'older' era of Galactic plane surveys came to an 'end' with the 10 GHz survey of Handa et al. (1987) at the highest radio frequency end of the spectrum and the 30 MHz northern sky survey of Kassim (1988) at the lowest frequency.

The mapping of polarized radio emission began in Cambridge (Wielebinski et al. 1962; Wielebinski and Shakeshaft, 1964) and Dwingeloo (Westerhout et al. 1962: Berkhuijsen and Brouw, 1963) almost at the same time. This work was taken up by Mathewson and Milne (1965) in the southern sky. Various Galactic regions were mapped by several teams but a complete all-sky survey of radio polarization does not exist. Large sections of the northern sky were mapped in polarization by Spoelstra (1972) and Brouw and Spoelstra (1976).

Maps of the Galactic plane in polarization have not been made until very recently. This is despite the fact that polarized emission clearly allows the separation of the thermal H II regions (which are unpolarized) from the supernova remnants and the foreground nonthermal emission. The first such survey was by Junkes et al. (1987) which showed that the emission could be traced to the inner galaxy.

2 The High-Frequency Surveys with Balloons and Satellites

One of the immediate north–south combinations is offered by the surveying satellites like IRAS, COBE, or ROSAT. The COBE satellite produced maps at 31.5, 53, and 90 GHz. Balloon-borne receivers with starts in the northern and southern hemispheres were also used to produce a map at 19 GHz (Boughn et al. 1990).

3 Molecular Line Surveys

The most abundant molecules (like CO) are so widely distributed that they can be mapped over a considerable extent of the Galactic plane. A composite survey of the whole Milky Way was presented by Dame et al. (1987) based on observations using two 1.2-m telescopes, one in New York, the other in Chile. This CO(1–0) line data is used to determine the distribution of molecular hydrogen in the Galaxy.

4 The More Recent Surveys

The excellent 11 cm Galactic plane survey of Reich et al. (1984) was followed by extensions (Reich et al. 1990b; Fürst et al. 1990). Also two 21 cm surveys of the same area were made by Reich et al. (1990a), Reich et al. (1997). A southern survey of Duncan et al. (1995) can be considered to be the complement of the northern data. The polarization data for the 2.4 GHz survey was published by Duncan et al. (1997). The polarization of the northern surveys at 2.7 GHz is soon to become available. The mapping of large areas of the Galaxy but with satisfactory angular resolution required massive data storage and a lot of telescope time. Nevertheless a medium-latitude 21 cm continuum survey (with polarization) has been started in the north (Uyaniker 1997; Uyaniker et al. 1998). This survey shows the details of the magnetic field structure which was anticipated by the work of Wieringa et al. (1993). A southern continuation of this 21 cm medium latitude survey is very important.

On the all-sky survey side, progress has been made with the completion of an absolutely calibrated 1400 MHz map combining the Reich (1982) and Reich & Reich (1986) data with a southern survey of Testori et al. (1998). A possible 'first' in the combination of north and south surveys can be reported: the southern 45 MHz survey of Alvarez et al. (1997) is waiting for its northern counterpart.

High angular resolution H I surveys exist only of limited parts of the sky. The use of multibeam systems offers the possibility to map the whole sky in the H I line with some 10 arcmin resolution.

References

Allen, C.W., Gum, C.S. (1950): Aust. J. Phys. (A) **3**, 224
Altenhoff, W.J. et al. (1970): A&AS **1**, 319
Altenhoff, W.J., Downes, D., Pauls, P., Schraml, J. (1979): A&AS **35**, 23
Alvarez, H., Apariaci, J., May, J., Olmos, F. (1997): A&AS **124**, 315
Berkhuijsen, E.M., Brouw, W.N. (1963): Bull. Astr. Inst. Ned. **17**, 465
Boughn, S.P. et al. (1990): Rev. Sci. Instr. **61**, 158

Brouw, W.N., Spoelstra, T. (1976): A&AS **26**, 129
Cane, H.V. (1978): Aust. J. Phys. **31**, 561
Colomb, F.R., Pöppel, W.G.L., Heiles, C. (1980): A&AS **40**, 47
Dame, T.M., Ungerechts, H., Cohen, R.S. et al. (1987): ApJ **322**, 706
Dröge, F., Priester, W. (1956): Zeitschr. f. Phys **40**, 236
Duncan, A.R., Stewart, R.T., Haynes, R.F., Jones, K.L. (1995): MNRAS **277**, 36
Duncan, A.R., Haynes, R.F., Jones, K. L., Stewart, R.T. (1997): MNRAS **291**, 279
Fürst, E., Reich, W., Reich, P., Reif, K. (1990): A&AS **85**, 691
Goss, W.M., Day, G.A. (1970): Aust. J. Phys. Suppl. No.**13**
Handa, T. et al. (1987): PASJ **39**, 709
Haslam, C.G.T., Salter, C.J., Stoffel, H., Wilson, W.E. (1982): A&AS **47**, 1
Haynes, R.F., Caswell, J.L., Simons, L.W. (1978): Aust. J. Phys. Suppl. No. **45**
Hill, E.R. (1968): Aust. J. Phys. **21**, 735
Jonas, J.L., de Jager, G., Baart, E.E. (1985): A&AS **62**, 105
Jones, B.B., Finlay, E.A. (1974): Aust. J. Phys. **27**, 687
Junkes, N., Fürst, E., Reich, W. (1987): A&AS **69**, 451
Kassim, N.E. (1988): ApJS **68**, 715
Kerr, F.J., Westerhout, G. (1964): in Blaauw and Schmidt (eds.) *Galactic Structure* Vol. 5, Chap. 8, Chicago University Press
Landecker, T.L., Wielebinski, R. (1970): Aust. J. Phys. Suppl. No. **16**
Mathewson, D.A., Milne, D.K. (1965): Aust. J. Phys. **18**, 635
Manchester, B.A. (1969) : Aust. J. Phys Suppl. No. **12**
Reich, P., Reich, W. (1986): A&AS **63**, 205
Reich, P., Reich, W., Fürst, E. (1997): A&AS **126**, 413
Reich, W. (1982): A&AS **48**, 219
Reich, W., Fürst, E., Steffen, P., Reif, K., Haslam, C.G.T.(1984): A&AS **58**, 197
Reich, W., Reich, P., Fürst, E. (1990a): A&AS **83**, 539
Reich, W., Fürst, E., Reich, P., Reif, K. (1990b): A&AS **85**, 633
Shain, C.A., Komesaroff, M.M., Higgins, C.S. (1961): Aust. J. Phys. **14**, 508
Spoelstra, T. (1972): A&AS **5**, 205
Testori, J.C. et al. (1998): in preparation
Thomas, B. MacA., Day, G.A. (1969): Aust. J. Phys. Suppl. No. **11**
Uyaniker, B. (1997): PhD Thesis, Bonn University
Uyaniker, B., Fürst, E., Reich, W., Reich, P., Wielebinski, R. (1998): A&A (submitted)
Westerhout, G. (1958): Bull. Astr. Inst. Ned. **14**, 215
Westerhout, G., Seeger, Ch.L., Brouw, W.N., Tinbergen, J. (1962): Bull. Astr. Inst. Ned. **16**, 187
Wielebinski, R., Shakeshaft, J.R. (1964): MNRAS **128**, 19
Wielebinski, R., Shakeshaft, J.S., Pauliny-Toth, I.I.K.(1962): Observatory **82**, 158
Wieringa, M.H. et al. (1993): A&A **268**, 215

Discussion

Tsvetanov: Have you tried to correlate the high polarization features above and below the galactic plane with galactic fountains, SN loops, etc? If such correspondence exists, it would have serious implications for the galactic dynamo theory you just mentioned.

Wielebinski: The dynamo works presumably in the plane of the galaxy. However, we do expect vertical structures, especially in nuclei of galaxies. The polarization maps allow indeed the detection of new SNRs. The Galactic fountains are sometimes thermal emission expelled from the plane and require multi-frequency studies. Some polarized structures resemble Parker Loops. The individual stock-taking of features has only now started. Correlations will follow.

A Study of Low/Intermediate-Redshift Radio Sources and Clues to the Nature of High-Redshift Objects

R. Morganti[1], C. Tadhunter[2], M. Villar-Martin[2] and R. Dickson[2]

[1] Istituto di Radioastronomia, Bologna, Italy
[2] University of Sheffield, Sheffield, UK

Understanding Active Galactic Nuclei (AGN) requires not only large samples but also detailed multiwaveband studies. At present, the latter can be carried out only for small/medium size samples. However, the new techniques, instruments and facilities (e.g. discussed in this workshop) will make it possible to extend these studies to much larger samples. Here we summarize the study done so far on a small sample of radio-selected AGN. Observations in a number of different bands together with the statistics have been used to answer a number of critical questions related to the AGN phenomenon. The results obtained for our sample (redshift limited $z < 0.7$) can be relevant for comparison with what obtained for high-z objects. Our study is based on a complete sample of radio sources formed by 88 objects (68 radio galaxies, 18 QSR and 2 BL Lacs) and selected for having $S_{2.7GHz} > 2Jy$, $\delta < +10°$ and $z < 0.7$. For this sample we have collected optical, radio and X-ray data, both ROSAT and SAX, (see Tadhunter et al. 1993, 1998; Morganti et al. 1997 and ref. therein).

1) Testing the AGN Illumination Model

By using spectroscopic data (Tadhunter et al. 1998) we have tested in our sample how well the emission line properties of the powerful radio galaxies can be explained in term of quasar illumination model, i.e. photoionization by EUV photons from a central illuminating quasar or AGN. In line with previous work, we find that significant correlations exist between the luminosities of the [OIII]λ5007, [OII]λ3727 and Hβ emission lines and the radio luminosity. However, we find that: 1) the scatter in the $L_{[OIII]}$ vs. L_{radio} correlation is significantly larger than in $L_{[OII]}$ vs. L_{radio} and $L_{H\beta}$ vs. L_{radio} correlations; 2) the ionization state deduced from the emission lines does not increase with radio power (or redshift) as predicted by the simple, constant ISM, photoionization model. Thus, the observed line luminosities are not easily reconciled with the idea that they are caused by the increase in the power of the photoionizing quasar as the jet power increases unless, e.g., the average ISM properties changing appreciably with redshift or radio power. This can have important implication for explaining some of the differences observed between low and high-z radio galaxies.

2) The Nature of the UV Excess

In our sample we find that the incidence of large UV polarization (Tadhunter et al. 1997) — a signature of scattered AGN light — is lower than reported in previous studies which were biased towards the brightest and more spectacular objects. Thus, the scattered AGN is not the main cause of the UV excess in most powerful radio galaxies and UV-continuum is multi component in nature. It comprises to varying degrees of: scattered AGN light, direct AGN light, the light from young stellar populations and nebular continuum (Dickson et al 1995). For this analysis is essential a careful modelling of high-quality spectropolarimetric data. It is now well-established that the scattered quasar component makes a significant contribution in many high redshift ($z > 0.4$) radio galaxies (Cimatti et al. 1996), but is not clear how the strength of the scattered component depends on redshift/radio power. When this analysis will be completed for the sources in our sample it will allow us to compare the polarization properties of high and low redshift radio galaxies in the same rest wavelength range (the near-UV).

3) Interaction Between Radio Plasma & ISM: How Common?

The illumination model fails to explain some of the properties observed at high-z while in these objects there is evidence of strong interaction between the radio jet and the ISM (McCarthy et al. 1987; Best et al. 1996). From this follows the importance of understanding the jet-cloud interaction phenomenon. It is known that this effect is not as strong at low-z. Nevertheless, our sample includes some interesting objects that could be good laboratories to study this phenomenon. So far we have studied in detail three of them: PKS 2152–69 (Fosbury et al. 1998), PKS 2250–41 (Clark et al. 1997) and PKS 1932–46 (Villar-Martin et al. 1998). They all show morphological evidence of interaction between the radio plasma and the ISM but only in the case of PKS 2250–41 this interaction and the shocks produced could be the dominant mechanism for ionizing the gas. Both PKS 1932–46 and PKS 2250–41 look very similar to these high-z objects suggesting that they are systems at low/intermediate redshift where we see situations very similar to the high redshift systems. The similarity must be in the environment. High-z radio galaxies show an excess of companion galaxies detected along the axes of the radio source (Röttgering et al. 1996): interestingly, both PKS 2250–41 and (possibly) PKS 1932–46 have a companion.

References

Best P., Longair M., Röttgering H. 1996, MNRAS, 280, L9
Clark N., Tadhunter C., Morganti R., et al 1997, MNRAS 286, 558
Cimatti A., Dey A., van Breughel W., Antonucci R., Spinrad H. 1996, Apj 465,145
Dickinson R., Tadhunter C., Shaw M., Clark N., Morganti R. 1995, MNRAS 273, L29
Fosbury R.A.E., Morganti R., Wilson E., Ekers R. 1998, MNRAS in press
McCarthy P. Spinrad H. et al. 1987, ApJ 319, L39
Morganti R., Oosterloo T., Reynolds J. et al. 1997, MNRAS 284, 541
Röttgering H., West M., Miley G., Chambers K. 1996, A&A 307, 376
Tadhunter C.N., Morganti R., di Serego Alighieri et al. 1993, MNRAS 263, 999
Tadhunter C.N., Morganti R., Robinson et al. 1998, MNRAS in press
Tadhunter C.N. et al., 1997, in "Quasar Hosts", Clements et al. (eds.) in press
Villar-Martin M., Tadhunter C.N., Morganti R. et al., A&A 1998 in press

The Local Galaxy Population from the HIPASS Survey

L. Staveley-Smith[1], R.L. Webster[2], G. Banks[3], V. Kilborn[2], B. Koribalski[1] and M. Putman[4]

[1] Australia Telescope National Facility, CSIRO, P.O. Box 76, Epping, NSW 2121, Australia
[2] School of Physics, University of Melbourne, Parkville, Victoria 3052, Australia
[3] University of Wales College of Cardiff, P.O. Box 913, Cardiff CF2 5YB, U.K.
[4] Mount Stromlo and Siding Spring Observatories, ANU, Private Bag, Weston Creek PO, ACT 2611, Australia

Abstract. A new, sensitive HI survey is underway with a multibeam receiver on the Parkes telescope. The aim of the HI Parkes All-Sky Survey (HIPASS) is to blindly survey the southern sky out to a velocity of 12700 km s^{-1}. In this paper, we describe the parameters of the survey, summarise the current status, and present a collection of preliminary results on the Magellanic Clouds, new dwarf galaxies discovered in the Centaurus A group, and the HI mass function. These results have implications for our understanding of the population of nearby galaxies.

1 Introduction

The gaseous content of nearby galaxies depends on their past average rate of star-formation and the mass of baryonic material contained in the initial protogalaxy. In general, more massive systems collapse earlier and contain the least gas relative to their baryonic mass at the present epoch. Understanding the evolution of galaxies and their gas consumption rate, and knowing the epoch of their formation is a fundamental goal in modern high-redshift astronomy. Present models (e.g. Pei & Fall 1995) are poorly constrained at high and low redshifts owing to uncertainties in gas content and star-formation rates. Some of these uncertainties arise from poorly constrained parameters at high redshifts, e.g. the correction for dust obscuration (Giavalisco et al. 1996, Calzetti 1997). However, there are significant uncertainties in local quantities such as the value for the space-averaged star-formation rate ($\dot{\Omega}_{\rm HI} = -3.7 \times 10^{-3}$ per Hubble time, Gallego et al. 1995) and the local gas density ($\Omega_{\rm HI} = 1.9 \times 10^{-4} h^{-1}$, Rao & Briggs 1993). Whereas local values for the star formation rate come from wide-area spectroscopic surveys, the local value for the gas density is dependent on targetted observations of optically-selected samples, and therefore may be biased. Moreover, the present cosmic star-formation rate is unsustainable for more than 0.5 Gyr unless there exists copious amounts of presently undetected HI or H$_2$ gas.

Blind surveys of neutral hydrogen (HI) in small regions of sky have been conducted and indicate some consistency with optically-selected samples (e.g.

Table 1. Summary of HIPASS parameters.

Parameter	Value
Sky Coverage	$\delta \leq 0°$
Integration time per beam area	500 s
Average system temperature	20 K
Central beam efficiency	63%
Central FWHP beamwidth	14$'$.0
Velocity coverage (cz)	-1200 to 12700 km s^{-1}
Channel separation	13.2 km s^{-1}
Velocity resolution[a]	18.0 km s^{-1}
3-σ Positional accuracy[b]	3$'$.0
3-σ detection limit[c]	30 mJy beam^{-1}
3-σ HI mass limit[d]	$10^6 d_{\mathrm{Mpc}}^2 M_\odot$

[a] After application of on-line Tukey 25% smoothing.
[b] Barnes (1998b); inversely proportional to signal-to-noise ratio.
[c] Measured value based on $r = 6'$ median gridding function.
[d] For $\Delta V = 150$ km s^{-1}.

Zwaan et al. 1997). However, there are strong selection effects operating against low column density galaxies which may not be optically prominent (e.g. Briggs 1990), and several surveys (e.g. Barnes et al. 1997) give only weak limits to the HI volume density of optically invisible galaxies. However the advent of the HIPASS survey, which is a sensitive all-sky survey of HI in the local Universe, will help to reduce some of the worst selection effects.

In this paper, we describe the parameters of HIPASS, and summarize a few preliminary results relevant to the population of nearby galaxies.

2 HIPASS Parameters

The HI Parkes All-Sky Survey (HIPASS) is being conducted with a new multibeam receiver (Staveley-Smith et al. 1996) located at the prime focus of the Parkes telescope. Table 1 summarises the observational parameters of the survey which commenced on 1997 February 27. The sky is surveyed in 8° declination strips. The width of each observed strip on the sky is about 1°.7. However, this strip is slightly undersampled in Right Ascension, so strips are overlapped so that after five scans, each of the 13 beams makes a well-sampled survey of the sky. At the end of 1997, the survey was approximately 27% complete, with the sky coverage shown in Figure 1. Software reduction strategies are discussed in Barnes (1998a), and bandpass removal techniques are discussed in Barnes et al. (1998).

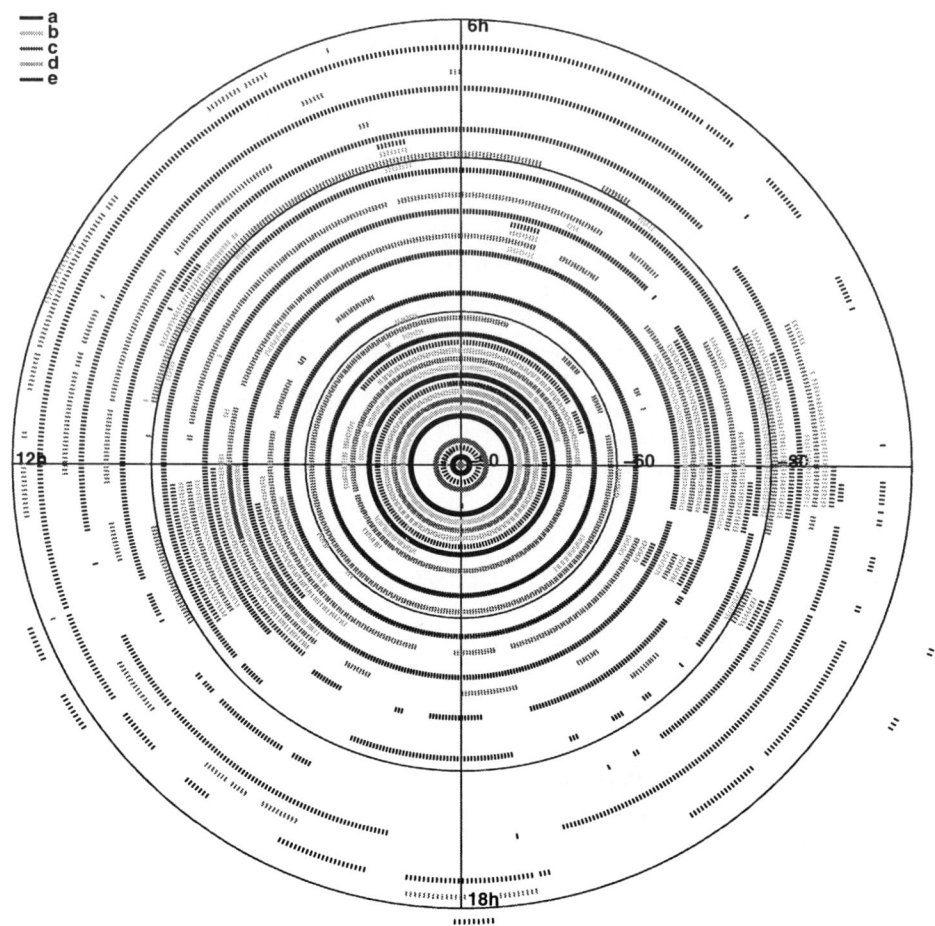

Fig. 1. A polar projection of HIPASS sky coverage obtained during 1997. Each of the five a-e scans is actually 8° long in declination (and not 1°.5 as plotted) and fully covers the sky.

3 Milky Way Dynamics

As can be seen from Figure 1, one of the best-sampled areas to date is the Southern Celestial Cap. This region contains our nearest gas-rich neighbours, the Magellanic Clouds. Although the velocity resolution and bandpass calibration schemes are not ideal for nearby extended objects, a large area around the Clouds has never been mapped with the full 14′ angular resolution of the Parkes telescope.

A brightness temperature image of an area 2400 deg^2 in extent and centred on declination −90° is shown in Figure 2 (Putman et al. 1998). This mosaic nicely shows the nature of the interaction between the LMC and SMC as well as the beginning of the Magellanic Stream which appears to peel off from the SMC at the left-hand edge of the figure.

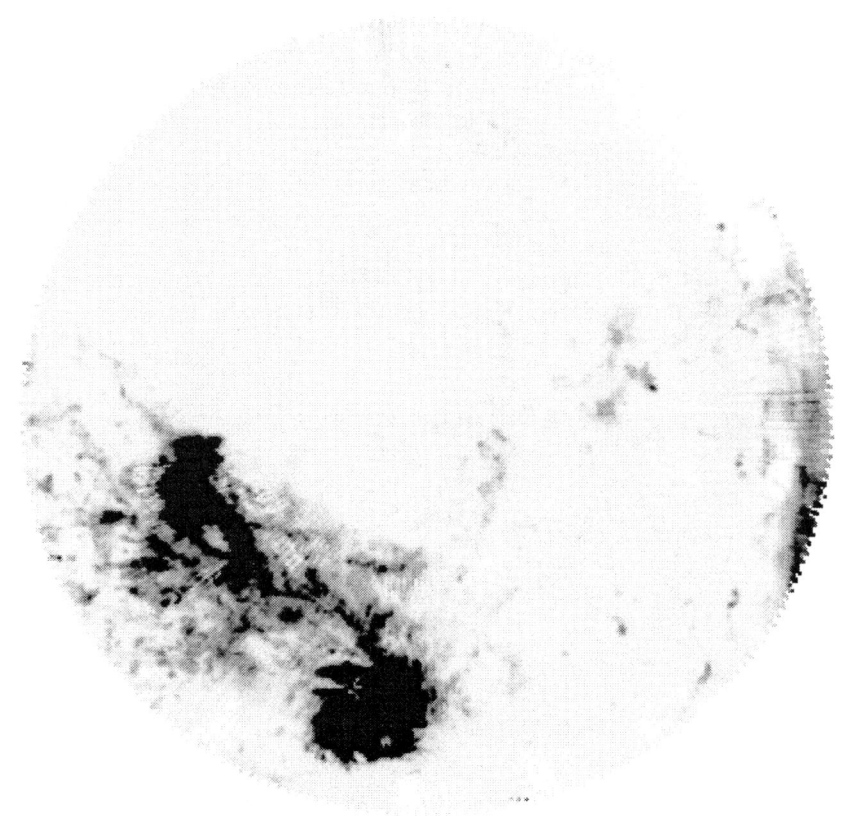

Fig. 2. A HIPASS view of the Southern Celestial Cap (Putman et al. 1998). This view contains the Large Magellanic Cloud (at the bottom), the Small Magellanic Cloud (towards the left) and the Bridge of gas connecting them. The southern Galactic Plane can be seen on the right-hand side. Many clouds of high velocity gas occupy the region between the Galaxy and the Clouds. The stripes in declination are scanning artefacts.

The wealth of new information apparent in this data is currently being used to re-examine the relationship between the LMC, SMC and the Galaxy in the context of various models describing their mutual interaction.

4 New Centaurus A Neighbours

The Centaurus A group of galaxies (as opposed to the more distant Centaurus cluster) is the second most nearby group in the southern hemisphere (Côté et al. 1997) and allows a quick look at whether the mix of galaxy morphological types being detected by HIPASS is different from optically-selected samples, or not. Again, as Figure 1 shows, this group (at right ascension 13^h, declination $-35°$) was one of the early targets of the HIPASS survey.

Preliminary results (Banks et al. 1998) reveal a substantial number of previously unrecognised group members (around 10 with HI masses greater than 10^7 $M_\odot$). Although some of these are objects with previously unknown or incorrect velocities, a large fraction appear to be faint dwarf irregular galaxies. An example, HIPASS1351-47, is shown in Figure 3. Such galaxies are expected to be more prominent in HI-selected samples since, in general, they appear to have greater HI mass-to-luminosity ratios.

Implications for the luminosity and mass distribution of Cen A galaxies, and the morphological mix of Cen A galaxies are discussed in Banks et al. (1998).

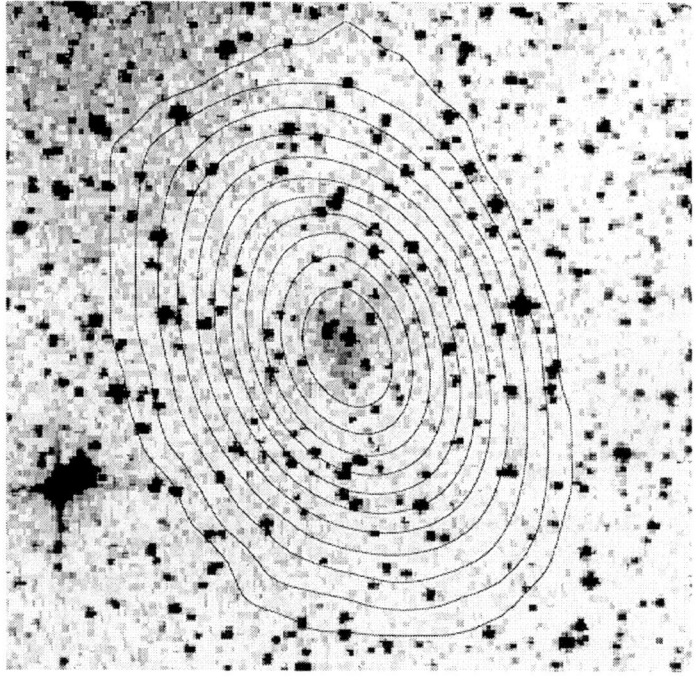

Fig. 3. Contours of HI column density overlayed on the Digitized Sky Survey for HIPASS1351-47, a new Centaurus A group member. The HI contours are from follow-up observations with the 375-m array of the Australia Telescope Compact Array taken on 1997 October 3.

5 Field HI Mass Function

A more quantitative description of the density of galaxies in an HI-selected survey is given by the field HI-mass function. At the time of writing, the numbers of galaxies uniformly selected and measured from HIPASS data cubes in a systematic way is only ~ 200. However, this already exceeds existing surveys (e.g. Zwaan et al. 1997). Based on a flux-limited sub-sample of objects at declination $-74°$, Kilborn et al. (1998) have produced the HI mass function shown in Figure 4. The data points are overlayed with Schechter functions, with the best-fitting faint-end slope being -1.35. This is somewhat steeper than the slopes (~ -1.0) generally found in traditional optically-selected samples (e.g. Loveday et al. 1992), and indicates the greater contribution of low-mass galaxies to the cosmic density of gas.

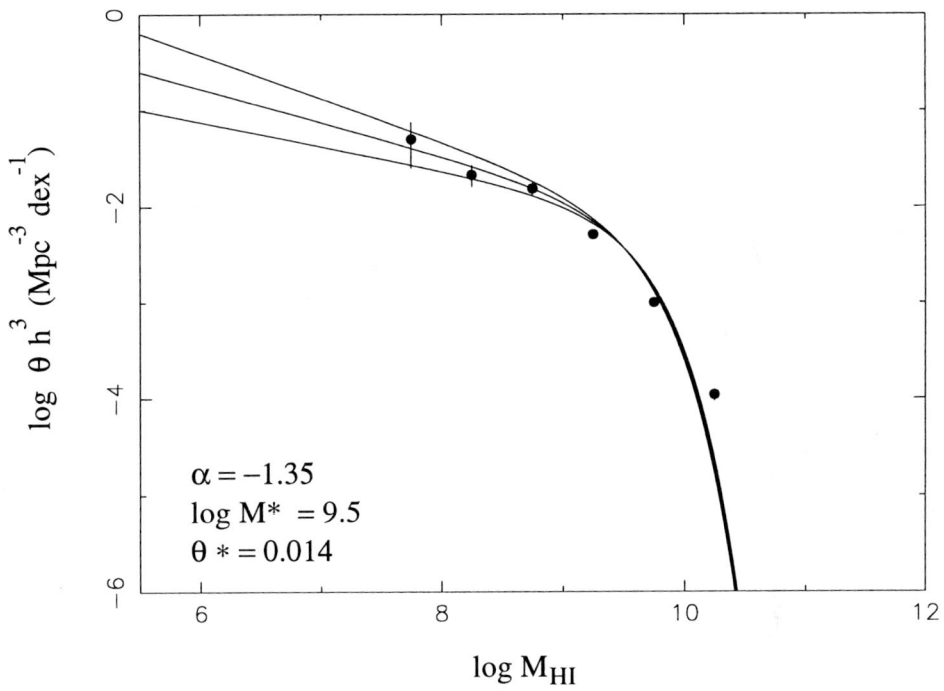

Fig. 4. A preliminary HI mass function from HIPASS using 99 galaxies from all HIPASS cubes in the $-74°$ declination zone. Three Schechter functions with faint-end slopes of -1.25, -1.35 and -1.45 are plotted.

Recent optical studies (e.g. Loveday 1997) suggest a possible upturn in the field luminosity function faintward of $M_B = -14$ mag. For an HI mass-to-luminosity ratio (in solar units) of unity, this corresponds to an HI mass of

6×10^7 M$_\odot$. Since this is approximately the lowest HI mass plotted in Figure 4, it will be interesting to see if there is any corresponding evidence for an upturn in the HI mass function as the amount of HIPASS data increases. Such an upturn would have dramatic consequences for our understanding of galaxy formation.

Acknowledgements

We acknowledge other members of the HIPASS team who have been enormously helpful in helping obtain the preliminary results quoted in this paper, and we thank the staff of the Parkes observatory for their support.

References

Banks, G. et al. 1998, (in preparation)
Barnes, D.G. 1998a, ADASS VII (San Francisco: ASP)
Barnes, D.G. 1998b, Ph.D. Thesis (University of Melbourne)
Barnes, D.G., Staveley-Smith, L., Ye, T., & Oosterloo, T. 1998, ADASS VII (San Francisco: ASP)
Barnes, D.G., Staveley-Smith, L., Webster, R.L., & Walsh, W., 1997, MNRAS, 288, 307
Briggs, F.H. 1990, AJ, 100, 999
Côté, S., Freeman, K.C., Carignan, C., & Quinn, P.J. 1997, AJ, 114, 1313
Calzetti, D. 1997, AJ, 113, 162
Gallego, J., Zamorano, J., Aragon-Salamanca, A., & Rego, M. 1995, ApJ, 455, 1
Giavalisco, M., Koratkar, A., & Calzetti, D. 1996, ApJ, 466, 831
Kilborn, V. et al. 1998, (in preparation)
Loveday, J. 1997, ApJ, 489, 29
Loveday, J., Peterson, B.A., Efstathiou, G., & Maddox, S.J. 1992, ApJ, 338
Pei, Y.C., & Fall, S.M. 1995, ApJ, 454, 69
Putman, M. et al. 1998, Nat, submitted
Rao, S., & Briggs, F. 1993, ApJ, 419, 515
Staveley-Smith, L. et al. 1996, PASA, 13, 243
Zwaan, M.A., Briggs, F.H., Sprayberry, D., & Sorar, E. 1997, ApJ, 490, 173

Discussion

Lahav: Does the HIPASS blind search detect a new population of galaxies?

Staveley-Smith: The majority of galaxies we find (away from the Galactic plane) have optical counterparts. About 10% of galaxies are not catalogued. So in that sense, no significant new population has been found so far.

Bland-Hawthorn: What are the limiting column densities you are reaching in your survey?

Staveley-Smith: The 3-σ sensitivity is 7×10^{17} atoms/cm^2 in a 13 km/s wide channel.

Mortlock: Is there any evidence, from the observations, of any tidal streaming between us and the Magellanic Clouds?

Staveley-Smith: Too early to say as the Magellanic Cloud data are still preliminary and distances are difficult to measure. The existing data (Putman et al. in preparation) only reveal tidal streaming from the Clouds.

Tsvetanov: If you assume some generic HI/dark matter ratio how much of the dark matter could be contained in the dwarf galaxies detected in HI by HIPASS?

Staveley-Smith: The HIPASS 2% mass function could be consistent with the bulk of the dark matter being contained in dwarf galaxies, in this case. I emphasise the very preliminary nature of the results.

The MNRF Upgrade to the Australia Telescope

Ray P. Norris

CSIRO Australia Telescope National Facility,
PO Box 76, Epping, NSW 2121, Australia

Abstract. I describe the recently-funded $11M upgrade to the Australia Telescope and the Australian Long Baseline Array. The upgrade will enable the Compact Array to become one of the world's leading millimetre arrays for a window of about 8 years, enabling high-resolution millimetre studies of galactic and extragalactic molecular line and continuum sources. It will also upgrade both the Compact Array and the Australian Long baseline Array to 22 GHz, both to increase resolution and to enable observations of objects such as water megamasers. For these and other parts of the upgrade, I describe the scientific drivers, the technical challenges and the current status.

1 Background

The CSIRO Australia Telescope National Facility (ATNF) consists of four instruments (the Compact Array at Narrabri, the 64-m Parkes telescope, the 22-m Mopra telescope, and the Long Baseline Array) which are equipped for radio-astronomical observations at centimetre wavelengths. It is open to all astronomers from all countries, and proposals are awarded time on the basis of scientific merit, as measured by a peer-review system.

In 1997, the ATNF was awarded $11M for an upgrade under the Major National Research Facilities (MNRF) program of the Australian Federal Government. The prime target of the upgrade was to enable operation at millimetre wavelengths, but it also included a number of other enhancements both to ATNF telescopes and to telescopes operated by the University of Tasmania. The upgrade will be complete in 2002-2004.

2 The MNRF Upgrade

The primary goal of the upgrade is to extend both the wavelength range and angular resolution of the AT by a factor of 10, and to upgrade the Australian Long Baseline Array (ALBA) both by equipping the Ceduna antenna for VLBI operation and by equipping the entire array for operation at 22 GHz (12 mm). The main elements of the upgrade are:

- equipping the Compact Array for 3.5 mm (85 - 100 GHz),
- extending the solid surface of five antennas, and adding extra stations,
- equipping all ATCA and Mopra antennas for 1 cm (12 - 25 GHz),
- upgrading the Ceduna antenna for VLBI,
- other technical developments (LO upgrade, atmospheric phase correction),

- strategic research (interference excision, focal plane arrays),
- international collaboration funding.

In addition, there are a number of related developments, which effectively supplement the MNRF funding, but are separately funded:

- MMIC receiver development, funded for $2.4M by CSIRO (in collaboration with CTIP)
- construction of a North ATCA spur, funded for $1.2M by CSIRO,
- SEST correlator development, funded for $340k by ATNF and ARC,
- Mopra surface extension, funded for $150k by UNSW and CSIRO.

3 The Upgrade of the Compact Array

3.1 Background

The Australia Telescope Compact Array consists of six 22-m antennas mounted on a 6-km East-West rail track. Each antenna is equipped with four cryogenic receivers and feeds covering four bands over a frequency range of 1.3 -10 GHz (corresponding to a wavelength range of 21 - 3 cm), and is able to image objects over an angular scale of 1 arcsec to 30 arcmin. Each antenna consists of a 15-m diameter inner solid inner section which is designed for use at wavelengths as short as 3 mm, and a perforated outer section which is designed for use at wavelengths down to 7 mm. The array is particularly optimised for spectral line operation and polarisation purity.

3.2 The Millimetre Upgrade

Millimetre arrays such as IRAM and BIMA have demonstrated in the last few years how much can be learnt from millimetre synthesis observations in studying the astrophysics of objects such as

- Galactic continuum sources (e.g. stellar disks, dust and free-free emission in star formation regions, SNR)
- Extragalactic continuum sources (e.g. AGN cores, IRAS galaxy dust tori)
- Galactic line sources (e.g. protostellar disks, young stellar objects)
- Extragalactic line sources (e.g. redshifted CO, CO mapping of AGN)

The Narrabri site of the ATCA is not a particularly good mm site, and performance at 115 GHz will be limited by the strong atmospheric absorption at that frequency). We therefore intend to concentrate on frequencies below 100 GHz, where atmospheric absorption is much less of a problem. Fortunately, a glance at the literature shows that the most exciting science at these wavelengths has moved from the traditional CO work at 115 GHz to the high-density tracer molecules, such as HCN, many of which have transition frequencies below 100 GHz. However, the ATCA will still be capable of observing redshifted CO from galaxies at redshift > 0.1.

The performance of the ATCA at 3 mm is compared with other synthesis arrays in Table 1. The ATCA will perform at least as well as the world's best arrays at 3 mm. Furthermore, it is the only millimetre array capable of observing the Southern hemisphere, and given the major optical facilities (such as the VLT and Gemini) being constructed in the Southern hemisphere, there is clearly a strong need for such an array in the South.

The ATCA also has the longest baseline of any existing array. This is important as observations have shown many exciting objects to be unresolved with existing baselines, resulting in an move to longer baselines. Whether the ATCA (or any other array) will in fact be able to make effective use of these long baselines will depend on the extent to which we are successful in solving the phase correction problem (see below). At present, we plan that only five of the ATCA antennas will be upgraded to 3 mm, giving a maximum baseline of 3 km. If there is sufficient scientific and engineering justification, the sixth antenna could also be upgraded in the future to give a 6-km maximum baseline.

Japan, Europe, and the US each have proposals in place for the construction of a large Southern millimetre array in Chile, by around 2010. It is likely that at least one of these proposals will go ahead, and the resulting instrument will be an enormous step forward. At that stage, the ATCA millimetre observations will no longer be able to compete, and will presumably close down or turn to niche projects. Thus there exists a window of opportunity of perhaps 7-8 years in which the ATCA can be a world-leader at millimetre wavelengths.

Table 1. How do we compare with other mm arrays?

Array	Completion Date	Wavelength Range (mm)	Sensitivity at 3mm (Jy)	max baseline (km)
BIMA (9*6m)	operational	3.0 (1.3)	0.7	1.4
OVRO (6*10m)	operational	3.0, 1.3	0.5	0.3
IRAM (5*15m)	operational	3.0, 1.5	0.3-0.8	0.4
Nobeyama (6*10m)	operational	3.0, 2.0	1.7	0.36
CfA SMA (6*6m)	1998?	1.6-0.35	-	0.47
AT (5*22m)	2002	3.0	0.5?	3.0 (6.0?)
US MMA (40* 8m?)	2010?	10.0-0.35	0.04?	10.0
European LSA (50*16m?)	2010?	3.0, 1.3 (7, 2, 0.8, 0.65, 0.45, 0.35?)	0.02?	10.0
Japanese LMSA (50*10m)	2010?	3.5 - 0.35	0.03?	10.0

Sensitivity is the rms continuum sensitivity at 100 GHz to a point source observed for 8 hours

3.3 The Technical Challenges of the Millimetre Upgrade

There are clearly many technical challenges associated with the upgrade. Here I highlight three key ones which require innovative solutions.

Atmospheric Phase Correction At 3 mm, variations in the refractive index of the atmosphere, primarily due to water vapour, cause phase variations and image smearing, which are equivalent to "seeing" at optical wavelengths, and could in principle limit the maximum baseline to tens or hundreds of metres. This problem must be solved for successful operation of any of the long-baseline arrays listed in table 3. A number of approaches are being tried at different observatories, mostly revolving around using a measurement of the water vapour content along the line of sight, to correct the phase of the signal in software. This is currently an active research project at the ATNF, and we hope to solve it by positioning our 12 mm receivers sufficiently close to the 3 mm receivers. The 12mm receivers can then be used, while observing at 3 mm, to measure emission from the 22 GHz water vapour line in the atmosphere, and thus derive the appropriate phase correction.

Receiver Development The existing SiS 3 mm receiver design used at Mopra is too bulky and expensive to be used to equip the Compact Array. We have embarked on a collaborative project with CTIP (CSIRO Telecommunications and Industrial Physics) to develop MMIC (Monolithic Millimetre Integrated Circuit) receivers, fabricated using InP which will enable us to produce inexpensive, high-performance, cryogenic low-noise millimetre receivers.

Pointing At present the ATCA antennas can be pointed to an rms accuracy of about ten arcsec. Since the primary beam width at 3 mm is 30 arcsec, the pointing accuracy needs to be improved by a factor of about three. We are exploring a number of solutions, including better pointing models, modelling antenna deformation, and making use of reference pointing observations. The final solution will doubtless consist of a number of these elements.

3.4 The 12-mm Upgrade

As well as the 3-mm receivers, the MNRF funding also provides for the construction of 22 GHz (12 mm) receivers for all compact array antennas. This will enable the ATCA to offer routine imaging with a resolution of 0.4 arcsec. However, this upgrade is important not just for the extra spatial resolution, but also because of the rapidly growing importance of extragalactic water megamasers. Since the discovery of a rotating Keplerian disk in NGC4258, a discovery which was widely hailed as being the first definitive evidence of the existence of black holes, these megamasers are seen as a key tool for probing the kinematics of the nuclear region of active galaxies.

Other science goals are:

- high-resolution observations of all the continuum objects currently being studied with ATCA (e.g. SN1987A, active galaxies, radio galaxies, supernova remnants, etc.),
- H_2O masers in star formation regions,
- Ammonia and other molecular lines in star formation regions and active galaxies.

4 The Australia Telescope Parkes Observatory

The Parkes telescope consists of a single 64-m diameter antenna, which is used both for single-dish work and as a member of the Australian VLBI network (see below). It has a frequency range of 0.3 - 43 GHz (corresponding to a wavelength range of 100 - 0.7 cm) and is equipped with a number of low-noise cryogenic receivers for all major radio-astronomical observing bands in this range.

Recently, the Parkes telescope was used intensively for a period of over a year to help recover the signals from NASA's Galileo spacecraft. With the funds earned from this contract, the telescope was recently upgraded with a larger focus cabin which enables greater versatility in scheduling receivers, and had also enabled the construction and installation of an innovative 13-beam receiver for studying neutral hydrogen from distant galaxies. This multibeam receiver is now being used in a major project to survey the entire Southern sky for HI (Staveley-Smith et al., elsewhere in these proceedings).

As part of the MNRF upgrade, we are also upgrading the backend of the telescope to enable greater flexibility and higher reliability.

5 The Australia Telescope Mopra Antenna

The Mopra telescope is a 22-m diameter antenna, located near Siding Spring observatory, which is used for single-dish millimetre observations and also as part of the Australian VLBI network. It has a frequency range of 1.3 - 10 GHz, and 85 - 115 GHz (corresponding to 100 - 0.7 cm, and 3 mm). Like the ATCA antennas, only the inner 15-m diameter surface is useable at 3 mm.

Although the cryogenic SiS dual-polarisation receiver is state-of-art, unfortunately the site is not! An atmospheric absorption line limits the performance at 115 GHz, the frequency of CO, although at lower frequencies (90-100 GHz, where many high-density molecular tracers such as HCN, are found) the performance rivals the best millimetre-telescopes in the world.

Under an agreement currently being negotiated, the University of New South Wales (UNSW) are funding an upgrade of the telescope to give it a solid 22-m surface useable at 3 mm, in return for observing time on the telescope being made available to UNSW staff and students in addition to any time won through the normal peer-reviewed time allocation process.

6 Australian Long Baseline Array

The Australian Long Baseline Array (ALBA) is an array of telescopes which use the technique of Very Long Baseline Interferometry (VLBI) to achieve a resolution of milliarcsec. It consists of the following telescopes:

- The tied AT Compact Array - equivalent to a 53-m antenna
- ATNF Parkes 64-m antenna
- ATNF Mopra 22-m antenna
- NASA/JPL Tidbinbilla 70/34-m antennas
- University of Tasmania Hobart 26-m antenna
- University of Tasmania Ceduna 30-m antenna
- Occasional use of Perth 15-m (ESA,UWA) and MOST (U. Sydney) antennas

The ALBA makes use of S-2 recording systems, giving 128 Mbit/s recording bandwidth, and the data are correlated at the ATNF Headquarters in Sydney. In addition, the ALBA frequently conducts international VLBI observations, particularly with South Africa, Japan, and China. Since last year, the ALBA has also been an active contributor to the Japanese VSOP space-VLBI project.

The Ceduna antenna is a communications antenna which was generously donated by Telstra to the University of Tasmania, but needed considerable modifications to equip it for VLBI use. Several of these modifications, including the S2 recorder, a hydrogen maser, new receivers, and an upgrade to the antenna to enable low-frequency operation, are funded under the MNRF program.

One of the prime goals of the MNRF upgrade was to equip the ALBA for operation at 22 GHz. This is important both to extend the resolution of the ALBA, and, perhaps more importantly, to enable it to study water megamasers, discussed above in connection with the 22 GHz upgrade of the ATCA.

As a result, the MNRF funding provides for the design and construction of 22 GHz receivers for Mopra, Narrabri, Hobart and Ceduna, and the provision of hydrogen maser clocks (necessary for adequate phase stability at this frequency) for all the antennas which do not yet have one.

7 Conclusion

The MNRF upgrade of the Australia Telescope is well under way and fully funded. Specifications are being fixed, science issues are being addressed, technology development has started, and we expect most of the upgrade will be available to users in about 2002-3. We very much welcome views and input from users and prospective users.

In this paper it has not been possible to give more than an overview of a selection of aspects of the MNRF upgrade. For more details, please see http://www.atnf.csiro.au/mnrf/mnrf.html.

The Square Kilometer Array Radio Telescope

R. Ekers

Australia Telescope National Facility, Australia

Abstract. Plans are being made to build the next generation radio telescope with sensitivity one to two orders of magnitude greater than any existing telescope. The sensitivity of this telescope will make it an excellent instrument to study the evolution of both neutral gas and star formation rate in galaxies at high redshift in a manner which will greatly extend and complement observations planned for future instruments at other wavelengths.

1 Introduction

It is well established that most scientific advances follow technical innovation (e.g. Harwit 1981). De Solla Price (1963), reached the same conclusion from his application of quantitative measurement to the progress of science. His analysis also showed that the normal mode of growth of science is exponential. A plot of the sensitivity of telescopes used for radioastronomy since the discovery of extra-terrestrial radio emission in 1940 shows this exponential character (Figure 1) with an increase in sensitivity of 10^5 since 1940, doubling every three years. To maintain the extraordinary momentum of discovery of the last few decades a very large new radio telescope will be needed.

An increase in sensitivity of this order cannot be achieved by improving the electronics of receiver systems, but only by *increasing the total effective collecting area of radio telescopes to about a million square meters*. The project has therefore acquired the appellation, the Square Kilometer Array.

New technologies involving large scale integration of transistor amplifiers into complex systems which can be duplicated inexpensively, and the rapidly increasing capability to apply digital processing at high bandwidth now make it possible to obtain this aperture at an affordable cost. This will require unconventional techniques involving the large scale replication of individual elements. The time frame during which a new radio facility is needed to complement other planned instruments will be in the years around 2010.

Under the auspices of the Large Telescope Working Group, established in 1993 by the International Union of Radio Science (URSI), a large scientific community is now beginning to cooperate to discuss the technical research required to the realization of the project. A group of 6 countries (Netherlands, U.S., China, Canada, Australia, India) are at present actively working on the technology study program. An internet site (`http://www.nfra.nl/skai/`) provide up-to-date information about the project (including the latest news, scientific and technical documents and the list of participant institutes). Much of the following material has been developed in the Large Telescope Working Group and

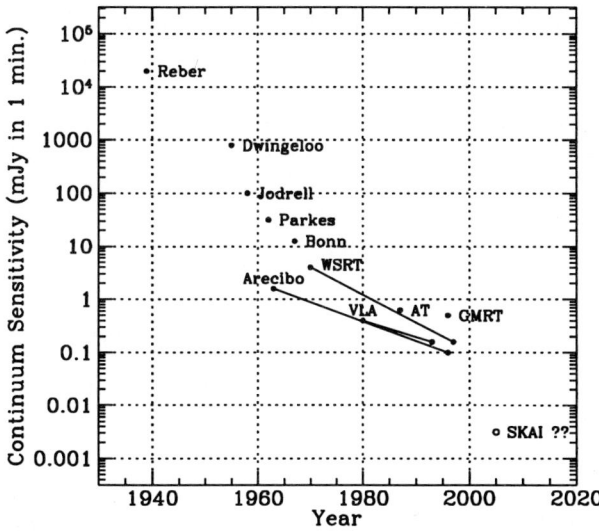

Fig. 1. Radio telescopes sensitivity

Parameter	Design Goal
Total Frequency Range	0.1 – 10 GHz
Imaging Field of View	1 square deg. @ 1.4 GHz
Angular Resolution	0.1 arcsec @ 1.4 GHz
Surface Brightness Sensitivity	1 K @ 0.1 arcsec (continuum)
Instantaneous Bandwidth	$0.5 + \nu/5$ GHz
Number of Spectral Channels	10^4
Number of Instantaneous Pencil Beams	100

Table 1. Instrumental Design Goals

will be available in full detail in the science case document now being developed (Braun & Taylor 1998).

2   Square Kilometer Array: the Concept

Extensive discussion of the envisioned science drivers and of the evolving technical possibilities has led to a set of design goals for the Square Kilometer Array. Some of the basic system parameters are summarized in Table 1.

By looking at the required instrumental parameters listed in the table, it is clear that the Square Kilometer Array will be the world's premier astronomical imaging instrument. It has a spatial resolution better than the Hubble Space Telescope (< 0.1 arcsec), a large field of view (~1 square degree), a spectral coverage of more than 50% ($\nu/\Delta\nu < 2$), a spectral resolution sufficient for kinematic studies ($\nu/d\nu > 10^4$), and all at a sensitivity which is about 100 times that which is now available.

By comparison, the largest optical integral field units which are now being considered for construction on 8-m telescopes would only provide a field of view of perhaps 1 arcmin on a side with 10% spectral coverage at a comparable resolution, while the next generation of millimeter arrays is envisaged to provide a field of about 40 arcsec with perhaps 10% spectral resolution.

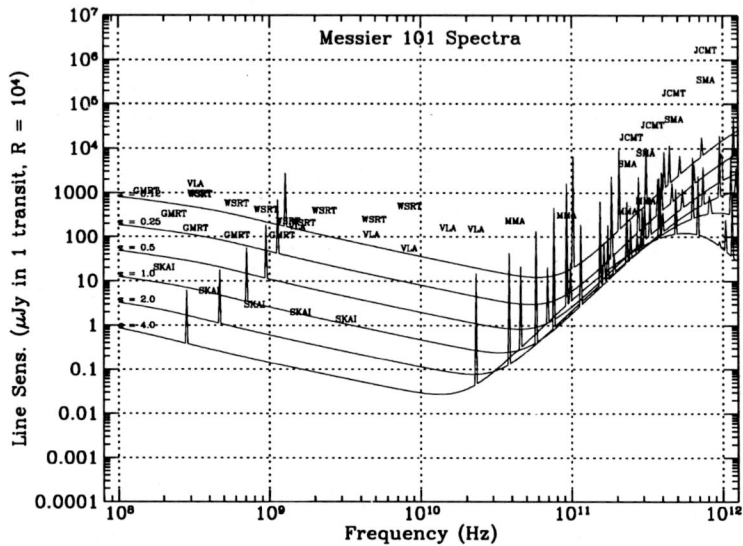

Fig. 2. Simulated spectra of the spiral galaxy M101 are shown for frequencies between about 10^8 and 10^{12} Hz after being red-shifted to $z = 0.12, 0.25, 0.5, 1, 2, 4$, under the assumption of no spectral evolution. Instrumental sensitivity (1σ) of existing and planned instruments are overlaid for spectral line observations.

The system sensitivities which follow from the design goals are very interesting. At 1.4 GHz, for the broad-band continuum ($\sim$600 MHz) and moderate resolution spectral line observations ($\nu/d\nu = 10^4$, corresponding to 30 km s^{-1}), the system sensitivities are 23 nanoJy and 1.6 μJy respectively after 8 hours observation. Braun (1995) has made an informative comparison of the spectral line sensitivity compared with those of many existing and planned facilities in Fig. 2 using a simulated spectra of the spiral galaxy M101 as it would appear at different red-shifts under the assumption of no spectral evolution. The striking result is that even a normal galaxy like M101 (with an atomic mass $M_{\rm HI} = 2 \times 10^{10} M_\odot$ and molecular gas mass $M_{\rm H_2} = 3 \times 10^9 M_\odot$) could be detected efficiently at arbitrary red-shift. The HI line can be tracked to red-shifts beyond 4, while the CO lines also become easily accessible at red-shifts of 4 and larger. Similarly the continuum emission of normal galaxies can be detected out to basically arbitrary distances and gives a direct measurement of the star formation rate.

3 Science with the Square Kilometer Array

The proposed telescope will have a sensitivity 100 times greater than that of the VLA. That corresponds to a 10,000 times increase in observing speed, enough to make projects which are currently completely impractical because they would take years of integration time, possible in a few hours. It is the steps of this magnitude which allow new types of phenomena to be investigated and which will result in new scientific advances.

Perhaps the most exciting of the scientific drivers is the opportunity to study the large scale structure of the universe by observing both the red shifted neutral hydrogen and the non-thermal emission from star forming regions in normal galaxies, like our own, at cosmological distances. The velocity field can be used to measure the rotation of these Galaxies probes the mass directly. It is interesting to note that, just as for the Hubble deep field, an image at the sensitivity of the square kilometer array is dominated by normal galaxies - the radio galaxies and quasars have already been seen to the beginning of the universe at higher flux levels and no longer dominate the faint radio sky.

3.1 Galaxy Evolution

The fascinating observations obtained with the HST, in particular the ongoing analysis of the few thousand galaxies in the Hubble Deep Field, show that there is already significant evolution detectable in the comoving star formation rate (SFR) density by looking back to a redshift of 1 (Madau 1998). Already looking back between $z = 0.5$ and $z = 1$ (3-4 h^{-1} Gyr) there should be a noticeable increase of a factor 2 - 3 in SFR density. The SFR density appears to peak around $z = 1.5$ with the most vigorous evolution between $z = 1$ and $z = 3$. Models indicate that the continuum images will be dominated by this population of galaxies. We will be able to detect hundreds of galaxies in this redshift range in an area the size of the HDF and we will be able to use this radio continuum measurement to estimate the star formation rate through the FIR-radio correlation. In fact with no absorption corrections needed this is likely to be the most accurate and straightforward indicator of SFR at high redshift.

Little is known about the evolution of the HI in galaxies out to redshifts of 1 and beyond, because present day instruments lack the sensitivity and resolution to direct measure the HI in galaxies at these redshift. Damped Lyα studies (Lanzetta et al. 1995) indicate that the comoving HI mass density is roughly 5 - 10 times the present beyond $z = 1$ and out to $z = 3$. This is also the period during which metal-rich gaseous halos appear, confirming that this is an era of strong evolution, where it is imperative to have a good insight into the evolution of the HI content in galaxies.

The Square Kilometer Array will be able to measure HI in galaxies back to redshifts of $z \sim 3$ and will revolutionize this are of research.

3.2 Large Scale Structure Studies

The Square Kilometer Array will be able to perform surveys covering a large area of the sky with very uniform sensitivity. In 3 months of observing time one could cover 1000 square degrees and be able to detect L_* galaxies out to redshift of $z = 1.3$ (or about 75% of the age of the universe). Assuming Zwaan et al. (1997) HI mass function one expects to detect 10^7 galaxies in a volume of 3×10^8 Mpc^3. This is an order of magnitude more than in optical surveys, such as the Sloan Digital Sky Survey (see e.g. Lahav, these proceedings) and the AAO 2dF Survey (Colless, these proceedings). Following the methodology of Lahav we have included the HI surveys in Fig. 3.

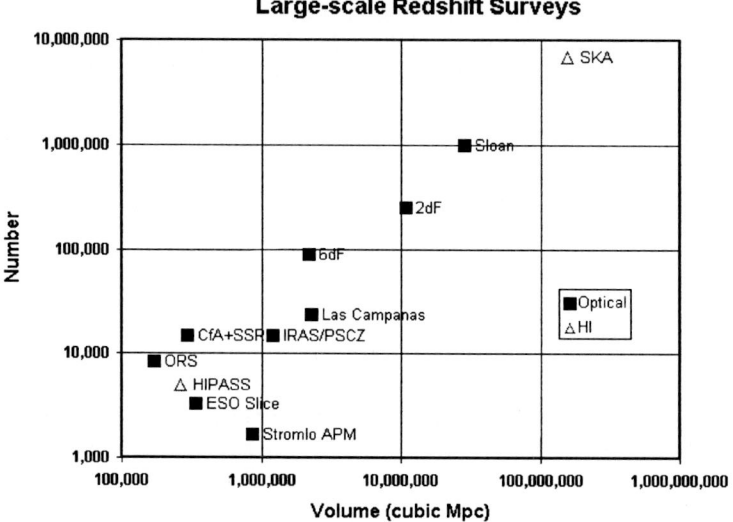

Fig. 3. Effective volume and number of galaxies in various planned and completed surveys from Lahav (these proceedings)

3.3 Probing Dark Matter with Gravitational Lensing

The technique of weak gravitational lensing has proven to be an important tool to study mass distribution in the universe. The projected mass distribution of foreground gravitational structure distorts the images of the faint background galaxies. As a result, gravitational lensing provides a direct measurement of the projected mass density (e.g. Kaiser & Squires 1993; Schneider, these proceedings).

The impact of the Square Kilometer Array on the field of weak lensing will be particularly profound for a number of reasons. With the sensitivity given above, we expect detected source densities of between 100 and 400 arcmin^{-2} in an 8 hours integration. At this flux level the radio source population is dominated by normal star forming galaxies with angular sizes in the range which is very accessible to the Square Kilometer Array.

While these source densities are comparable to those of the HST, there are two very significant differences. Firstly, the Square Kilometer Array PSF is both compact (smaller than about 0.1 arcsec) and extremely well-defined (to about one part in a million) over the entire field of view. Secondly, the enormous instantaneous field of view is sufficient to probe scales of some Mpc on a side (at $z = 0.3$) per pointing. This should enable clean measurement of cluster mass surface densities as well as routine detection of the weak lensing signature due to large-scale structure.

A special case of considerable interest becomes possible with the sensitivity and resolution of the Square Kilometer Array. It is possible to see the strong lensing effects caused by the core of a cluster of galaxies on a background galaxy seen in HI. By scanning in redshift the background galaxy's HI field provides a moving source of emission which can map out the 2D structure in the lensing mass distribution.

3.4 Dark Sunyaev-Zeldovich Clusters

At this meeting we have heard reports of the detection of hot gas by the Sunyaev-Zeldovich (S-Z) effect (Liang & Birkinshaw, these proceedings) even though no galaxies can be seen. At the upper frequency (10GHz) the individual stations of the Square Kilometer Array (dense packed arrays of elements covering an area of about one hectare) will have sufficient brightness sensitivity to detect the S-Z effect with sensitivity more than an order of magnitude better than any existing telescope. Normally one would push to higher frequencies to observe the S-Z effect in order to avoid contamination with weak discrete continuum sources in the field, but since the Square Kilometer Array will have excellent simultaneous sensitivity to discrete sources it will be able to subtract them with high accuracy.

3.5 Gamma Ray Bursters

It is only in the last year that we have begun to understand the gamma ray bursts (GRB's). At least one GRB is known to be securely of extragalactic origin (Metzger et al. 1997). This revolution in our understanding is due to the accurate localization of GRB's by the Italian-Dutch BeppoSAX satellite and the discovery of the relatively long lived "afterglow" seen at X-ray, optical and radio wavelengths. At radio wavelengths this afterglow is long lasting but very weak and at the sensitivity limits of existing telescopes (Frail et al. 1997). The Square Kilometer Array not only has adequate sensitivity to follow these afterglows but has more than one instantaneous pencil beam available to do so. This means that it can conduct monitoring programs with μJy sensitivity on a number of GRB's without interfering with other imaging observations. In addition to the obvious value for detecting and locating the radio afterglow from GRB's, continuous monitoring at high sensitivity will show the deep modulation of the received signal on a time scale of hours caused by the interstellar scattering (e.g. Walker 1998). Frail et al. (1997) have already detected this modulation in GRB970508 and have used it to estimate the changing size of the fireball on the scale of a few micro arcseconds. There is no other known technology which can access this scale of sizes in extragalactic objects.

References

Braun, R. & Taylor, R., 1998, *"Square Kilometer Array Radio Telescope – The Science Case"*
Braun, R. 1995, in *"Cold Gas at High Redshift"*, eds. M. Bremer et al. (Kluwer), 437
Frail, D., Kulkarni, S., Nicastro, L., Feroci, M., & Taylor, G. 1997, Nature 389, 261
Kaiser, N., & Squires, G. 1993, ApJ 404, 441
Lanzetta, K., Wolfe, A., & Turnshek, D. 1995, ApJ 440, 435
Madau, P., 1998, in press, astro-ph/9707141
Metzger, M. et al. 1997, Nature 387, 878
Harwit, M., "Cosmic Discovery - The Search, Scope & Heritage of Astronomy", Basic Books, Inc., New York (1981)
Price, Derek J. de Solla "Little science, big science" 1963 Columbia University Press
Walker, M. 1998, Mon. Not. R. Astron. Soc. 294, 307

Prospects with Large Millimeter Arrays

Peter Shaver

ESO, Karl-Schwarzschild-Str. 2, 85748 Garching bei München, Germany

Abstract. The new generation of large millimeter arrays will make possible observations of the epoch of first galaxy formation above $z = 5$, and studies of star-forming galaxies at lower redshifts that may be invisible to the HST and VLT because of dust obscuration. Millimeter and submillimeter observations are crucial for a full understanding of the evolution of galaxies of all types. Europe, the U.S., and Japan are all working on large millimeter array projects in the southern hemisphere, and these efforts may well merge into one global collaboration.

1 High Redshift Star-Forming Galaxies

What is the Hubble Deep Field missing? From the redshifts obtained so far, the global star formation rate deduced from HDF galaxies appears to peak at about $z \sim 1 - 2$. The steep rise up to $z \sim 1$ and the decline beyond $z \sim 2 - 3$ is similar to that found in large complete quasar samples that are unaffected by dust obscuration (Shaver *et al.*, 1998). On the face of it, this may suggest that the galaxy samples are relatively complete too. However, there are certainly corrections that have to be applied for internal dust, and these can be large, particularly at the higher redshifts (bluer rest wavelengths). As a result, even the existence of a peak in the global star formation rate is uncertain.

Thus, the HDF may well be missing large numbers of star-forming galaxies at high redshifts. It could also be missing star-forming galaxies at lower redshifts with very high dust opacity. And of course it gives a very incomplete idea of the bolometric luminosity of the galaxies it does detect, as most of the luminosity of star-forming galaxies comes out in the far-infrared and (sub)millimeter wavebands. Large millimeter arrays can detect all of these, and are therefore an essential complement to the large optical and near-infrared telescopes.

The millimeter wavebands are particularly well suited to the detection of high-redshift star-forming galaxies. In the continuum, the huge far-infrared dust emission peak is redshifted to millimeter wavelengths, and galaxies out to $z = 20$ can be detected as a result of this large negative K-correction. This may well be the best way to find the first galaxies in the "dark ages" beyond $z = 5$. Also, the millimeter wavebands abound with spectral lines, and the "ladder" of molecular transitions essentially guarantees that a redshifted line will appear in one of the observing bands. Molecular lines such as CO can be detected with a large millimeter array from galaxies up to $z = 1 - 10$, and atomic lines such as [CII] 158μ can be detected up to $z = 5 - 10$. Thus, there is a wealth of possibilities with a millimeter array of adequate sensitivity.

At the moment, millimeter observations have just scratched the surface, and given tantalizing glimpses of what will be possible. Millimeter continuum and line observations have already been made of objects up to $z = 4.7$, near the present redshift limit (Omont et al., 1996a,b; Guilloteau et al., 1997). The possible detection of a far-infrared background with COBE (Puget et al., 1996) suggests the presence of a large number of sources at $z \sim 3 - 10$. The possible detection of high-redshift galaxies at 175μ with ISO (e.g. Rowan-Robinson et al., 1997) may provide evidence for strong evolution. And the detection of previously unknown high-redshift galaxies at 850μ with SCUBA on the JCMT (Smail et al., 1997) also indicates a large population and strong evolution. The density of such objects on the sky may be so large that single-dish millimeter telescopes will be confusion limited, and large arrays with high angular resolution will be required. In summary, the high-redshift Universe is just now being opened up to millimeter wavelengths, and we appear to be poised for a major breakthrough.

2 A Next Generation Millimeter Array

A next generation millimeter array will be a major step for astronomy, and will revolutionize millimeter astronomy. It will be the millimeter equivalent of the VLT, HST, and NGST, and is now the top-ranked ground-based astronomy project in many countries. The various concepts all call for high angular resolution (0.1" at 3 mm) provided by baselines up to 10 km, and high sensitivity provided by dozens of antennas with a total collecting area of about 5,000-10,000 m^2. Receivers would cover all wavebands accessible from the ground in the millimeter and submillimeter range. The site must be high, dry, large, and flat - a high Andean plateau in the Atacama desert is ideal, so interest has focussed on northern Chile.

A giant millimeter/submillimeter array with these parameters will have a major impact on all areas of astronomy, and will change the notion of millimeter astronomy as a "specialist" field forever. The main science drivers are of course the origins of galaxies (the epoch of first galaxy formation in the "dark ages" beyond $z = 5$) and the origins of stars and planets (all phases of star formation hidden away in dusty molecular clouds). But the array will go far beyond these major objectives (see the proceedings of the ESO Workshop on Science with Large Millimeter Arrays (Shaver, 1996)), and targets will include: gravitational lenses, quasar molecular absorption lines out to high redshifts, active galactic nuclei (free of synchrotron and dust obscuration), nearby galaxies (including nuclear accretion disks and kinematic detections of black holes), objects of all types in the Magellanic Clouds, the galactic center (free of obscuration), galactic molecular clouds, astrochemistry, protostars, young stellar objects (with their protoplanetary accretion disks and outflow jets), masers, thousands of stars over the entire H-R diagram, circumstellar shells around evolved stars (time-dependent chemistry, which can be observed across the Galaxy), planetary nebulae, supernovae (particularly 1987A) and supernova remnants, details in planetary atmospheres, asteroids and comets, and the search for extrasolar planets. With its broad scientific potential, such an array will be heavily used by the general astronomical community, and it will be a wonderful counterpart to the VLT in the southern hemisphere.

Three major millimeter array projects are under consideration around the world: the Large Southern Array (LSA) in Europe, the Millimeter Array (MMA) in the U.S., and the Large Millimeter and Submillimeter Array (LMSA) in Japan. The LSA and MMA projects have recently agreed to explore the possibility of a partnership; the combined array would comprise 64 × 12m antennas, giving a total collecting area of 7,000 m^2. It would be located at the 5,000m site of Chajnantor, near the town of San Pedro de Atacama. The LMSA will also be located near this site, and a merger of all three projects into one "world array" is presently being discussed. The time scales for all three projects are very similar: design studies until 2002-2003, and construction over the period 2003-2008. Operation need not await final completion, and could begin when just a part of the array is available. In view of the scientific potential outlined above, it is clear that a millimeter array of this kind will usher in a new era of astronomy.

References

Guilloteau, S., Omont, A., McMahon, R.G., Cox, P., Petitjean, P. (1997): *Astron. Astrophys.* **328**, L1

Omont, A., Petitjean, P., Guilloteau, S., McMahon, R.G., Solomon, P.M., Pécontal, E. (1996a): *Nature* **382**, 428

Omont, A., McMahon, R.G., Cox, P., Kreysa, E., Bergerion, J., Pajot, F., Storrie-Lonbardi, L.J. (1996b): *Astron. Astrophys.* **315**, 1

Puget, J.-L., Abergel, A., Bernard, J.-P., Boulanger, F., Burton, W.B., Désert, F.-X., Hartmann, D. (1996): *Astron. Astrophys.* **308**, L5

Rowan-Robinson, M. et al. (1997): *Mon. Not. R. astr. Soc* **289**, 490

Shaver, P.A. (ed.) (1996): *Science with Large Millimetre Arrays* (ESO Astrophysics Symposia; Springer-Verlag, Berlin Heidelberg)

Shaver, P.A., Hook, I.M., Jackson, C.A., Wall, J.V., Kellermann, K.I. (1998): in *Highly Redshifted Radio Lines* (eds. C. Carilli, S. Radford, K. Menten, and G. Langston; PASP: San Francisco) (in press) (also: astro-ph/9801211)

Smail, I., Ivison, R.J., Blain, A.W. (1997): *Astrophys. J.* **490**, L5

van der Werf, P.P., Israel, . F.P. (1996): in *Science with Large Millimetre Arrays* (ESO Astrophysics Symposia; ed. P. A. Shaver; Springer-Verlag, Berlin Heidelberg), p. 51

Discussion

R. Ekers: What is the evidence that the objects detected by SCUBA are at high redshift?

P. Shaver: Their broadband spectral properties are consistent with high redshift galaxies, and in some cases redshifts (as high as $z \sim 2$) have actually been measured.

M. Dickinson: If there are sources to be detected at $z \sim 20$, how will we measure the redshifts? Can it be done somehow with millimeter observations alone, and if so to what fluxes/luminosities?

P. Shaver: Redshifts can be measured at millimeter and submillimeter wavelengths using molecular and atomic spectral lines. There are many lines, so unique redshift determination may be as easy as it is in the optical. Lines will be detectable in starburst galaxies out to very high redshifts; van der Werf and Israel (1996) have estimated sensitivities.

M. Burton: What are the prospects for *sub*-millimeter interferometry? How far into the sub-mm regime do you think it will be possible to extend interferometric techniques?

P. Shaver: There are proven techniques for phase calibration at millimeter wavelengths, and our understanding of atmospheric physics indicates that we can do adequate phase calibration at submillimeter wavelengths also. The ultimate fall-back is fast switching.

C. Lineweaver: What will the angular resolution be?

P. Shaver: 0.1 arcsec at 3 mm, or about 10 milliarcsec at the shortest accessible submillimeter wavelengths.

Galileo & *d.o.lo.res.*

E. Molinari, P. Conconi, M. Pucillo, S. Monai

[1] Osservatorio Astronomico di Brera, via Bianchi 46, I-23807 Merate (LC), Italy
[2] Osservatorio Astronomico di Trieste, Trieste, Italy

Abstract. In the course of 1998 the Italian *Galileo* Telescope will begin operation. The first generation of instruments will include a low resolution spectrograph with multi object capabilities and full remote on-line control.

1 *d.o.lo.res.*

LSR *d.o.lo.res.* (acronym of device optimised for the low resolution) is an imager-spectrograph which will be mounted at Nasmyth focus B of the Italian National Telescope "Galileo" (TNG). Its full implementation includes multi-object on-line spectroscopy, photo- and spectro-polarimetry, long slit and echelle modes. Commissioning and test phase will start mid 1998 at the Observatory of Roque de los Muchachos in Canary Island (Spain).

The optical design consists of 11 elements including the spheric closing window for the cryostat. The image quality is better than 0.3″ for $D_{80\%}$ on the whole field. Foreseen filters are classical U, B, V, R, I, g, r, while the grisms have dispersions from 220 to 46 Å/mm covering wavelength range 3000-10000 Å. One low order echelle will allow a higher resolution on a limited wavelength range. For the echelle grism two cross dispersers will be available, one optimised for the red and the other for the blue.
The mechanical construction and assembly took place in Officine Castellini (Brescia, Italy) while the optical workshop was SESO (Aix-en-Provence, France). Final integration and test are in progress at the Observatory of Trieste.

2 Observing Modes

A series of (guided) observing mode will be accessible by final users.

IMG - The **imaging** mode will have a useful field of view of 10'×10'. LSS **Long slit spectroscopy** will use an adjustable long slit and a series of 8 grisms.
ECH - Also **echelle** spectroscopy will use the adjustable long slit an a choice of two cross dispersers.
MOS - Two kind of **Multiple Object Spectroscopy** will be available: (a) two series of on-board 19 slitlet frames will be controlled by on line software within an useful area of 10'×6'; (b) up to seven user supplied masks can be placed in the mask holders an positioned on the TNG focal plane. A punching machine is in completion phase.

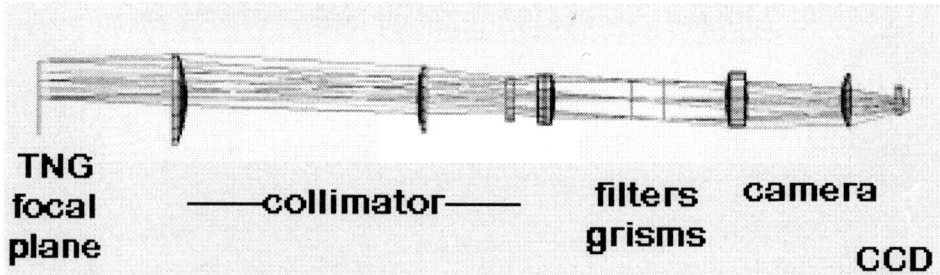

Fig. 1. The optical configuration of *d.o.lo.res.*. The useful area in the TNG focal plane (left) is about 10 by 10 cm. Different slit set-ups will be available. The collimated, parallel beam is 76 mm wide and the two filter and grism wheels will operate in this zone. The 30 by 30 mm CCD detector will have 2048^2 pixel $0.3''$ on a side.

POI - A Wollaston prism on the grism wheel can be used for **polarimetric imaging**. Before the focal plane a polarimetric plate can be inserted and a custom mask will allow direct imaging.

POS - **Polarimetric spectroscopy** can also choose between linear and circular polarisation.

CAL - **Calibration** can be used in imaging (lamp flat fielding) and in spectroscopy: 6 lamps will send light in an integrating sphere which will be imaged on the focal plane with same f/# of main telescope.

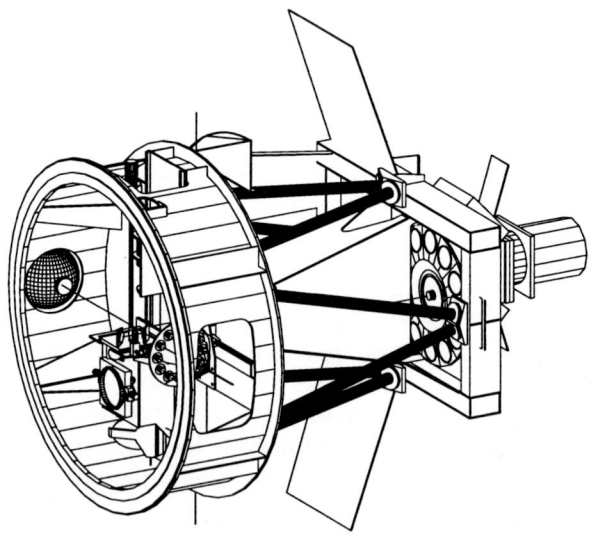

Fig. 2. The mechanical design of LRS *d.o.lo.res.* is shown. Its length is 2 meter, excluding the detector dewar, and its weight is 1.2 tons. All the movements are controlled by a network of transputers

The Future of Observations of the Sunyaev-Zel'dovich Effect

Haida Liang and Mark Birkinshaw

University of Bristol, Physics Department, Tyndall Avenue, Bristol BS8 1TL, UK

Abstract. We present an overview of the current status and the future of observations of the Sunyaev-Zel'dovich effect. In particular we examine the capabilities of an upgraded Australia Telescope for detecting the SZ effect.

1 Introduction

The Sunyaev-Zel'dovich effect (SZE) is the change in brightness temperature of the cosmic microwave background (CMB) radiation along the line of sight towards a dense cluster of galaxies (Sunyaev and Zel'dovich 1972). Photons from the cosmic microwave background radiation are scattered by the hot gas in a cluster. This causes a change in brightness of the CMB radiation. Two SZE components can arise: the thermal and the kinematic effects.

The thermal SZE is given by the Doppler effect due to the random thermal motion of the hot electrons in the intra-cluster medium (ICM) causing the CMB photons to change their frequencies after scattering off the electrons. If the thermal plasma is optically thin then the problem is reduced to the case of single scattering. Since scattering conserves the number of photons along the line of sight, and for $h\nu \ll kT_g$ scattering processes tend to increase the photon energy, the result is a decrease in intensity in the background radiation at low frequencies and an increase at high frequencies. Thus we expect a decrement in brightness temperature in the Rayleigh-Jeans part of the CMB radiation spectrum in the direction of a cluster and an increase in brightness temperature in the Wien part of the spectrum. In the non-relativistic limit, where the gas temperature $kT_g \ll 20$ keV, the thermal SZE is given by

$$\frac{\Delta T_r}{T_r} = \frac{\Delta I_\nu}{I_\nu}\frac{d\ln T_r}{d\ln I_\nu} = y(x\frac{e^x+1}{e^x-1} - 4) \quad (1)$$

where $x = h\nu/kT_r$, the Comptonization factor y is given by

$$y = \int \frac{kT_g(r)}{m_e c^2}\sigma_T n_e(r) dl \quad , \quad (2)$$

σ_T is the Thompson cross section, and $n_e(r)$ is the electron density.

If the cluster has a peculiar velocity, the Doppler effect due to the bulk motion of the ICM will also cause a change in the brightness temperature of the CMB

(Sunyaev & Zel'dovich 1981; Rephaeli & Lahav 1991). This kinematic SZE in the non-relativistic limit is given by

$$\frac{\Delta T_r}{T_r} = -\frac{V_r}{c} \int \sigma_T n_e(r) dl \qquad (3)$$

where V_r is the receding radial component of the peculiar velocity. Unlike the thermal effect, the sign of the kinematic effect is independent of frequency and is solely dependent on the direction of the radial component of the peculiar velocity. Figure 1 shows that the kinematic SZE reaches its peak when the thermal SZE is zero. Thus we can separate the kinematic and thermal effects and measure the radial component of the peculiar velocity of a cluster at any redshift even though the kinematic effect is usually an order of magnitude smaller than the thermal SZE. Although this will enable us to probe the deviation from the Hubble flow at high redshifts, the accurate (relativistic, multiple-scattering) expression for the thermal SZE is required in the spectral fitting to obtain V_r (Rephaeli 1995).

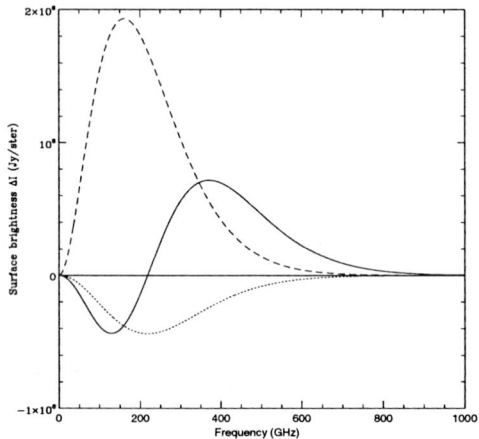

Fig. 1. Specific intensities of the thermal SZE as a function of frequency (solid curve) for optical depth $\tau_T = 0.03$ and $T_g = 7 \times 10^7$ K, and the kinematic SZE (dotted curve) for $V_r = 3000 \, \mathrm{km \, s^{-1}}$. The CMB blackbody (dashed curve) scaled down 200 times is plotted for comparison.

1.1 Classic Charms of the SZE

Ever since the suggestion by Silk & White (1978) and Gunn (1978) that H_0 and q_0 can be determined by comparing an SZE map with X-ray surface brightness and temperature maps, the determination of cosmological parameters has been a major focus of SZE work. Basically, the thermal SZE is proportional to the average electron density n_e along the line of sight and the X-ray surface brightness

S_x is proportional to the average of n_e^2. If the cluster is spherically symmetric, the line-of-sight linear size deduced from these data can be compared with the angular size, and the distance to the cluster can be determined. In principle a sample of high redshift clusters at $z > 0.5$ could be used to deduce q_0. Uncertainties caused by the non-sphericity of a cluster can be overcome by observing a randomly-oriented sample of clusters. Since $S_x \propto <n_e^2>$ and $\Delta T_r^2 \propto <n_e>^2$, the main theoretical limitation of the method is the assumption that the gas is not clumped on small scales, $<n_e^2(r)>=<n_e(r)>^2$. Even ignoring this intrinsic limitation, the uncertainty in the measurement of H_0 is still large compared with conventional methods because of the difficulties of the observations involved. While conventional determinations of H_0 in the local Universe (e.g., using the Tully-Fisher relation) yield a measurement with greater accuracy, the SZE/X-ray method allows us to measure H_0 at large distances and so eliminates systematic effects from local deviations from the universal expansion, while also avoiding any presumptions about the universality of "standard candles".

Apart from the classic charms of the SZE in providing direct measurements of cosmological parameters, it is also an important probe of the intracluster medium (ICM) since the SZE is essentially proportional to the pressure of the ICM along the line of sight (see section 3).

2 Current Status

Observations of the SZE in the radio spectrum use single dish radio telescopes or interferometer arrays. Early detections of the SZE were made with single dishes (e.g. Birkinshaw and Gull 1984) because of their higher brightness sensitivity relative to conventional interferometric arrays such as the VLA, which were designed for high resolution radio mapping. However, high sensitivity compact interferometer arrays with short baselines have enhanced brightness sensitivity and are ideal instruments for the detection of the SZE. Interferometers register only correlated signals, which greatly reduces the effects of atmosphere and interference. Moreover, interferometers provide a convenient way of subtracting radio sources; while the short baseline data that have high brightness sensitivities are used for the detection of the SZE, the long baseline data can be used to detect and subtract discrete radio sources. Such a procedure provides the most systematic-free subtraction of radio sources, since the long and short baseline data are collected simultaneously and with the same telescope.

Table 1. A comparison of interferometers used to observe the SZE

	Ryle	ATCA	VLA	OVMMA	BIMA
D (kλ)	1	0.85	1.5	1	0.8
ν (GHz)	15	8.8	8	28.5	28.5

For many years the SZE was only known in the three extremely X-ray luminous clusters A2218, A665 and CL0016+16. Since 1993, when the first reliable interferometric detection of an SZE was reported by the Cambridge group (Jones et al. 1993), several high X-ray luminosity clusters have been detected by the Ryle telescope. Shortly afterwards, an innovative implementation of high sensitivity 30 GHz receivers on mm interferometers with small dish sizes (the OVMMA and BIMA) greatly improved the efficiency of producing high sensitivity SZE maps (Carlstrom et al. 1996). More than 20 clusters now have interferometric SZE maps (Carlstrom and Joy; Jones; private communications). Table 1 gives a list of interferometers used for SZE detections with operating frequencies and shortest baselines in wavelength units.

In the radio regime, an increase of frequency ν means a decrease of astrophysical confusion by $\sim \nu^{0.3}$ to $\sim \nu$ since the majority of radio sources have steep radio spectra; it also means an increase of the specific brightness intensity ΔI_r by $\sim \nu^2$. However, for the same telescope, i.e. fixed baselines, an increase in ν means an increase in uv-spacing by a factor of ν and hence a decrease of brightness sensitivity by a factor of $\sim \nu^2$ or worse. Thus the ideal telescope would operate at high frequency on short baselines: this is partly why the BIMA and OVMMA have been highly effective at detecting the SZE.

At high frequencies ($\nu > 90$ GHz), bolometers have been used because of their high sensitivity. The SuZIE 3-element bolometer array has detected the SZE in several clusters (e.g., Holzapfel et al. 1997b). Currently a number of groups are observing the SZE using the newly-commissioned SCUBA array on the JCMT. This provides an array of 37 bolometers at 350 GHz. The advantage of these high frequency measurements is that they are more sensitive to the kinematic SZE. For instance, Holzapfel et al. (1997b) measured the SZE at 2.1, 1.4 and 1.1 mm, where the thermal SZE signal is negative, close to zero and positive respectively, for cluster A2163. Observations at these frequencies provide a check for the unique spectral features of the thermal SZE as well as separating the kinematic and thermal effects.

Figure 2 shows the current cluster Hubble diagram deduced from the SZE and X-ray data. It is difficult to place an uncertainty on the best fit for H_0 due to the effects of clumpiness, selection, and absolute flux calibration. The data came from 3 different SZE telescopes and 2 different X-ray satellites, an inaccurate absolute calibration of even one telescopes could substantially bias the result for H_0. Since the detection of the SZE is still relatively difficult, only the brightest and hottest X-ray clusters are currently detected: this can bias the sample towards clusters that are elongated along the line of sight. Small-scale clumpiness of the ICM will also bias the value of H_0, although X-ray spectroscopy places some constraints on the type of clumping that is permitted (e.g. Holzapfel et al. 1997a). For a detailed review of the subject see Birkinshaw (1998).

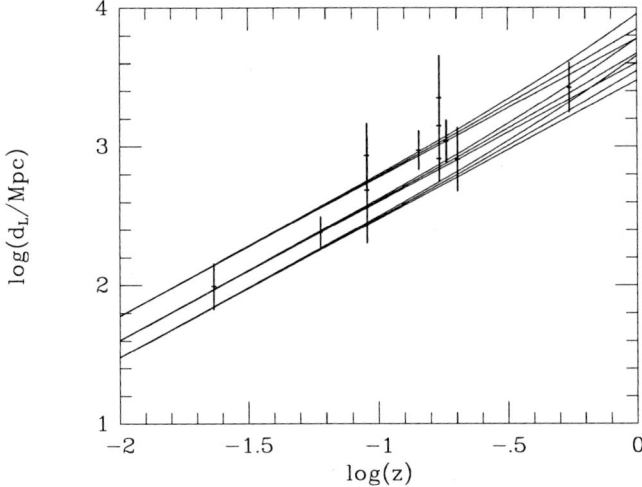

Fig. 2. A Hubble diagram based on distances measured for the clusters Abell 1656, 2256, 478, 2142, 1413, 2163, 2218 and 665 and CL0016+16 (Herbig *et al.* 1995; Meyers *et al.* 1997; Grainge *et al.* 1996; Holzapfel *et al.* 1997; McHardy *et al.* 1990; Birkinshaw & Hughes 1994; Jones *et al.* 1995; Birkinshaw *et al.* 1991; and Hughes & Birkinshaw 1998). Three values are shown for A2218. The theoretical curves are for H_0=50, 75 and 100 km s^{-1} Mpc^{-1} with q_0=0, 0.5 and 1. The best fit for H_0 is $\sim$ 60 km s^{-1} Mpc^{-1} with no strong constraint on q_0.

3 Future

An ideal instrument for the SZE would be a multi-band 2-D interferometer with many small dishes that can be packed close together to obtain short baselines (dish diameters < 10 m for frequencies > 15 GHz). Due to problems with radio confusion, we can only expect to detect the SZE efficiently at frequencies > 15 GHz. A scaled array with a wide range of frequencies will assist not only in identifying and removing confusing radio sources but also in separating the kinematic and thermal SZEs.

In order to deduce the cosmological parameters using the SZE, we also need high sensitivity, high resolution and low background X-ray telescopes for spectro-imaging observations to complement improved in SZE instruments.

To compare the mass distribution of the cluster components and test the validity of the assumption of hydrostatic equilibrium, we will also need large scale shear maps due to gravitational lensing of background objects by the cluster. The gas mass deduced from the SZE provides a more direct measure of the cluster baryonic fraction than that of the X-ray measurements when compared with the lensing mass. An incorporation of X-ray surface brightness, temperature maps, SZE maps, large scale weak lensing maps as well as multi-slit optical spectroscopy of cluster member galaxies will no doubt produce a wealth of information not only on H_0 but also on cluster dynamics and mass distribution.

Finally, the SZE ΔT_r provides an unique method of finding very high redshift clusters of galaxies given the redshift independent nature of the effect. In a model without evolution the detection of the SZE may even be easier at high than at low z, due to the diminished effects of confusion due to radio sources (especially radio halo sources, which may be practically impossible to separate from the SZE on the basis of structure).

3.1 Feasibility of an Upgraded ATCA as a Future SZE Instrument

History The Australia Telescope Compact Array (ATCA) has been used to observe the SZE. Five dishes were arranged in a close-packed configuration with minimum baseline 30 m. The detection experiments were conducted at the highest available frequency, 8.8 GHz, to minimise the effects of confusion from discrete radio sources. The ATCA proved to have a high brightness sensitivity (about 6 times more sensitive than the Ryle Telescope), however it failed to detect the SZE in the 7 clusters that were observed due to radio source confusion (Liang 1995). At least 3 of the clusters contain a radio halo source that masks the SZE. Three further clusters had either extended radio sources or blends of discrete radio sources. MS2137-23 was the only cluster where radio source emission did not affect the observation (Liang & Ekers 1994). A frequency of 8.8 GHz appears to be too low for the detection of the SZE since the effects of discrete radio sources are still appreciable.

AT Upgrade and the Future The planned upgrade of the AT will provide us with a receiver with operating frequencies centred at 16, 22 and 25 GHz as well as North-South baselines. The expected thermal noise at 16 GHz will be $35\,\mu\text{Jy beam}^{-1}$ over a 12-hour integration with 5 antennas. The increase in frequency provided by the new receiver will alleviate problems with radio source confusion. The addition of a north-south baseline will not only increase the number of short baselines but also improve uv-coverage in the 30 m to 200 m baseline range for sources with $\delta > -40°$ (for upgrade details see Norris, these Proceedings). On the other hand, the increase in frequency increases the uv-spacing. If the minimum baseline can be decreased to 26 m, which is still sufficient to avoid neighbouring dish collisions, then the loss from the increased uv separation will not be too severe.

We will examine the case of A2163 where the detection of the SZE was hampered at 8.8 GHz by the strong radio halo source in the cluster. All estimates will be based on the expected SZE signal from X-ray parameters for $H_0 = 50\,\text{km}\,\text{s}^{-1}\,\text{Mpc}^{-1}$. In a 60 hr observation, we expect an S/N of 13 at 8.8 GHz without the radio halo source (c.f. S/N = 15 for BIMA at 28.5 GHz); instead we found 1.1 mJy of extended emission. At 16 GHz we expect an S/N of 6 or 10 without the radio halo, if the minimum spacing is 30 m or 26 m respectively. The effect of the radio halo is reduced by a factor of ~ 3 by increasing the frequency to 16 GHz. However, if we move A2163 to a redshift $z = 3$, the expected SZE S/N is 10 or 16, if the minimum spacing is 30 m or 26 m respectively. The effect of the radio halo would be smaller by a further factor of ~ 60 and hence negligible. Thus the upgraded AT would be suitable for detecting high redshift clusters or objects like the VLA dark SZE cloud (Richards et al. 1997).

3.2 The Planned Next Generation of SZE Instruments

There are a number of ground based dedicated instruments for the detection of CMBR anisotropy planned in the southern hemisphere (in Chile or Antarctica). At least two of these instruments have enough angular resolution to be sensitive to the SZE as well: the CBI ($\sim 5'$ resolution) and the MINT ($\sim 3'$). Space missions for CMB anisotropy measurements, such as Planck, will have resolutions of $\sim 6'$ and will also be sensitive to the SZE. We would expect these experiments to provide all-sky surveys of clusters of galaxies at various redshifts, but not SZE maps. We will still need higher resolution instruments with sufficient sensitivity on scales between $\sim 0.5'$ and $\sim 6'$ (i.e. $0.5 - 7\,\mathrm{k}\lambda$) to obtain maps of the SZE, vital for the H_0 determination. For example at 30 GHz, such a telescope should have at least 20 dishes of diameter $< 5.5\,\mathrm{m}$, with wide band correlators. Currently available telescopes are only able to map the SZE in the most luminous and hottest X-ray clusters (i.e. $L_{[0.1-2.4\mathrm{keV}]} > 10^{45}$ ergs s^{-1} and $kT_g > 8$ keV). So far the only plan on improving the mapping sensitivities for the SZE is the upgrade of the correlator bandwidth from 800 MHz to 10 GHz (an increase by a factor of 3.5 in sensitivity) for BIMA, which is in the northern hemisphere. Perhaps modifications to the designs of currently-planned southern CMB anisotropy telescopes can be made to extend their high sensitivities to the smaller angular scales which will enable sensitive SZE mapping.

References

Birkinshaw, M., Gull, S. (1984), MNRAS **206**, 359
Birkinshaw, M., Hughes, J. (1994) ApJ **420**, 33
Birkinshaw, M., Hughes, J., Arnaud, K. (1991) ApJ **379**, 466
Hughes, J., Birkinshaw, M. (1998) ApJ in press
Birkinshaw, M. (1998), Physics Report, submitted
Carlstrom, J., Joy, M., Grego, L. (1996) ApJ **456**, L75
Gunn, J. (1978) *Observational Cosmology* (Geneva Observatory), 1
Jones, M., Saunders, R., Alexander, P., Birkinshaw, M., Dillon, N. et al. (1993) Nature **365**, 320
Grainge, K., Jones, M., Pooley, G., Saunders, R., Baker, J., et al. (1996) MNRAS **278**, L17
Herbig, T., Lawrence, C., Readhead, A., Gulkis, K., Kogut, A. et al. (1996) ApJ **449** L5
Holzapfel, W., Arnaud, M., Ade, P., Church, S., Fischer, M. et al. (1997) ApJ **480**, 449
Holzapfel, W., Ade, P., Church, S., Mauskopf, P., Rephaeli, Y. et al. (1997) ApJ **481**, 35
Liang, H., Ekers, R. (1994) *Clusters of Galaxies*, Proceedings of the XXIXth Rencontres de Moriond, 401

Liang, H. (1995), PhD thesis, ANU
McHardy, I., Stewart, G., Edge, A., Cooke, B., Yamashita, K. et al. (1990) MNRAS **242**, 215
Myers, S., Baker, J., Readhead, A., Leitch, E., Herbig, T. (1997) ApJ **485**, 1
Rephaeli, Y. (1995) ApJ **445**, 33
Rephaeli, Y., Lahav, O. (1991) ApJ **372**, 21
Richards, E., Formalont, E., Kellerman, K., Partridge, R., Windhorst, R. (1997) AJ **113**, 147
Silk, J., White, S. (1978) ApJ **226**, L3.
Sunyaev, R., Zel'dovich, Ya. (1972) Comm. Astrophys. Sp. Phys. **4**, 173
Sunyaev, R., Zel'dovich, Ya. (1981) Astrop. & Sp. Phys. Rev. **1**, Soviet Scientific Reviews, Section E, 1.

Discussion

Lineweaver: I'm worried about the preferential selection of radially elongated clusters in the sample used to derive H_0. How do you remove the selection effect?

Liang: Make sure we don't lose the low X-ray surface brightness clusters, i.e. a flux limited sample rather than a surface brightness limited sample. SZ upper limits should also be included in the fit from H_0 to avoid the selection effect.

Cosmology in a Nutshell + an Argument Against $\Omega_\Lambda = 0$ Based on the Inconsistency of the CMB and Supernovae Results

Charles H. Lineweaver

University of New South Wales, Sydney, Australia

Abstract. I present several simple figures to illustrate cosmology and structure formation in a nutshell. Then I discuss the following argument: if we assume that $\Omega_\Lambda = 0$ then the CMB results favor high Ω_m while the supernova results favor low Ω_m. This large inconsistency is strong evidence for the incorrectness of the $\Omega_\Lambda = 0$ assumption. Finally I discuss recent CMB results on the slope and normalization of the primordial power spectrum.

1 Cosmology in a Nutshell

The Big Bang model became the standard cosmological model soon after the discovery of the cosmic microwave background (CMB). The Big Bang model has a hot, dense early epoch (see Figure 1) when nucleosynthesis occurred. It also has an opaque surface that can naturally produce the Planckian spectrum of the CMB. The Steady State universe was not hotter in the past, has no epoch of Steady State Nucleosynthesis and has no opaque surface to produce the CMB.

Gravitational collapse is the leading model of structure formation (Figure 2). Slight over-densities are gravitationally unstable and collapse under their own self-gravity. In an alternative family of models, structure forms from topological defects. In gravitational collapse models CMB anisotropies larger than ~ 1 degree are acausal and rely on inflation to explain their existence. In defect models these large anisotropies are close by, causal and sub-horizon sized.

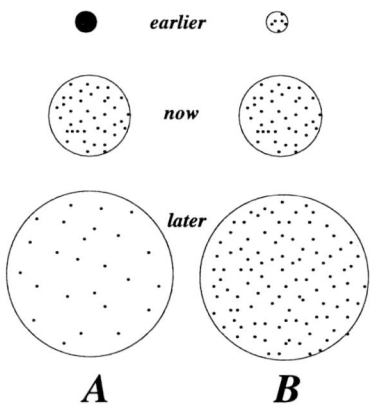

Figure 1. Big Bang vs Steady State. Expanding horizon volumes in the Big Bang (A) and Steady State (B) models. The dots represent the density of the Universe. At early times only the Big Bang model was dense and hot providing an oven for nucleosynthesis and later an opaque surface of last scattering which naturally produces the Planckian spectrum of the CMB.

Figure 2. Gravitational collapse vs topological defects. The leading model of structure formation is on the left: small over-densities are gravitationally unstable and collapse under their own gravity to form the structures we see around us. Topological defect models (right) are an alternative. When symmetry is broken in the early universe causally disconnected regions are occupied by different vacuum states (indicated by the direction of the lines in the figure). Large energy densities are present at the boundaries between such regions and this is where structure forms.

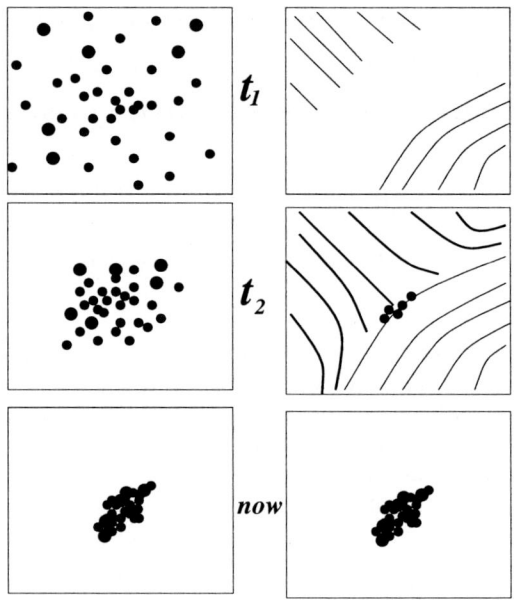

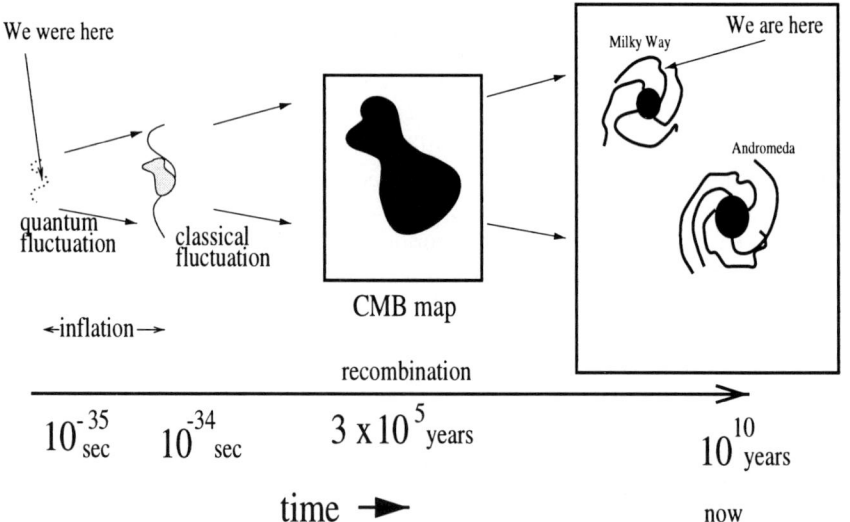

Fig. 3. Galaxies are CMB anisotropies are Quantum fluctuations. According to the inflationary scenario, quantum fluctuations of a scalar field are the origin of all structures. These quantum fluctuations are not caused by any preceeding event in the same sense as radioactive decay or quantum tunneling are not caused. They are non-deterministic prime movers. Inflation of the universe by a factor of more than 10^{26} transforms these quantum fluctuations into super-horizon classical density fluctuations. On their way to becoming galaxies we can monitor their progress by looking at CMB maps.

One of the most important questions in cosmology is: What is the origin of all the galaxies, clusters, great walls, filaments and voids we see around us? The inflationary scenario provides the most popular explanation for the origin of these structures: they used to be quantum fluctuations.

Figure 3 illustrates the metamorphosis of quantum fluctuations to CMB anisotropies to galaxies. Primordial quantum fluctuations of a scalar field get amplified and evolve to become classical seed perturbations and eventually large scale structure. This process can be monitored by CMB observations since matter fluctuations produce temperature fluctuations in the CMB: $\frac{\delta\rho}{\rho} \propto \frac{\Delta T}{T}$.

How does a particular fluctuation know whether it will become a spiral or an elliptical galaxy? Does the density and irregularity of its environment determine its morphology by controlling its angular momentum and the amount of merging? With a full understanding of galaxy formation we may be able to look at CMB cold spots and their neighborhoods and predict where they will end up in the Hubble tuning fork diagram of galaxy types. The distribution of morphological types at high redshift discussed by Driver in these proceedings would then be a derivable function of the characteristics of the CMB anisotropies.

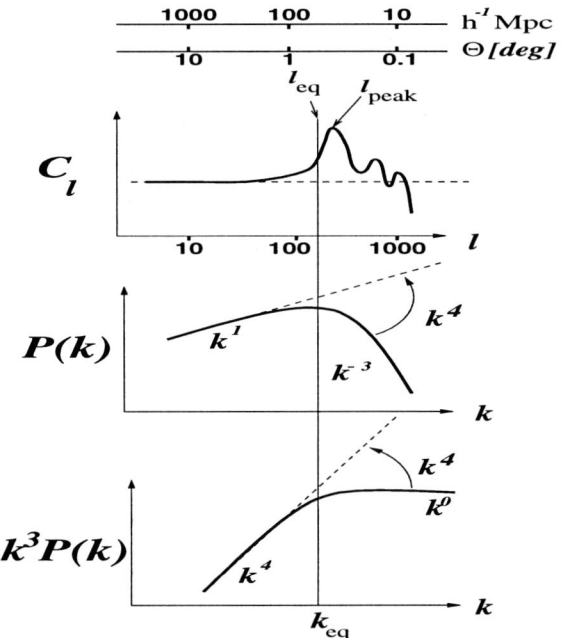

Figure 4.
CMB powerspectrum (C_ℓ, top) compared with the matter density powerspectrum ($P(k)$, middle and bottom). C_ℓ is a measure of the power in the spatial variations of the CMB as a function of the angular scale. The upper axes give the angular scales and comoving sizes corresponding to the Legendre polynomial index ℓ. In the bottom two panels the turnover in $P(k)$ at $\sim k_{eq}$ occurs at the scale ($L_{eq} = 2\pi k_{eq}^{-1}$) which just enters the horizon as the Universe changes from radiation dominated to matter dominated. Smaller sizes entered the horizon earlier during radiation domination and were unable to grow during this period. The k^4 growth suppression which they suffered is indicated.

1.1 There is no Scale Beyond which the Universe is Homogeneous

It has been claimed that some recent, deep, galaxy redshift surveys have reached the scale at which the Universe becomes homogeneous. Strictly speaking however there is no scale beyond which the universe is homogeneous. The amplitude of the density contrast $(\delta\rho/\rho \propto k^3 P(k)$ decreases for larger scales but is never zero. A more meaningful question is: Where is the turnover in the power spectrum? This turnover is due to a suppression of growth of a given k mode by k^4 relative to modes which enter the horizon during matter domination (assuming $\Omega_o = 1$). Thus, the horizon scale at matter-radiation equality is an important diagnostic of this fundamental scale. See Figure 4.

Lineweaver & Barbosa (1998) have used current CMB anisotropy measurements to determine the position of the adiabatic peak in the CMB spectrum under the assumption of open or critical density CDM dominated universes: $\ell_{peak} = 260^{+30}_{-20}$.

Figure 5 illustrates how harmonic sound bumps appear in the CMB power spectrum driven by the wells and valleys of the CDM potentials. The epoch when matter and radiation densities are equal has a redshift of z_{eq} while decoupling occurs at z_{dec}. The number of oscillations between z_{eq} and z_{dec} and thus the phase of the oscillations at z_{dec} is determined by i) the physical size of the potential well, ii) the speed of sound and iii) the time interval between z_{eq} and z_{dec}.

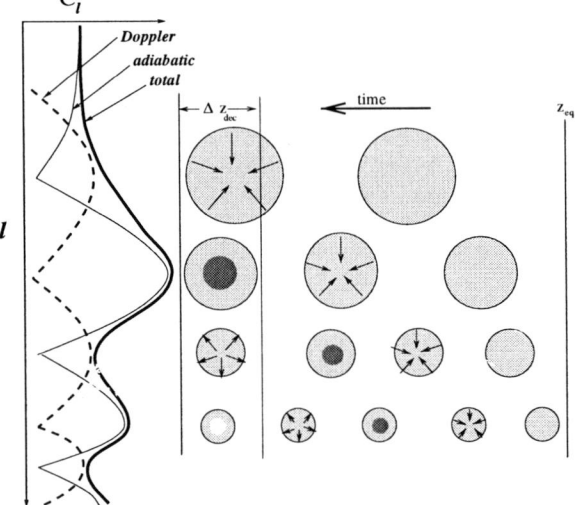

Figure 5.
Sound waves in the photon-baryon fluid create bumps in the CMB power spectrum. The grey spots are cold dark matter potential wells which initiate infall and then oscillation of the photon-baryon fluid in these wells. The Doppler and adiabatic effects make the sound visible in the radiation when the baryons decouple from the photons during the interval marked Δz_{dec}. These bumps are analogous to the standing waves of the resonant frequencies of a plucked string or of a good shower and may be the oldest music in the Universe. See Hu etal (1997) and Lineweaver (1997) for details.

2 An Argument Against $\Omega_\Lambda = 0$ Based on the Inconsistency of the Latest CMB and Supernovae Constraints on Ω_m

The CMB is already giving us useful constraints on cosmological parameters in popular but restricted families of CDM models. In Figure 6 the region of the $h - \Omega_m$ plane preferred by the CMB data is shown (since $\Omega_\Lambda = 0$, $\Omega_0 = \Omega_m$). The best-fit is indicated with an **X**. The values of the spectral index n which minimize the χ^2 values for a given (h, Ω_o) pair are indicated by the thin solid iso-n lines and are labeled with $0.8 - 1.3$. High values of h require high values of n.

Figure 6.
The dark grey banana-shaped region is the approximate 68% confidence level preferred by the current CMB anisotropy measurements. The thick solid lines are the approximate 2, 3 and 4 σ contours. The age interval shown is $10 - 18$ Gyr. The thin lines are contours of the spectral index n values which minimize the χ^2 for each pair (h, Ω_o). Note the monotonic relations: the higher the h value the higher the n value and the lower the Ω_o value. A favored open model $h \approx 0.70$ with $\Omega_o \approx 0.3$ is rejected at greater than $\sim 4\sigma$. The best-fit value is $h = 0.40$ and $\Omega_o = 0.85$. The corresponding n value is 0.91. Figure adapted from Lineweaver & Barbosa (1998).

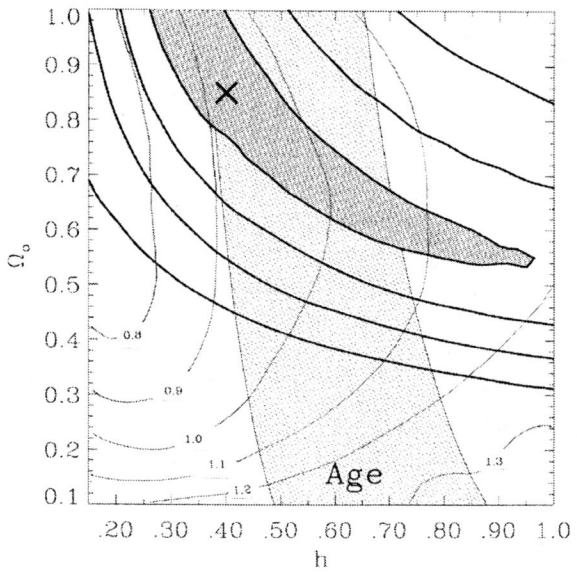

Under the assumption that $\Omega_\Lambda = 0$ (i.e., $\Omega_o = \Omega_m$) the most recent CMB results on the density of matter in the Universe yield $\Omega_m > 0.3$ at the $\sim 4\sigma$ confidence level (Figure 6). This result is independent of the value of Hubble's constant, of the spectral index n and of the normalization Q_{10}. In this same model the new supernovae results are $\Omega_m = -0.1 \pm 0.5$ (Garnavich et al. 1998) and $\Omega_m = -0.4 \pm 0.1$ (statistical) ± 0.5 (systematic) (Perlmutter et al. 1998). With additional supernovae Ω_m stays low and the error bars decrease making the inconsistency between CMB and supernovae results even stronger (Schmidt 1998 private communication). People who like open models with $\Omega_\Lambda = 0$ could argue that these supernovae results are consistent with some cosmological measurements which yield $0.1 \lesssim \Omega_m \lesssim 0.3$. However the strong inconsistency between the CMB results (which strongly exclude low values of Ω_m) and the supernovae which favor very low, even negative (and thus unphysical) values of Ω_m is strong evidence against $\Omega_\Lambda = 0$ models.

3 Results for the Slope and Normalization of the CMB Power Spectrum

The CMB solutions for h or n can be read from either Figure 6 or 7. CMB anisotropy measurements in open and critical CDM models yields $n = 0.91^{+.29}_{-0.09}$ (Lineweaver & Barbosa 1998). In Figure 7 the thin solid lines indicate the h values ($0.2 \leq h \leq 1.0$) which minimize the χ^2 for (n, Q_{10}) pairs. Conditioning on $n = 1$ or $h = 0.50$ changes the results (see Table 2 of Lineweaver & Barbosa 1998).

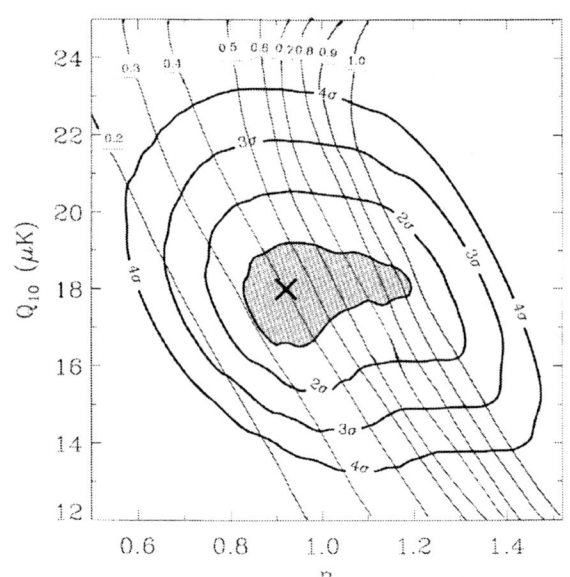

Figure 7. Contours in the plane of the slope n and normalization Q_{10} of the primordial power spectrum of CMB anisotropies. Notation is analogous to the previous figure except here the thin lines are contours of the h values which minimize the χ^2 for a given pair of (n, Q_{10}). Note the monotonic relation: for a given Q_{10} value, the higher the n value the higher the h value. The best-fit values of n and Q_{10} are $n = 0.91^{+0.29}_{-0.09}$ and $Q_{10} = 18.0^{+1.2}_{-1.5}$. To my knowledge, these are the tightest constraints on these parameters. Figure adapted from Lineweaver & Barbosa (1998).

References

Garnavich, P.M. et al. 1998 *Astrophysical Journal*, submitted, astro-ph/9710123
Hu, W., Sugiyama, N. & Silk, J (1997) Nature, 386, 37
Lineweaver, C.H. 1997, in "From Quantum Fluctuations to Cosmological Structures", edt. D. Valls-Gabaud, M.A.Hendry, P. Molaro & K. Chamcham, ASP conference series, 126, p 185-205
Lineweaver, C.H., Barbosa, D. (1998), *Astrophysical Journal*, 496, (in press), astro-ph/9706077
Perlmutter, S. et al. 1998, http://panisse.lbl.gov/public/papers/aasposter198dir/wwwposter2c2.jpg

Discussion

Bland-Hawthorn: What about galaxy fluctuations ($z = 0$–10) as you go to smaller angular scales (less than COBE)?

Lineweaver: Obvious point sources can be removed. A wide spread in the observation frequencies can be used to remove non-flat radio sources and far-infrared sources. Cross-correlation analysis with known galaxies can be used to measure the removing effect. We do what we can. The SZ effect clusters seems to be a kind of worry (cf. Hobson et al. MNRAS 1998). "Foreground separation methods for satellite observations of the CMB"

Tsvetanov: If you had a billion dollars and could build you favourite satellite, what would you go for? For example, higher angular resolution, much better limit on $\delta T/T$, or something else?

Lineweaver: I'd increase the angular resolution of Planck HFI by a factor of 2, not more. I'd separate the HFI and LFI into separate satellites and try to keep the angular radiation independent of frequency. This would mean a bigger satellite for the LFI but much simpler data analysis downstream. I'd pack in as much redundancy as I could and use the rest of the money to insert in ground-based interferometers and long-duration balloon-borne bolometric missions.

ASCA Results and Future Japanese X-Ray Missions

Koujun Yamashita

Department of Physics, Nagoya University,
Furo-cho, Chikusa-ku, Nagoya 464-8602, Japan

Abstract. The Japanese X-ray astronomy satellite, ASCA, has been operating quite well in orbit for five years. I present some of the observational results from ASCA for extragalactic objects, describe the status of ASTRO-E and outline future projects of Japanese X-ray missions.

1 Introduction

The first imaging spectroscopic observations in the energy region up to 10 keV were performed for a variety of astronomical objects with ASCA – launched on February, 1993, as the fourth Japanese X-ray astronomy satellite. In particular, X-ray spectra were observed with superior energy resolution in 0.5–10 keV. They gave new insight into understanding the physical state in extreme conditions.

I briefly mention some of the observational results of extragalactic sources, such as early-type galaxies, Seyfert 1, Seyfert 2, quasars, clusters of galaxies and cosmic X-ray background (CXB). The Fe-K emission line feature indicates direct evidence of a massive black hole in active galactic nuclei. The Hubble constant has been derived from Sunyaev-Zel'dovich clusters. A number of new faint sources were found by the large sky survey related to CXB.

ASTRO-E is the forthcoming satellite after ASCA, which will be launched in 2000 by the newly developed M-V rocket. We are now planning a future satellite probably coming up in 2006-7. These missions aim at looking deep in the universe.

2 ASCA Instrumentation

ASCA put on board four sets of multi-nested thin foil mirror X-ray telescopes (XRT) incorporated with two X-ray CCDs (Solid state Imaging Spectrometer, SIS) and imaging gas scintillation proportional counters (Gas Imaging Spectrometer, GIS) for each two sets. XRT has the outer diameter of 35 cm and focal length of 3.5 m. The effective area of 4 sets of XRT is 1200 cm^2 at 1.5 keV and 400 cm^2 at 8 keV. SIS is sensitive down to 0.5 keV, having superior energy resolution (2% at 6 keV) and small field of view (22 × 22 arcmin square), whereas GIS is more sensitive up to 10 keV with moderate energy resolution (8% at 6 keV), covering a wide field of view (50 arcmin in diameter). Both detectors are complementarily functioned. The angular resolution mostly depends on the point spread function of XRT with 30 arcsec (FWHM) and 3 arcmin (HPD). Maximum time resolution is 60 μsec and 16 sec for GIS and SIS, respectively.

3 Observational Results of ASCA

Spatially-resolved spectra were obtained for the following extragalactic objects with GIS and SIS. X-ray surface brightness distributions were observed for early type galaxies and clusters of galaxies with the angular resolution of 3 arcmin (HPD).

3.1 Early Type Galaxies

The chemical evolution and distribution of dark matter are interesting subjects for X-ray observations of early type galaxies. Thirty nearby galaxies at distances of 17–70 Mpc were analyzed by Matsushita (1997) and Matsumoto et al. (1997). X-ray spectra comprised a soft component, originating from the hot interstellar medium confined in the gravitational potential, and a hard component arising from the sum of low mass binaries. The soft component was fitted by a thermal emission curve with temperature of 0.3–1 keV and several emission lines of O, Ne, Fe, Mg, Si and S. Fe-L lines around 1 keV were important to derive Fe abundance. The hard component was fitted with a power law of slope 1.8 and thermal spectra of kT = 10 keV. X-ray luminosities were obtained to be in the range $10^{40} - 10^{43}$ erg/sec. 10^{41} erg/sec is a critical value to distinguish properties of these galaxies. In general they are classified into two categories, one is X-ray extended, high gas density at the center and solar abundance, the other is compact, low density and low abundance (Matsushita 1997). The former is confined in the superposed potential of groups of galaxies, whereas lower abundance in the latter is explained by the escape of hot gas from the galaxy potential.

NGC4636, located in the Virgo cluster, was observed for about 200 ksec. Its X-ray surface brightness profile is more extended than that of the galaxy itself, which was fitted with a double β-model. The thermal spectrum is expressed with kT = 0.76 keV and 0.5 solar abundance.

3.2 Seyfert 1 Galaxies

Eighteen Seyfert 1 galaxies with redshift 0.002–0.047 were analyzed by Nandra et al. (1997a,b). Fe-K_α lines were detected from fourteen of them. X-ray luminosities were obtained to be in the range of $10^{42} - 10^{44}$ erg/sec in the 2–10 keV range. X-ray spectra were expressed with average photon index of 1.91 and broad Fe-K_α emission lines around 6.4 keV with line width of 0.43 keV. The line profile is interpreted to reflect the gas motion and gravitational redshift around a massive black hole. A typical example is MGC-6-30-15 (Tanaka 1995). The mean profile for the sum of 12 samples consists of a narrow line at 6.4 keV and a broad line at the energy centroid of 6.1 keV (Nandra 1997b).

3.3 Seyfert 2 Galaxies

Seyfert 2 galaxies are less luminous than Seyfert 1, since the central source is obscured by the surrounding thick absorbing gas. Ueno (1997) analyzed twelve Seyfert 2 galaxies in the distance of 10–300 Mpc. Those spectra comprised a hard component with photon index of 1.5–1.7 incorporated with heavy absorption with column density of $10^{22} - 10^{24}$ cm^{-2} and soft excess emission. Prominent Fe-K emission lines were detected with an equivalent width of 80–1600eV. X-ray luminosities were obtained to be $10^{40} - 10^{44}$ erg/sec at 2–10 keV after correcting for absorption.

3.4 Quasars

Ten high-redshifted ($1.3 < z < 3.9$) radio-loud quasars were observed. The most distant one, Q1745+624 ($z = 3.9$), has a photon index of 1.68 and $N_H = 6.1 \times 10^{22}$ cm^{-2} (Kubo 1997). The spectral indices are larger than that of CXB, which are compared with low-redshift quasars ($z < 1$).

3.5 Clusters of Galaxies

ASCA has already observed nearly 200 clusters of galaxies which were classified as group of galaxies, poor clusters, nearby rich clusters and distant clusters. Spatially-resolved spectra were obtained for most of them within the angular resolution of ASCA up to a redshift of 0.69. Spectral analysis gives plasma temperatures and chemical abundances of the hot intracluster medium, which were found to be 1–10 keV and 0.3 times solar values, on average. The temperature and abundance distributions were obtained by analysing the cluster region by region or central core and envelope. This analysis makes clear the evidence of cooling flows as well as merging of substructure. The patchy temperature distribution was obtained for the Coma cluster with kT = 4–12 keV (Honda 1996).

Gravitational lensing clusters (A370, CL0500-24, CL2244-02) were analyzed to derive the mass distribution which was smaller than the optical value (Ota 1997). Sunyaev-Zel'dovich clusters (A665, A2218, CL0016+16) were analyzed to derive the Hubble constant which, when combined with the radio observations, was found to be 68 ± 22 km/sec/Mpc, as shown in figure 1 (Furuzawa 1997).

3.6 Cosmic X-Ray Background

The large sky survey (LSS) was done via 76 pointed observations with total exposure of 515 ksec in the area of 6 degrees square near the North Galactic Pole (Ishisaki 1997, Ueda 1997). A total of 119 sources were detected with intensities higher than 1.5×10^{-13} erg/sec/cm^2 at 2–10 keV. The slope of the Log N – Log S relation was derived to be 1.5 on the extrapolation of previous observations. The spectrum of the CXB was found to have a photon index of 1.5 at 2–10 keV and 6.4 at 0.5–2 keV. The contribution of point sources is estimated to be 30%.

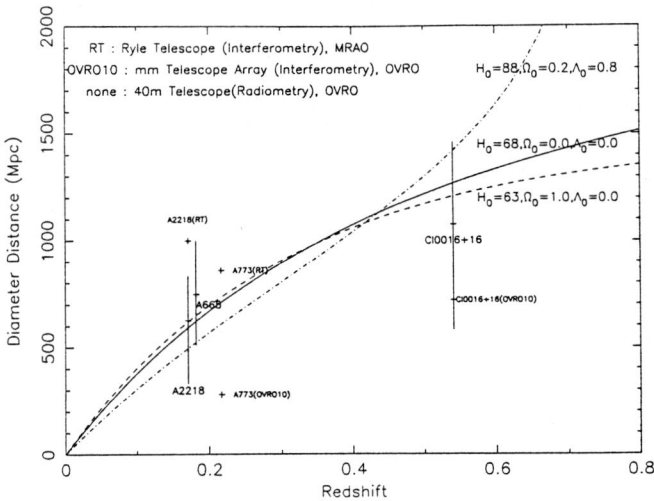

Fig. 1. Diameter distance vs. redshift for Sunyaev-Zel'dovich clusters to derive the Hubble constant (Furuzawa 1997).

4 Future Japanese X-Ray Missions

Japanese scientific satellite missions are conducted by the Institute of Space and Astronautical Science (ISAS). We have had four X-ray astronomy satellites in orbit and the fifth one, ASTRO-E, will be launched in 2000. The sixth one, temporarily called ASTRO-G, is being planned by the X-ray astronomy group, with an aimed launch in 2006-7.

4.1 ASTRO-E

The preparation of ASTRO-E is now going to meet its launch schedule of 2000. ASTRO-E is four times bigger than ASCA, in which 5 sets of multi-nested thin foil mirror X-Ray Telescopes (XRT) are put on board, incorporated with 4 sets of CCD cameras (Imaging Spectrometer, XIS), and and one set of microcalorimeters (X-Ray Spectrometer, XRS) at the focal plane. XRS has superior energy resolution of 12 eV in 0.5–10 keV, which is cooled down to 60 mK by solid Ne, Liquid He and ADR (Kelly 1997). Therefore the life is limited to two years. Moreover, the Hard X-ray Detector (HDX) made of Si-PIN + GSO crystal is also on board to cover the energy region of 10–700 keV. ASTRO-E is a complementary mission to AXAF in 1998 and XMM in 1999.

The effective area is 2200 cm^2 at 1.5 keV and 1400 cm^2 at 6 keV for XIS, and 550 cm^2 at 1.5 keV and 140 cm^2 at 6 keV for XRS. The angular resolution of XRT is 1.5 arcmin (HPD). XRS is the most powerful spectrometer in the X-ray region ever operated in orbit. We expect a lot of exciting results for all classes of astronomical objects.

4.2 X-Ray Mission after ASTRO-E

We are now discussing to propose an X-ray mission after ASTRO-E. It will be expected to be launched by M-V rocket in 2006-7. Its total weight is similar to ASTRO-E. Present and planned X-ray telescopes only cover the energy region up to 10 keV. Therefore we intend to extend the energy region up to at least 40 keV using a multilayer supermirror. The development is going on at Nagoya University (Yamashita 1997). They have already made a prototype supermirror telescope and confirmed the reflectivity and image quality by measuring 10–50 keV X-rays. The hard X-ray reflectivity is shown in Figure 2. At the focal plane a combination of X-ray CCD and CdZnTe solid state detectors is mounted as an imaging detector to cover the energy region of 0.1-40 keV. This instrument's objectives are to investigate the high energy phenomena occurring in SNR, AGN, and clusters of galaxies with a S/N superior by three orders of magnitude. It is expected to find new types of astronomical objects obscured by thick absorbing gas. It will also be able to measure the polarization of X-ray sources.

Another instrument under discussion is one that will be capable of high resolution spectroscopy using microcalorimeter or multilayer gratings with an energy resolution of 1000. These developments are under way. It is obvious that X-ray spectroscopy is an essential tool for astronomical observations.

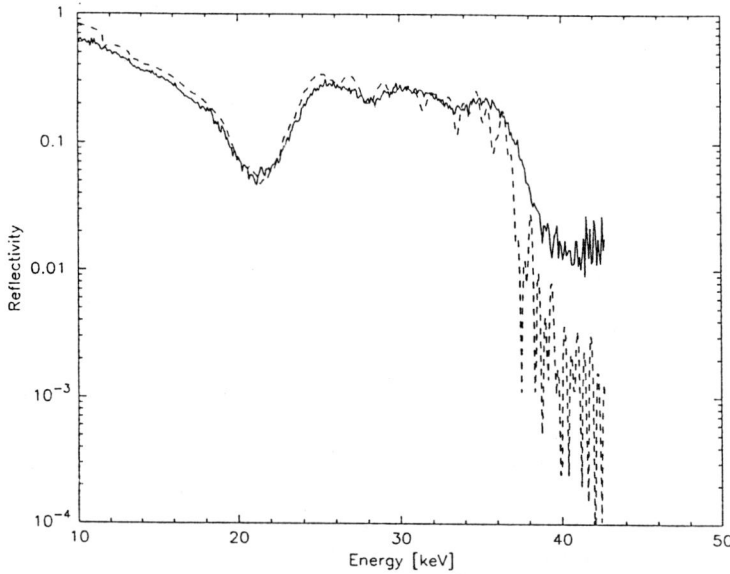

Fig. 2. Reflectivity of Pt/C multilayer supermirror in the hard X-ray region at incidence angle of 0.3 deg.

5 Summary

ASCA's imaging spectroscopic observations have made substantial contributions to the study of extragalactic objects in the 0.5–10 keV range and have provided a deeper understanding of high energy astrophysical phenomena. Succeeding missions are expected to make it possible to explore fainter and more distant objects out to the edge of the universe using high resolution spectroscopy and hard X-ray imaging observations.

References

Furuzawa, A. (1997): The Global Value of the Hubble Constant by Observations of Sunyaev-Zel'dovich Clusters of Galaxies with ASCA, ISAS Res. Note No. **615**

Honda, H. et al. (1996): Detection of a Temperature Structure in the Coma Cluster of Galaxies with ASCA, ApJ., **473**, L71-L74

Ishisaki, Y. (1997): Spectra and Large Scale Isotropy of the Cosmic X-ray Background from ASCA Observations, ISAS Res. Note No. **613**

Kelly, R.L. (1997): High Resolution X-Ray Spectroscopy using Microcalorimeter, *The Next Generation of X-Ray Observatories* (eds. M.J.L. Turner and M.G. Watson, University of Leicester), pp. 81-87

Kubo, H. et al. (1997): Observation of the High Redshift Quasar Q1745+624 ($z = 3.9$) with ASCA. ISAS Res. Note No.**609**

Matsumoto, H. et al. (1997): X-ray Properties of Early-Type Galaxies as Observed with ASCA, ApJ., **482**, 133-142

Matsushita, K. (1997): X-ray Study of Hot Interstellar Medium in Early Type Galaxies, Ph D thesis, University of Tokyo

Matsushita, K. et al. (1997): New Measurement of Metal Abundance in Elliptical Galaxy NGC4636 with ASCA, ApJ., **488**, L125-L128

Nandra, K., et al. (1997a): ASCA Observations of Seyfert 1 Galaxies. I. Data Analysis, Imaging, and Timing. ApJ. **476**, 70-82

Nandra, K. et al. (1997b): ASCA Observations of Seyfert 1 Galaxies. II. Relativistic Iron K_α Emission. ApJ. **477**, 602-622

Ota, N. et al. (1997): ASCA Observations of the Lensing Clusters, CL0500-24, CL2244-02 and A370, ISAS Res. Note No. **625**

Tanaka, Y., Inoue, H., and Holt, S.S. (1994): The X-ray Astronomy Satellite ASCA, PASJ, **46**, L37-L41

Tanaka, Y., et al. (1995): Gravitationally redshifted emission implying an accretion disk and massive black hole in the active galaxy MCG-6-30-15. Nat. **375**, 659-661

Ueda, Y. (1997): ASCA Studies of Faint X-ray Sources and the Relation to the Cosmic X-ray Background, ISAS Res. Note No. **621**

Ueno, S. (1997): X-Ray Study of Type 2 Seyfert Galaxies, ISAS Res. Note No. **619**

Yamashita, K. (1997): Multilayer X-ray Optical Systems for Future X-ray Astronomy Missions, *The Next Generation of X-Ray Observatories* (eds. M.J.L. Turner and M.G Watson, University of Leicester), pp. 115-121

Discussion

Tsarevsky: What is ASCA's time resolution? and with such a fairly good time resolution, is ASCA involved in gamma-ray burst science?

Yamashita: A few milliseconds for the G.I.S. instrument and two observations have been done following the information of gamma-ray bursts. ASCA is not involved in the direct observations of gamma-ray bursts.

High-Redshift X-Ray Clusters

Isabella M. Gioia[1,2]

[1] Istituto di Radioastronomia del CNR, Via Gobetti 101, 40129 Bologna, Italy
[2] Institute for Astronomy, 2680 Woodlawn Drive, Honolulu, Hawai'i 96822, USA

Abstract. Despite their extreme importance for cosmology, the number of known high redshift clusters is rather small. A few clusters at high redshift have been discovered through their X-ray emission. Follow-up observations of these clusters with X-ray telescopes, and deep imaging from the ground and from HST, have shown that massive bound structures at high redshift are not as rare as was once believed. Weak lensing techniques have provided masses of the order of $10^{14} - 10^{15}$ M$_\odot$. I will present results obtained on high-z, massive clusters extracted from the *Einstein* Medium Survey and the *ROSAT* NEP survey. The study of these clusters can provide stringent constraints on theories of large-scale structure formation.

1 Introduction

In hierarchical clustering theories, clusters of galaxies form from the high peaks in the original density field; thus they provide crucial constraints on the shape, amplitude, and temporal evolution of the primordial mass fluctuation spectrum. Despite their importance, the statistics on the abundance of high-z (>0.5) clusters are poor since they are so difficult to find. Optical surveys of distant galaxy clusters are well known to have serious statistical shortcomings such as the effects of superpositions of unvirialized systems (see, amongst others, Frenk *et al.* 1990; Van Haarlem *et al.* 1997). In addition, at high redshifts it becomes extremely difficult to detect enhancements in the galaxy surface density against the overwhelming field galaxy population. Therefore, the selection of high-z clusters by means of their X-ray emission is one of the cleanest ways to avoid sample contaminations and waste of efforts, and it is based on simple physical quantities such as X-ray luminosity and temperature. As observational capabilities become more accurate, the number of distant clusters is steadily increasing and it might come to be in severe conflict with various cosmological models.

In Table 1, five examples of the Einstein Medium Sensitivity Survey (EMSS; Gioia *et al.* 1990, Stocke *et al.* 1991) clusters with average $z = 0.66$, and $z_{max} = 0.826$ are listed. In addition to their high X-ray luminosity, there is other compelling evidence that these clusters are genuinely massive, like the presence of lensed arcs or the detection of a strong shear signal, which allow an estimate of the mass in the clusters. The letter s in Table 1 indicates the presence of giant arcs in the cluster core, and thus strong lensing, while the letter w indicates that weak lensing analysis has been performed. L_X is given in the 0.3–3.5 keV band in units of $\times 10^{44}$ erg s^{-1}, T_X is in the 2–10 keV band and N$_{0.5}$ gives an indication of the richness of the cluster.

Table 1. EMSS Clusters with $z > 0.5$

Name	redshift	lensing	L_X (10^{44} erg s^{-1})	T_X (keV)	$N_{0.5}$	σ (km/s)
MS0015.9+1609	0.546[a]	w	14.6[b]	8.4[c]	66±6[d]	1324[a]
MS0451.6−0305	0.550[e]	sw	20.0[b]	10.4[f]	47±5[d]	1371[g]
MS1054.5−0321	0.826[b]	w	9.3[b]	12.3[h]	82±10[i]	1360[h]
MS1137.5+6625	0.782[b]	sw	7.6[b]		56±6[d]	
MS2053.7−0440	0.583[e]	sw	5.8[b]		20±5[d]	

[a]Dressler & Gunn 1992 [b]Gioia & Luppino 1994 [c]Furuzawa et al. 1994 [d]Luppino & Gioia 1995 [e]Maccacaro et al. 1994 [f]Donahue, 1996 [g]Carlberg et al. 1994 [h]Donahue et al. 1998 [i]Luppino & Kaiser, 1997

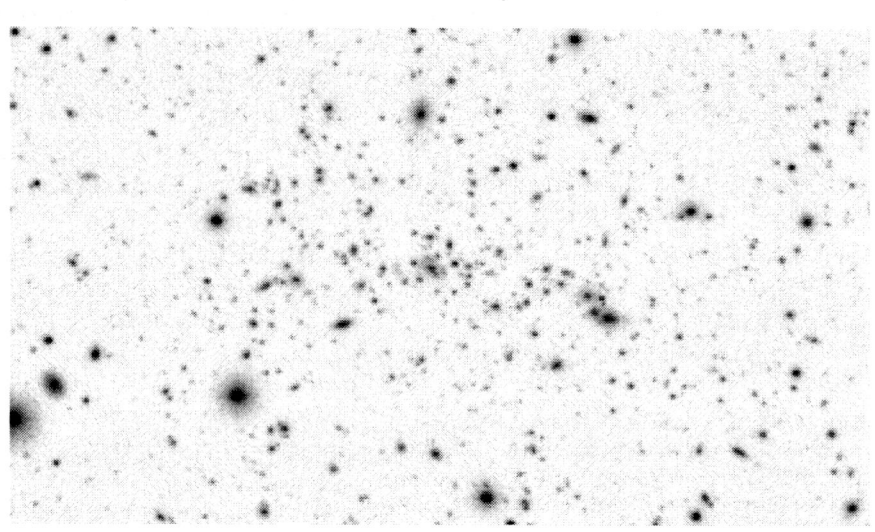

Fig. 1. Subarray I image of MS1054−03 taken with the Tek 2048 CCD camera at the UH2.2m.

2 The Medium Survey $z > 0.5$ Clusters

MS0015.9+16, also known as CL0016+16 (Koo 1981), has been the archetypal rich, X-ray luminous, distant cluster since its discovery. It has a linear structure elongated in the NE-SW direction. Smail et al. (1994) reported detection of weak shear, which has been reconfirmed by the analysis performed by Doug Clowe (PhD thesis, in preparation) using Keck R-band deep images with a larger field of view, and a different technique (Kaiser and Squires 1993; Squires and Kaiser 1996).

MS0451−03 is the most X-ray luminous cluster in the EMSS and among the brightest clusters known. ASCA data by Donahue (1996) show a hot cluster at

10.4 keV with iron abundance of 15% solar and a total mass (within $1h_{50}^{-1}$ Mpc) of $9.7^{+3.8}_{-2.2} \times 10^{14}$ M⊙ . The shear signal detection, performed by Clowe using two hours of R-band Keck data, confirms the presence of a large mass.

MS1054−03 is the most distant cluster of the EMSS and among the most distant X-ray selected clusters known. In Fig. 1 the filamentary morphology of the cluster is evident; there is an elongation of galaxies in the E-W direction. The cluster is extremely rich and quite hot at $12.3^{+3.1}_{-2.2}$ keV, as obtained by ASCA (Donahue et al. 1998). A strong shear signal at the 6σ level was detected by Luppino & Kaiser (1997) and confirmed by Clowe on deeper Keck images. The total mass (within 1 Mpc) from X-rays and from weak lensing are consistent ($2-6\times 10^{14} h_{50}^{-1}$ M⊙ vs $3-30\times 10^{14} h_{50}^{-1}$ M⊙).

MS1137+66: deep images have been collected of this cluster with Keck and with the UH 2.2m. From the reduced images a large arc has been discovered close to the cluster center (Clowe et al. 1998). In contrast to MS1054−03 and MS0015.9+16, this cluster is compact and concentrated with no filamentary structure. A weak lensing analysis, performed by Clowe on a deep R-band Keck exposure, using the I band 2.2m data as a color selection to remove cluster galaxies, finds a nice centrally-concentrated mass peak falling exactly on the BCG with a $M/L_V = 270$. The mass from weak lensing comes out to be 2.45×10^{14} h_{50}^{-1} M⊙ at 500 h_{50}^{-1} kpc (assuming the background galaxies lie at $z = 2$).

MS2053-04 is a compact, optically-poor cluster with an arc system close to the core. Weak lensing analysis of Keck data also detects shear signal in this cluster. A recent X-ray observation has been performed with the BeppoSAX satellite by Scaramella et al. (in preparation). The data have not been fully reduced yet but a preliminary analysis shows this cluster to be much cooler than MS1054−03 or MS0451−03.

3   RXJ1716+66 from the ROSAT NEP Survey

We are in the process of completing the optical identification of all the sources in the $9° \times 9°$ region of the North Ecliptic Pole (NEP) survey. The NEP region has the largest exposure time of all the ROSAT All-Sky Survey (Trümper 1991), totalling about 10 ks. The principal derivative is a statistically complete sample of galaxy clusters appropriate for a better characterization of the X-ray luminosity function evolution (Mullis et al. 1998). We have recently discovered a cluster at $z = 0.81$, (**RXJ1716+66**, Henry et al. 1997), which is among the most distant X-ray selected clusters known together with MS1054−03 and the X-ray clusters detected by Rosati et al. (1998) in the RDCS (ROSAT Deep Cluster Survey) and by Ebeling et al. (1998) in the WARPS (Wide Angle ROSAT Pointed Survey). As with MS1054−03, it is unlikely that RXJ1716+66 is in virial equilibrium. The galaxies are in an inverted S-shaped filament running northeast to southwest (Fig. 2) with a distance from the top to the bottom of the 'S' of about 1.5 h_{50}^{-1} Mpc. It is intriguing that the morphology of these two clusters is filamentary with the X-rays coming from the center. The initial formation of

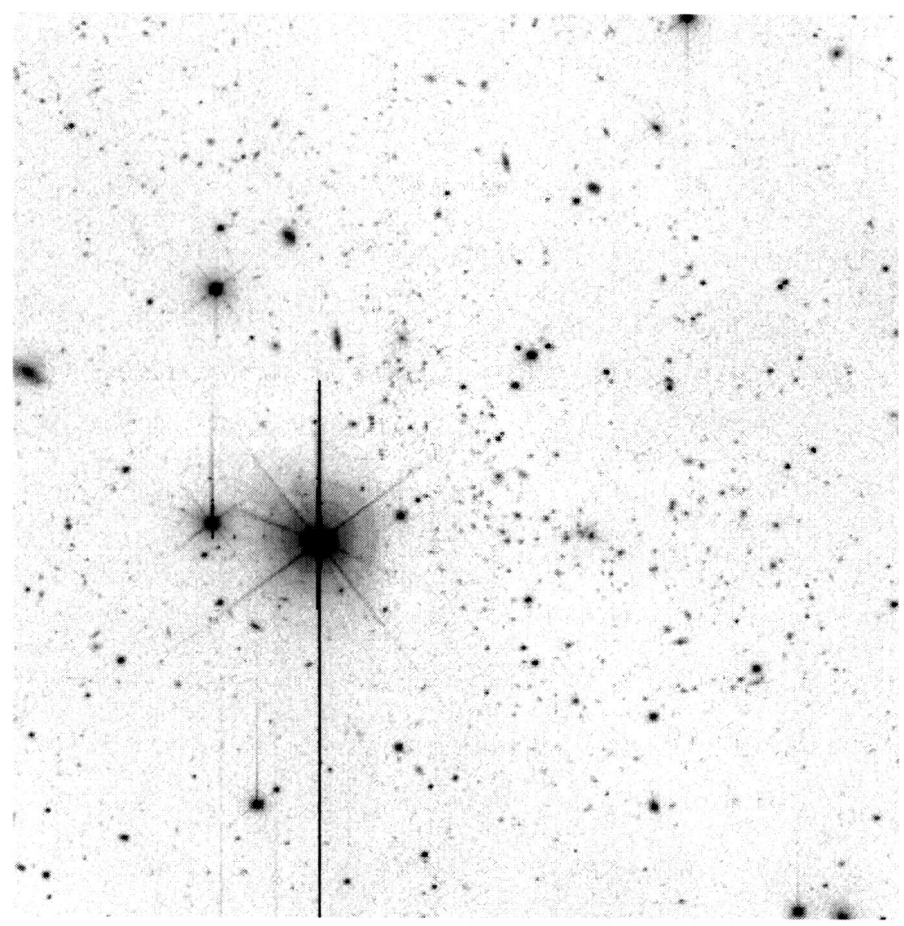

Fig. 2. Subarray I image of RXJ1716+66 taken with the UH 8K×8K CCD mosaic camera on the CFHT prime focus. North is at the top, east to the left.

protoclusters is often described as matter flowing along filaments (Bond et al. 1996) with the X-rays generated at the impact point of the two colliding streams of matter. From the weak lensing analysis, a mass peak is visible just east of the brightest cluster galaxy, and an arm of mass to the NE, neatly following the line of galaxies NE of the cluster core, is detected (Clowe et al. 1998). Both the optical lightmap and weak lensing massmap have two spatially distinct massive sub-clusters, as well as a long filamentary structure.

4 Conclusions

Identifications of high-z clusters provide useful observational constraints on cosmological models. Cluster abundances have been cited as one of the strongest pieces of evidence against the standard CDM model, normalized to reproduce the microwave background anisotropies seen by the COBE satellite. The number of high-z massive clusters predicted by standard CDM, or other mixed Dark Matter models, is too low with respect to the number of clusters observed. Massive clusters should be much rarer at epochs earlier than $z = 0.5$ if $\Omega=1$, contrary to the observations. Donahue et al. (1998) and Bahcall & Fan (1998) show that the existence of even the single most distant cluster in the EMSS at $z = 0.83$, MS1054.5−0321, with its large gravitational lensing mass, high temperature, and large velocity dispersion is sufficient to establish powerful constraints.

Acknowledgments

I am grateful to P. Henry, C. Mullis, N. Kaiser, G. Luppino, M. Donahue, D. Clowe for stimulation, help and advice. This work has received partial financial support from NASA-STScI grant GO-5402.01-93A and GO-05987.02-94A, from NSF AST95-00515, from NASA grant NAG5-2594 and ASI grants ARS-94-10 and ARS-96-13.

References

Bahcall, N.A. & Fan, X., astro-ph/9803277, ApJ, in press.
Bond, J.R., Kofman, L. & Pogosyan, D., 1996, Nature, 380, 603.
Carlberg, R., Yee, H. & Ellingson, E., 1994, ApJ, 437, 63
Clowe, D.I., Luppino, G.A., Kaiser, N., Henry, J.P. & Gioia, I.M., 1998, ApJL, in press.
Donahue, M., 1996, ApJ, 468, 79.
Donahue, M., Gioia, I.M., Luppino, G., Hughes, J.P. & Stocke, J.T., 1998, ApJ, in press.
Dressler, A. & Gunn, J., 1992, ApJS, 78, 1
Ebeling et al. , 1998, ApJ, submitted.
Frenk, C.S., White, S.D., Efstathiou, G. & Davis, M., 1990, ApJ, 351, 10.
Furuzawa, A., Yamashita, K., Tawara, Y., Tanaka, Y. & Sonobe, T., 1994, in "New Horizon of X-ray astronomy", eds F. Makino & T. Ohashi, (Universal Academy Press: Tokyo), pag 541.
Gioia, I.M. et al. , 1990, ApJS, 72, 567.
Gioia, I.M. & Luppino, G.A., 1994, ApJS, 94, 583.
Henry, J.P. et al. , 1997, AJ, 114, 1293.
Kaiser, N. & Squires, G., 1993, ApJ, 404, 441.
Koo, D.C., 1981, ApJ, 251, L75.
Luppino, G. A.& Gioia, I.M., 1995, ApJ, 445, L77.
Luppino, G. A. & Kaiser, N. 1997, ApJ, 475, 20.
Maccacaro et al., 1994, Astr. Lett.& Comm. , Gordon and Breach, Publ., 29, 267.
Mullis, C.R., Gioia, I.M. & Henry, J.P., 1998, to appear in "The Hot Universe", IAU symposium 188.

Rosati, P., Della Ceca, R., Norman, C. & Giacconi, R., 1998, ApJ, 492, L21.
Smail, I., Ellis, R., Fitchett, M. & Edge, A., 1994, MNRAS, 270, 245.
Stocke, J.T. *et al.* , 1991, ApJS, 76, 813.
Squires, G. & Kaiser, N., 1996, ApJ, 473, 65.
Trümper 1991, Adv. Spce Res., 2, 241.
Van Haarlem, Frenk, C. & White, S., 1996, MNRAS, 287, 817

Discussion

Illingworth: Have you been able to get redshifts for the arcs, particularly the radial arc, in MS2137-23?

Gioia: Unfortunately not. We tried a couple of times but we only get 1–2 nights of Keck telescope time a year, and if the weather is not co-operative ...

The Deep X-Ray Radio Blazar Survey (DXRBS)

Paolo Padovani[1,2,3], Eric Perlman[1], Paolo Giommi[4], Rita Sambruna[5], Laurence R. Jones[6], Anastasios Tzioumis[7] and John Reynolds[7]

[1] Space Telescope Science Institute, 3700 San Martin Drive, Baltimore, MD 21218, USA
[2] Affiliated to the Astrophysics Division, Space Science Department, European Space Agency
[3] On leave from Dipartimento di Fisica, II Università di Roma "Tor Vergata", Italy
[4] SAX Science Data Center, ASI, Viale Regina Margherita 202, I-00198, Italy
[5] Pennsylvania State University, Department of Astronomy, 525 Davey Lab, University Park, PA 16803
[6] School of Physics & Astronomy, Univ. of Birmingham, Birmingham B15 2TT, UK
[7] Australia Telescope National Facility, CSIRO, PO Box 76, Epping NSW 2121, Australia

Abstract. We have undertaken a survey for blazars by correlating the ROSAT WGA-CAT database with publicly available radio catalogs, restricting our candidate list to serendipitous flat-spectrum sources ($\alpha_{\rm r} \le 0.7$, $f_\nu \propto \nu^{-\alpha}$). We discuss here our survey methods, identification procedure and first results. Our survey is found to be $\sim 95\%$ efficient at finding blazars, a figure which is comparable to or greater than that achieved by other radio and X-ray survey techniques. DXRBS provides a much more uniform coverage of the parameter space occupied by blazars than any previous survey. Particularly important is the identification of a large population of flat-spectrum radio quasars with ratios of X-ray to radio luminosity $\gtrsim 10^{-6}$ ($\alpha_{\rm rx} \lesssim 0.78$) and of many low-luminosity flat-spectrum radio quasars. Moreover, DXRBS fills in the region of parameter space between X-ray selected and radio-selected samples of BL Lacs.

1 Introduction

Blazars are the most extreme variety of Active Galactic Nuclei (AGN) known. Their signal properties include irregular, rapid variability; high optical polarization; core-dominant radio morphology; apparent superluminal motion; flat ($\alpha_{\rm r} \lesssim 0.5$) radio spectra; and a broad continuum extending from the radio through to gamma-rays (e.g., Urry & Padovani 1995). The broadband emission from blazars is dominated by non-thermal processes (most likely synchrotron and inverse-Compton radiation). Blazar properties are consistent with relativistic beaming, that is bulk relativistic motion of the emitting plasma at small angles to the line of sight (as originally proposed by Blandford & Rees in 1978), which gives rise to strong amplification and collimation in the observer's frame. It then follows that an object's appearance depends strongly on orientation. Hence the need for "Unified Schemes", which look at intrinsic, isotropic properties, to unify fundamentally identical (but apparently different) classes of AGN.

The blazar class includes flat-spectrum radio quasars (FSRQ) and BL Lacertae objects. The main difference between the two classes lies in their emission lines, which are strong and quasar-like for FSRQ and weak or in some cases outright absent in BL Lacs. As a consequence of their peculiar orientation with respect to our line of sight, blazars represent a very rare class of objects, making up considerably less than 5% of all AGN (Padovani 1997).

2 Why a Blazar Survey?

A blazar survey is needed essentially for two reasons: number statistics and limiting fluxes. All existing blazar samples, in fact, are relatively small and at relatively high fluxes. The largest radio-selected BL Lac sample, the 1 Jy sample (Stickel et al. 1991), includes 37 objects with $f_{\rm 5GHz} > 1$ Jy. In the X-rays, there are a few available samples, made up of $40 - 70$ objects (the EMSS sample [Maccacaro et al. 1994], the Slew sample [Perlman et al. 1996], plus various ROSAT-based samples [Laurent-Muehleisen et al. 1997; Bade et al., in preparation]) but the deepest sample is still the EMSS one, which reaches $f_{\rm x} \sim 2 \times 10^{-13}$ erg cm^{-2} s^{-1}. (See, however, Wolter et al. 1997 for an ongoing deeper survey.)

Moving to FSRQ, until very recently the only sizeable sample with complete redshift information was the one extracted from the 2 Jy sample (Wall & Peacock 1985), which includes 52 sources. (The FSRQ in the 1 Jy and S4 samples reach lower fluxes [1 Jy and 0.5 Jy respectively] and are more numerous but the identification of the two samples is still not complete.) Drinkwater et al. (1997) have recently published the PKS 0.5 Jy sample, which includes 323 flat-spectrum radio sources with $f_{\rm 2.7GHz} > 0.5$ Jy, 86% of which have a measured redshift. Finally, Shaver, Hook, and collaborators have just put together a sample of more than 400 flat-spectrum objects down to $f_{\rm 2.7GHz} = 0.25$ Jy (Hook, these proceedings).

There is then clearly a need for a deeper, larger blazar survey, to address many open questions of blazar research, namely:

- BL Lac evolution and luminosity functions; unified schemes
- Flat-spectrum radio quasar evolution and luminosity functions; unified schemes
- The relationship between BL Lacs selected in the radio and X-ray band
- The relationship between flat-spectrum radio quasars and BL Lacs
- What is a BL Lac?
- The relationship between physical parameters (e.g., X-ray and radio spectral indices, equivalent width, line luminosity, continuum luminosity, etc.)

3 The DXRBS

The basic idea behind our Deep X-ray Radio Blazar Survey (DXRBS) is quite simple: blazars are relatively strong X-ray and radio emitters so selecting X-ray and radio sources with a flat radio spectrum (one of their defining properties) should be a very efficient way to find these rare sources. We adopt a spectral index cut $\alpha_{\rm r} \leq 0.7$. This will: (1) select (by definition!) flat-spectrum radio quasars; (2) select basically 100% of BL Lacs; (3) exclude the large majority of radio galaxies.

We have then cross-correlated WGACAT (White, Giommi & Angelini 1995), the publicly available database of ROSAT PSPC sources (restricting ourselves to sources having quality flag ≥ 5 to avoid problematic detections) with a number of publicly available radio catalogs. North of the celestial equator, we used the 20 cm and 6 cm Green Bank survey catalogs NORTH20CM and GB6 (White & Becker 1992; Gregory et al. 1996), while south of the equator, we used the Parkes-MIT-NRAO catalog PMN (Griffith & Wright 1993). All sources with radio spectral index $\alpha_r \leq 0.7$ at a few GHz were selected as blazar candidates. Note that WGACAT reaches $f_x \sim 10^{-14}$ erg cm^{-2} s^{-1} (although its flux limit varies widely on the sky), while the flux limits of the radio catalogs are the following: $f_{5GHz} \sim 20$ mJy (GB6), $f_{1.4GHz} \sim 100$ mJy (NORTH20CM), and $f_{5GHz} \sim 40$ mJy (PMN). (See Fig. 1 and section 4.)

For objects north of the celestial equator, $6 - 20$ cm radio spectral indices were obtained directly from the cross-correlation of the GB6 and NORTH20CM catalogs. For sources at southern declinations, the lack of a comparably deep radio survey at a second frequency required a different strategy[1]. We then conducted a snapshot survey with the Australia Telescope Compact Array (ATCA) at 3.6 and 6 cm, to get also radio spectral indices unaffected by variability (which will be a problem for our northern sample).

As a result of the correlations between the X-ray and the radio catalogs we have obtained a list of about 200 blazar candidates (with $|b| > 10°$), to which we add 88 previously known, serendipitous (i.e., not ROSAT targets) blazars (77 FSRQ and 11 BL Lacs), for a total of about 300 sources.

We note that, as the original catalogs included tens of thousands of objects, our search strategy has narrowed down the number of candidates by more than two orders of magnitude. This kind of approach is extremely important for surveys, like ours, that look for rare objects in large catalogs and will be vital with the advent of even larger and deeper catalogs, foreseen in the near future.

3.1 The Identifications

Accurate positions to pinpoint the optical counterparts were obtained from either the NVSS (Condon et al. 1997) or our ATCA survey. Magnitudes for all X-ray/radio sources with counterparts on the POSS and UKST plates which comprise the Digitized Sky Survey were obtained from the Cambridge APM and Edinburgh COSMOS projects (Irwin et al. 1994; Drinkwater et al. 1995). Most X-ray/radio sources without counterparts on the survey plates were imaged at either the KPNO 0.9m or the CTIO 0.9m telescopes. This allowed identification of optical counterparts to $R = 23$. The magnitude distribution of the blazar candidates peaks around 18. Spectroscopic observations were conducted at the KPNO 2.1 m, MMT, Lick 3 m, ESO 2.2 m and 3.6 m, and CTIO 1.5 m telescopes.

---

[1] The NVSS survey (Condon et al. 1997) was not available when we started this project. Moreover, it is still not 100% completed and covers the sky north of $-40°$.

Table 1. DXRBS Identifications

Class	Newly identified	Previously known	Total
Radio Quasars	86	77	163
BL Lacs	26	11	37
Radio Galaxies	4	10	14
Total	116	98	214

The breakdown of the identifications at the time of writing (February 1998) is given in Table 1. So far we have identified ∼ 50% of our candidates and 97% of them are blazars. 90% of the previously known objects are also blazars[2]. Our method is then indeed very efficient (93.5%) at identifying blazars.

4 First Results

The 163 DXRBS FSRQ we have so far identified span the redshift range of 0.1 to 3.8, with a mean value of 1.25. Figure 1a shows the greatly improved coverage of the radio/X-ray flux plane for FSRQ provided by DXRBS, as compared to previously available samples (note that both the 1 Jy and S4 samples are not completely identified and X-ray data are available only for ∼ 60% of their FSRQ). DXRBS FSRQ go about an order of magnitude deeper in radio flux and a factor of ∼ 4 deeper in X-ray flux. Moreover, while very few previously known FSRQ had relatively large X-ray-to-radio flux ratios ($f_x/f_r > 10^{-11.5}$ in the units of Fig. 1: see the dashed line), many DXRBS FSRQ are quite "X-ray bright."

This is better seen in Fig. 2, which plots radio versus X-ray luminosity for FSRQ. About 25% of DXRBS FSRQ have $L_x/L_r \gtrsim 10^{-6}$ (or, alternatively, $\alpha_{rx} \lesssim 0.78$). Only nine 1 Jy/S4 FSRQ had such luminosity ratios. Based on the overall spectral energy distribution of these sources and extrapolating from the situation in BL Lacs, these "X-ray bright" FSRQ probably have the peak of their synchrotron emission at UV/X-ray energies, unlike the other, more common, FSRQ, which peak in the IR/optical band. Due to their lower radio fluxes, DXRBS FSRQ are also reaching relatively low radio luminosities, approaching what should be the lower end of the FSRQ luminosity function according to unified schemes ($L_r \approx 10^{24.5}$ W Hz^{-1}: Urry & Padovani 1995). In particular, more than 20% of them have $L_r < 10^{26.5}$ W Hz^{-1}, as compared to only 3% for the 1 Jy and S4 samples.

---

[2] Note that it can be difficult to distinguish between a BL Lac and a radio galaxy for border-line sources. Therefore, while we are confident that the classification criteria have been applied consistently for our newly discovered blazars, this might not be the case for the previously known sources.

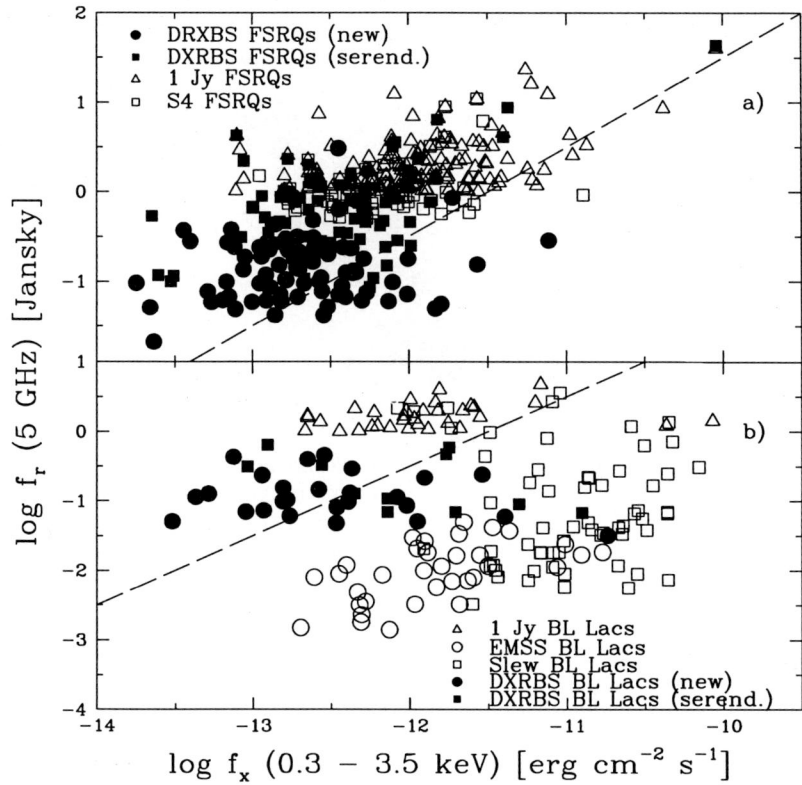

Fig. 1. The radio/X-ray flux plane for FSRQ (top panel) and BL Lac (bottom panel) samples. Note how DXRBS is sampling previously unexplored regions of parameter space. The dashed lines correspond to $\log f_x/f_r = -11.5$.

As regards BL Lacs, DXRBS is again exploring uncharted territory, as it is finding BL Lacs which cover a previously unexplored region of the radio/X-ray plane, and go deeper (by almost an order of magnitude) in X-ray flux than currently available samples (see Fig. 1b). Note also how the DXRBS BL Lacs are intermediate in their X-ray-to-radio flux ratios as compared to the "classical" radio and X-ray selected samples. A more complete description of the DXRBS first results is given by Perlman et al. (1998).

References

Blandford, R., D., & Rees, M. J. (1978): *Pittsburgh Conference on BL Lac Objects*, A. N. Wolfe ed., 328
Condon, J. J., et al., (1997): preprint (http://www.nrao.edu)
Drinkwater, M. J., Barnes, D. G., & Ellison, S. L. (1995): PASA **13**, 12

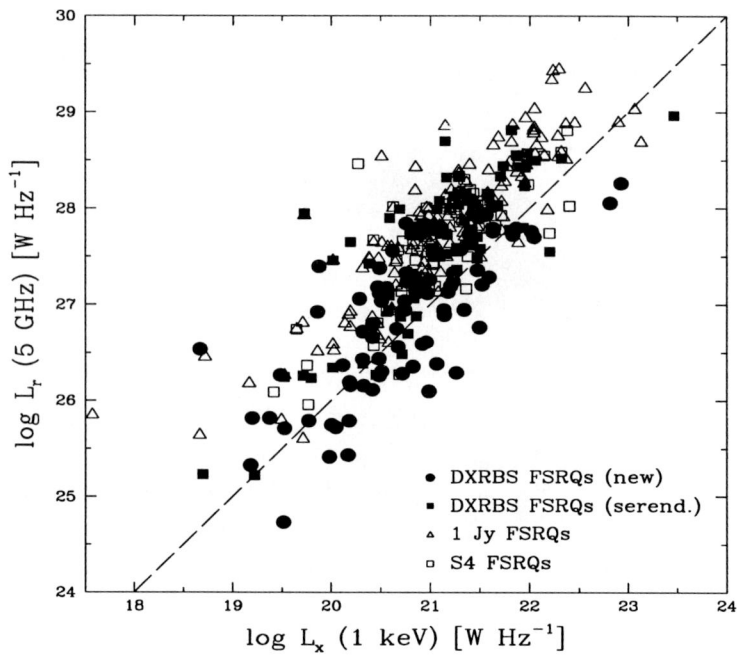

Fig. 2. The radio and X-ray luminosities of FSRQ.

Drinkwater, M. J., et al. (1997): MNRAS **284**, 85
Gregory, P. C., Scott, W. K., Douglas, K., & Condon, J. J. (1996): ApJS **103**, 427.
Griffith, M. R., & Wright, A. E. (1993): AJ **106**, 1095
Irwin, M., Maddox, S. & McMahon, R. (1994): Spectrum **2**, 14
Laurent-Muehleisen, S. A., Kollgaard, R. I., Ciardullo, R., Feigelson, E. D., Brinkmann, W., & Siebert, J. (1997): preprint (astro-ph/9711268)
Maccacaro, T., et al. (1994): Astrop. Lett. & Communications **29**, 267
Padovani, P. (1997): *Very High Energy Phenomena in the Universe*, Y. Giraud-Héraud, J. Trân Thanh Vân eds., Editions Frontieres, 7
Perlman, E. S., et al. (1996): ApJS **104**, 251
Perlman, E. S., Padovani, P., Giommi, P., Sambruna, R., Jones, L. R., Tzioumis, A., & Reynolds, J. (1998): AJ **116**, in press
Stickel, M., Padovani, P., Urry, C. M., Fried, J. W., & Kühr, H. (1991): ApJ **374**, 431.
Urry, C. M., & Padovani, P. (1995): PASP **107**, 803
Wall, J. V., & Peacock, J. A. (1985): MNRAS **216**, 173
White, R. L., & Becker, R. H. (1992): ApJS **79**, 331
White, N. E., Giommi, P., & Angelini, L. (1995):
  http://lheawww.gsfc.nasa.gov/users/white/wgacat/wgacat.html
Wolter, A., Caccianiga, A., Della Ceca, R., Gioia, I. M., Maccacaro, T., & Ruscica, C. (1997): *From the Micro- to the Mega-Parsec*, A. Comastri, T. Venturi, M. Bellazzini eds., Mem. Soc. Astron. Ital. **68**, 147

Discussion

Tsvetanov: From the newly identified BL Lacs, how many have emission lines in their spectra?

Padovani: That is a complicated issue. We are finding that the radio-galaxy / BL Lac / radio quasar divisions are fuzzier than previously found for samples with higher flux limits so the answer to your question is not straightforward. (One of the aims of our survey is also to try to answer the question "What is a BL Lac?".) In any case, at present I can say that a fair number (say 20–30%) of BL Lacs in our sample have (weak) emission lines.

Wagner: Do you carry out/plan to carry out any optical or radio studies of the polarisation properties of this sample?

Padovani: We might carry out an optical polarisation study for the brightest sources. Regarding radio polarisation, we will either carry out a dedicated program or use the polarisation information available in the NVSS survey.

Tsarevsky: By the initial definition, the BL Lacertae-type objects are highly variable (in the optical and radio). Are they prominently variable in X-rays too?

Padovani: They certainly are. As far as our sample is concerned, we have X-ray exposures (2–3) only for a small fraction of the sources so we might be able to study X-ray variability in detail.

La Franca: Have you already been able to compute the X-ray luminosity function?

Padovani: Not yet. We are working on the sky coverage of the X-ray catalog we are using (UGACAT). Once that is done, we should be able to get a preliminary X-ray luminosity function even before the completion of the optical identifications.

X-Ray Surveys and Their Follow-Up

I.J. Danziger

Osservatorio Astronomico di Trieste, Via G.B. Tiepolo 11, 34131 Trieste, Italy

Abstract. Published results of optical identifications in major X-ray surveys are briefly reviewed. How the more recent deeper observations originating from ROSAT differ from earlier work is pointed out. It is shown that deeper radio observations combined with X-ray work are providing more statistically significant results. A review is given of recent deep ROSAT X-ray identification work and reasons for apparently different results are discussed.

1 Introduction

The main reasons for pursuing observations of X-ray sources identified in surveys may be summarised briefly. a. To determine the origin of the X-ray background radiation where a large sample is required to demonstrate if unresolved sources constitute all of this radiation. Although we now know that 70–80 percent of the background consists of discrete AGN sources the remaining 20–30 percent is still in contention. b. To determine if new types of X-ray sources are contributing at the faint limits, an as yet unresolved problem. c. To study the evolution in the Log N – Log S diagram for different classes of objects, again to separate distinct classes of objects. d. To study particular objects which usually are the very strong X-ray emitters to determine whether this strength of emission gives rise to other observable effects. e. To use the sample to look for correlations of physical properties, and in particular to test unified theories of AGN's.

This talk will concentrate only on extragalactic aspects of X-ray emission in the relatively soft X-ray range defined by EINSTEIN and ROSAT observations. Some of the results constitute parts of on-going projects, and there will be comparisons made between published and unpublished/about to be published results.

2 Completed Surveys

The largest completed survey which set the standard for what followed is the EINSTEIN EMSS (0.3–3.5 keV) summarized by Gioia et al. (1990) and Stocke et al. (1991). It covered 778 square degrees of sky embracing 835 serendipity sources. Classification of sources requires several different parameters to separate stellar sources from AGN's for example, to separate clusters of galaxies from others, to subdivide AGN types. A close reading of identification papers reveals that almost all workers have employed the same criteria, specifically stated and quantified, or tacitly assumed. Statistical results expressed in percentages of the extragalactic

component are summarised as follows.
AGN - 73; Clusters of galaxies - 18; BL Lac - 6; Normal galaxies - 2. In total there were 220 stellar sources.

The EINSTEIN Pavo Deep Field involving pointed HRI + IPC observations, with a new unpublished reduction by F. Primini, have been the subject of a complete identification by Danziger and Gilmozzi (1997). Of the possible 21 X-ray sources the identifications may be summarized in percentages as follows.
AGN - 80-95; Clusters of galaxies - 5; BL Lac - 0; Normal galaxies - 0. Only 1 star was found. The optical colour information strongly suggested a large range of AGN luminosity from QSO's to low luminosity AGN's embedded in galaxies. It also sets a firm lower limit to the density of AGN's contributing to the X-ray background at a sensitivity of 1.5×10^{-14} ergs/cm^2/sec (0.5–2.0 keV).

3 New Surveys

The same Pavo Field has been the subject of a deep (25080 sec) pointed ROSAT PSPC observation (a collaboration involving Gilmozzi, Danziger, Zimmermann, Hasinger, Macgillivray). There exist probably more than 80 bona fide X-ray detections of which, including the EINSTEIN results, 41 objects have been identified with AGN's, 4 with clusters, $\sim$ 4 with BL Lacs. Thus with a better (and deeper) sample, the proportions begin to resemble those of the EMSS. This work is continuing but requires a larger telescope.

Radio observations (a collaboration involving Anderson, Ekers, Danziger, Gilmozzi) with ATCA at 1.42 GHz provide a limiting (5σ) sensitivity of 0.23 mJy. A total of 81 radio sources have been detected of which as many as 12 and certainly 6 are associated with an X-ray source. This statistic is similar to that reported later for the Lockman Field, and significantly increases the detection rate in Pavo compared to previous attempts.

Another ongoing ROSAT project is the southern ALL-SKY Survey (PI's Danziger and Trumper) which for identification work covered 575 square degrees of sky divided into 4 distinct fields. This work is not yet complete. Out of 688 X-ray sources there are so far 521 identifications. The breakdown in percentages of extragalactic objects is as follows.
AGN - 74; Clusters of galaxies - < 14; BL Lac - 4; Normal galaxies - $\sim$ 6. So far there are 166 stars. Not strikingly different from the EMSS result.

Radio follow-up with the ATNF and VLA at 4.8 GHz reported by Anderson, Ekers, Danziger shows a likely association of radio emission with 76 AGN's, < 41 clusters, 15 BL Lac's and $\sim$ 21 normal galaxies.

The deepest ROSAT observations have been made in the Lockman Field by Hasinger et al. (1993). These observations using 143.7 ksec have been combined with observations of 26 shallower fields to produce 1176 sources with a sample complete to 2.5×10^{-15} ergs/cm^2/sec. With this data they have reported a change in slope (flattening) of the differential source count at 2.5×10^{-14} ergs/cm^2/sec. At the faintest limiting flux reported here 60 percent of the background is resolved into discrete sources. Moreover a fluctuation analysis extending beyond

2.5×10^{-15} ergs/cm^2/sec suggests that < 25 percent can be truly diffuse and a reasonable extrapolation to zero fluxes could easily account for 100 percent of the background. A resteepening could saturate the background at limits conceivably detectable with AXAF. It should be noted that this study deals with X-ray sources not X-ray sources optically identified as AGN's. Therefore in discussing Log N – Log S one has to allow for a stellar component.

An important follow-on to these results has been presented by Hasinger et al. (1998) who have analysed a total of 207 ksec of PSPC data and 1.32 Msec of HRI data for the Lockman Field producing a Log N – Log S reaching 1×10^{-15} ergs/cm^2/sec. They report that simulations for PSPC fields with exposures > 50 ksec suggest that such surveys are confusion limited a factor of 2 above the 4σ threshold. This may have important implications for some of the differences reported later.

Radio follow-up of the Lockman Field with the VLA at 1.492 GHz by de Ruiter et al. (1997) revealed 149 sources with S > 120 microJy (4σ). Approximately 30 percent are associated with optical objects, and there are 16 possible radio/X-ray associations, not so different from the result for Pavo.

4 Comparisons of Optical Identification Work

Apart from the as yet incomplete work in Pavo there are 5 different reported results from deep ROSAT observations which may be usefully compared and contrasted. They are as follows:

CRSS – Boyle et al. (1995). Serendipity pointed (PSPC) with $S > 2 \times 10^{-14}$. Of the 123 X-ray sources, 68 are QSO's and 12 NELG's (1/2 Seyfert 2, 1/2 other).

DRS – Georgantopoulos et al. (1996). 5 fields 21–49 ksec. $S > 3 \times 10^{-15}$. Of the 194 X-ray sources, 107 are AGN's (including QSO's), 19 galaxies (no discrimination among types), 1 cluster (and 12 stars).

NEP – Bower et al. (1996). 1 field 79.1 ksec. $S > 1.0 \times 10^{-14}$. Of 20 X-ray sources, 13 AGN's, 0 galaxies.

UKDS – McHardy et al. (1998). 1 field 115 ksec. $S > 1.6 \times 10^{-15}$. Of 70 X-ray sources ($S > 2 \times 10^{-15}$), 32 QSO's, 18 NELG's (without secure separation of Seyfert 2's from others), 6 clusters.

RDS – Schmidt et al. (1998). Complete sample Lockman Field. $S > 5.5 \times 10^{-15}$. Of 50 X-ray sources, 39 AGN's, 3 groups, 1 galaxy, 3 stars. This last paper contains a careful discussion and quantification of criteria for classification. Also included is a finding that 79 percent of field galaxies near the identified object are NELG's (Narrow Emission Line Galaxies). In the following table I summarize 2 important percentages from each of the surveys, the percentages of AGN's and of galaxies.

CRSS	AGN 55-60	Galaxy 10-5
DRS	AGN >55	Galaxy <10
NEP	AGN 65	Galaxy 0
UKDS	AGN >46	Galaxy <26
RDS	AGN 78	Galaxy 2

Clearly the RDS identification reveals a significantly higher percentage of AGN's than the others. The earlier Pavo result also showed a much higher percentage of AGN's.

So the question is whether NELG's are contributing to source counts at faint limits, and therefore whether one is seeing the effects of a new component affecting the Log N – Log S plot. Schmidt et al. (1998) argue that because of the high surface density of faint emission line galaxies in the field it is very possible that incorrect identifications have been assigned to X-ray sources simply because of the presence of narrow emission lines. That is why the paper by Hasinger et al. (1998) concerning X-ray source confusion is important and relevant.

Another less apparent problem also exists in the statistics provided by these 5 surveys. There are in total > 265 AGN's identified but no BL Lac objects. Statistics from earlier surveys would suggest that there should be ∼ 13 BL Lac objects. The simplest explanation is that (also suggested by Brian Boyle here) a low s/n spectrum, generally characterizing optical identification work, of a featureless continuum typical of BL Lac spectra, might be quickly discarded in favour of emission line objects that are nearby, but not a correct association. Thus we see on 2 counts that narrow emission line galaxies should be regarded with special suspicion in the presence of finite error bars on positions of X-ray sources.

5 Conclusions

From the above discussion some obvious and hardly original points can be made concerning future approaches to surveys.

What are required are:

1. Deep AXAF observations of several fields with sizes ∼ 0.5 square degrees. The considerably improved imaging and pointing will help with the first problem above. The several fields of this size should provide good statistics and a measure of how much the surface density of X-ray sources varies from place to place, a reality as evidenced by some of the ROSAT data.
2. Deep radio (6 cm) observations made before optical identification work begins. The reasonable (25 percent) association of radio with X-ray sources could save large optical telescope time.
3. Various colour imaging photometry with good seeing will help to discriminate low luminosity AGN's. The ratio of visual to X-ray flux will be a useful tool particularly when objects are too faint for spectroscopy.

4. Low resolution high s/n flux calibrated spectra, the flux calibration being necessary to measure emission line ratios for discriminating Seyfert 2's from star forming galaxies with HII regions. The high s/n will greatly assist in resolving the second problem of BL Lac objects.

References

Bower, R.G. et al. 1996, MNRAS 281, 59
Boyle, B. et al. 1995, ApJ 272, 462
Danziger, I.J., Gilmozzi, R. 1997, A&A 323, 47
De Ruiter, H. et al. 1997, A&A 319, 7
Georgantopoulos, I. et al. 1996, MNRAS 280, 276
Gioia, I. et al. 1990, ApJS 72, 567
Hasinger, G. et al. 1993, A&A 275, 1
Hasinger, G. et al. 1998, A&A 329, 482
McHardy, I.M. et al. 1998, AN 319, 51
Schmidt, M. et al. 1998, A&A 329, 495
Stocke, J. et al. 1991, ApJS 76, 81

Cosmic Rays and the Structure of the Local Universe

Roger W. Clay[1] and Bruce R. Dawson[1]

University of Adelaide, South Australia, 5005

Abstract. The next generation of extremely high energy cosmic ray arrays will record significant numbers of events with energies above 10^{20} eV. At those energies, protons should retain directional information about their sources which are likely to be closer than about 100Mpc due to attenuation by the CMB. We have compared the actual cosmic ray directional distribution from the Haverah Park cosmic ray array with predicted sky maps using a number of extragalactic astronomical catalogs to examine the feasibility of such a procedure for interpreting directional data. Even with very few events, it appears that the technique yields useful results.

1 Modelling Cosmic Ray Sky Distributions

In travelling to us through intergalactic space, the highest energy cosmic rays interact with the 3K Cosmic Microwave Background (CMB) and lose energy through photopion production. Such particles also scatter in any intergalactic magnetic fields. Lampard et al. (1997) have examined these processes in detail. Clay et al. (1997) have calculated the diffusive scattering and provide an expression for the mean angular deviation (in degrees):

$$\Delta\phi = 4 \left(\frac{2.10^{20}eV}{E}\right) \left(\frac{d}{10Mpc}\right)^{0.5} \left(\frac{L}{100kpc}\right)^{0.5} \left(\frac{B}{0.1\mu G}\right)$$

(observational energy, E: displacement, d: largest field turbulence scale, L: magnetic field strength, B)

We have taken the six highest events from the Haverah Park air shower array (Wada (1980)) to investigate how the directions of such events might be studied when much larger datasets accumulate with the operation of the High Resolution Fly's Eye and the Pierre Auger Observatory. We took the measured distribution of arrival directions and compared it with sky distributions which we generated using available astronomical catalogs.

We first took the Center for Astrophysics redshift catalog of galaxies and estimated the relative strength of the cosmic ray flux from each object assuming that the flux was proportional to the visual brightness but reduced by attenuation in the CMB, based on the calculations of Lampard et al. (1997) (assuming a particular mean intergalactic field strength (0.1, 0.3 or 1.0μ G) and turbulence structure (100kpc maximum scale length)). We then determined an angular uncertainty for protons from the object, based on intergalactic scattering and the instrumental angular uncertainty (Wada (1980)). A Monte Carlo sky distribution was derived and weighted by the Haverah Park declination distribution.

We thus obtained a sky distribution (in 3° pixels) which would be a long term distribution of events such as were recorded by the Haverah Park array.

We took the directions of the actual Haverah Park highest energy events and counted the number of sky distribution events in the pixels to which they corresponded. This was repeated by selecting six random directions from the appropriate declination distribution one million times. We determining how many times the observed directions had accumulated pixel counts below the simulated ones. This gives a probability that the Haverah Park events are not related to the catalogued galaxies. These data are shown in table 1, as are data based on the IRAS galaxy sample for brightnesses at 12,25,60 and 100μm.

Table 1. The probability (%) that random event directions are more consistent with expected cosmic ray sky distributions than the Haverah Park highest energy events based on a number of extragalactic source catalogues.

Catalogue	Magnetic Field (μ G)		
	0.1	0.3	1.0
Visual	2.4	0.6	0.04
12 micron	2.4	2.2	0.3
25 micron	4.6	2.3	0.07
60 micron	1.9	0.02	0.00
100 micron	0.1	0.01	0.00

Table 1 demonstrates that, given any of the catalogues we used, the Haverah Park data fit the sky distribution better than most of the randomly generated data sets. There is thus evidence that the observed cosmic rays are from the galaxies themselves. This stronger than noting a clustering about the supergalactic plane (Stanev et al. (1995)). Also, the cosmic ray intensity is best related to the infrared intensity at 60 or 100 microns. It is well known that there is a strong correlation between the far infrared flux from galaxies and their radio brightness and this could be an extension of the same effect.

The data in table 1 indicate that more scattering is needed than that which we assumed for a 0.1μG intergalactic field but this may either be due to an underestimate of the strength of that field or due to local scattering in the vicinity of our galaxy.

References

Clay, R., Cook, S., Dawson, B., Smith, A., Lampard, R. (1997): U. of Adel. preprint.
Lampard, R., Clay, R., Dawson, B. (1997): Astroparticle Physics **7**, 213-218
Stanev, T., et al. (1995): Phys. Rev. Lett. **75**, 3056
Wada, M. (1980): *Catalogue of Highest Energy Cosmic Rays No. 1* (World Data Center C2 for Cosmic Rays, Institute of Physical and Chemical Research, Tokyo)

Looking Deep from the South Pole: Star Formation in the Thermal Infrared

Michael G. Burton, John W.V. Storey and Michael C.B. Ashley

Joint Australian Centre for Astrophysical Research in Antarctica,
School of Physics, University of New South Wales, Sydney, NSW 2052, Australia

Abstract. The Antarctic Plateau provides the pre-eminent conditions on the Earth for wide-field imaging at thermal infrared wavelengths. We describe a project to equip the 60 cm SPIREX telescope at the South Pole with a large format (1024 × 1024) IR array camera (Abu) to demonstrate this potential. With it we aim to survey the Large Magellanic Cloud for sites of massive star formation at 3.5μm at the 1.4" diffraction limit of the telescope. We also discuss the potential for studying extra-galactic star formation through a deep survey of the Hubble Deep Field–South in this band. We compare the sensitivity of such surveys from Antarctica with those from mid-latitude sites, and propose a 2 m-class telescope, SPIRIT, which would be able to achieve unique new science at low cost compared to the new generation of 8 m-class telescopes now under construction.

1 Introduction

Understanding star formation is an active research area in contemporary astrophysics. Of interest is the study of processes which occur as a single star forms, their collective effects during group and cluster formation, and the birth of entire galaxies. We seek to understand the dependence of these events on the local environment of the molecular clouds in which they occur. This endeavour is very much the realm of infrared and millimetre astronomy, both to peer into the murky depths of the clouds to see the young stars, and to measure the properties of the natal gas from which they came. A key goal of these studies is to undertake a complete population census of regions of star formation in order to determine the number and types of stars that form in them, and how this varies between different complexes. For this, observations in the thermal infrared ($\lambda > 3\,\mu$m) are necessary. These wavelengths not only penetrate to the depths of cloud cores, but also allow us to distinguish between the embedded population and background stars. In simple terms, young stellar objects are surrounded by warm (few hundred K) disks which emit strongly at $\lambda > 3\,\mu$m, and thus are readily distinguished in infrared colour-colour diagrams (*e.g.* [1.65μm − 2.2μm]/[2.2μm − 3.8μm]) from reddened stars (*e.g.* see Figure 1). Near–IR colour-colour diagrams (*e.g.* [1.25μm − 1.65μm]/[1.65μm − 2.2μm]), while relatively easy to construct due to the better sensitivities available, show only small IR excesses due to disks. These are readily confused with reddening vectors, and also fail to include the most deeply embedded sources.

The problem has been that at 3.8μm sensitivities are typically 4–5 magnitudes worse than at 2.2μm from most observing sites, thus limiting the work

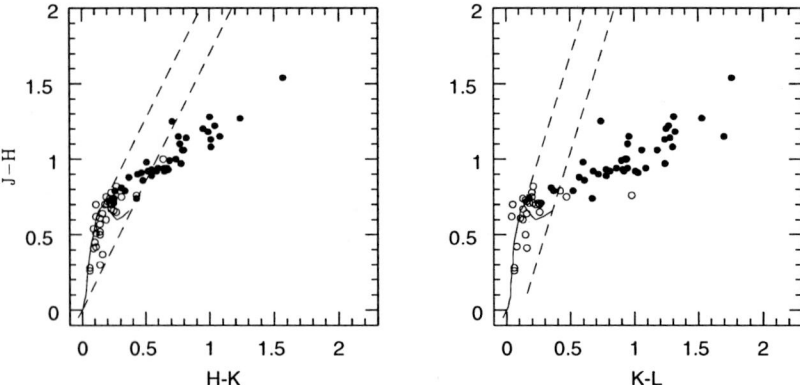

Fig. 1. Near–IR colour-colour diagrams for the nearby Taurus–Auriga molecular cloud (Strom et al. 1989). *Left*, J–H as a function of H–K; *right*, J–H as a function of K–L. Filled circles represent pre-main sequence stars with spectroscopic evidence for accretion disks; open circles represent stars without. In both diagrams, the solid line corresponds to the main sequence and the dotted lines denote the reddening band. It is readily apparent how much clearer the separation between stars with and without disks is using the K–L colour index compared to the H–K colour (J=1.25μm, H=1.65μm, K=2.2μm and L=3.6μm).

that has been done at this wavelength. Needed are deep, wide-field surveys of comparable sensitivity to those conducted at 2.2μm in order to determine the complete stellar membership of a star formation region. Such an opportunity is afforded a telescope on the Antarctic Plateau in view of the vastly reduced thermal background at these wavelengths over temperate sites. In this paper we describe a project now underway at the South Pole which aims to demonstrate that such surveys are feasible, with an initial objective being a 3.8μm survey of the Large Magellanic Cloud for massive star formation.

2 The Antarctic Plateau

The Antarctic plateau provides unique conditions on the Earth for the conduct of observational astronomy. The air is thin, dry and cold and the weather stable, attributes all offering gains to the observational astronomer. These conditions are quite different to those experienced at Antarctic coastal locations, which are frequently subject to violent storms.

The plateau is over 3,000m in elevation, rising to over 4,000m at Dome A. An average year-round temperature of −50° C, falling to −90° C at times, vastly reduces the thermal background in the near–IR. The precipitable water vapour content of the air is typically around 250μm and can fall below 100μm, opening up new windows in the infrared and sub-millimetre regimes to ground-based observation. The lack of a diurnal temperature cycle and the low wind speeds on the highest parts of the Antarctic plateau provide conditions of extraordinarily stability, benefiting a wide range of observational programs (Burton et al. 1994).

In the near–IR, from 2.2 to 5 μm, the thermal emission from sky and telescope at the South Pole has been measured to be 20–100 times less than at Mauna Kea (Ashley et al. 1996, Nguyen et al. 1996). However, while the greatest gains occur at 2.4μm, it is only in a relatively narrow (0.2μm wide) window. The entire L–band (3–4μm) experiences a background reduction of 20–40 times that of Mauna Kea, opening up the possibility of deep observations in the thermal IR that simply cannot be conducted from any other ground based site. In the mid–IR measurements also show reductions of up to 20 in the background, a result of a reduced atmospheric emissivity in addition to the temperature drop. There is also exceptional stability in the DC level of the sky (Smith & Harper, 1997).

The stable atmospheric conditions, tenuous air and absence of jet streams combine to produce conditions of superb clarity, or 'super-seeing'. Mitigating against these positive attributes, however, is the presence of a strong inversion layer which occurs on the most stable days of winter, when the air temperature can rise by 10° C in a few metres. This produces relatively poor ice-level seeing, $\sim$ 1.5″. However it occurs almost entirely in the lowest $\sim$ 100 m of the atmosphere; there is virtually no contribution from above this height (Marks et al. 1996, 1998). The seeing is thus of a quite different nature to that encountered at temperate latitudes, with a much larger isoplanatic angle and coherence time for the seeing cells. The depth of the inversion layer decreases with elevation on the Plateau, and on top of the ice Domes may be small enough that a telescope on a raised platform will be above it entirely.

The US NSF has funded the Center for Astrophysical Research in Antarctica (CARA) to establish an astronomical observatory at the Amundsen–Scott South Pole station. Building upon a series of isolated experiments, construction of the observatory is now well advanced. Three major experiments have wintered over, an infrared camera (on the 60-cm SPIREX telescope), a sub-millimetre spectrometer (the 1.7-m AST/RO telescope), and a microwave background anisotropy experiment (COBRA). An extensive site-testing program has been conducted, with plans to extend it to the high points of the plateau, Domes A and C, using an automated observatory (the 'AASTO', Storey 1998).

3 SPIREX/Abu

SPIREX, the 'South Pole InfraRed EXplorer', is a 60-cm telescope built in 1994 by CARA in order to conduct observations from 2.4 μm from the South Pole to determine whether the postulated 'cosmological window' on the Universe exists (Harper 1989, Hereld 1994). The window indeed existed, but the background turned out to be $\sim$ 5 times higher than anticipated (though still some 40 times better than at temperate sites). However, as a result of parallel measurements using the IRPS (Ashley et al. 1996), the original IR spectrometer of the AAT, it was demonstrated that the entire L–band (3–4 μm) has even greater potential as a new window, especially while only small aperture telescopes are available at the site.

An agreement between the Universities of Chicago and Ohio State, NASA Goddard, the US National Optical Astronomy Observatory and the University of New South Wales, has resulted in a project to upgrade the SPIREX telescope and install a state-of-the-art IR focal plane array; NOAO's "Abu" camera, incorporating a 1024 × 1024 InSb Aladdin array, with broad band and line imaging filters. UNSW's contribution includes the addition of a CCD fast guider and a tip-tilt secondary mirror, to improve the tracking and pointing of SPIREX. Automated observing and reduction protocols have been designed by Ohio State to allow continuous observing when conditions permit. This upgrade has taken place during the 1997/98 summer season at the Pole, with Australian participation funded by the Major National Research Facilities Program.

The system has been optimised for the thermal infrared L–band, around 3.5μm. Expected sensitivities are given in Table 1 through the 6 filters. These include both summer and winter time figures as in the thermal IR day time operation will be possible with only slightly reduced sensitivities. Abu has an 11' circular field of view with 0.7" pixel scale, and will fully-sample at the diffraction limit of 1.4". With a Fried parameter of $\sim$ 50 cm at 3.5μm, tip-tilt optics should allow us to obtain near-diffraction limited images even in poor seeing conditions.

4 Plans

There are many projects that SPIREX/Abu will be well suited to. These include:

- Line imaging of embedded HII regions along the southern galactic plane in the Brα, PAHs and H_2 filters, to study the environment of the ionized, neutral and molecular mediums of massive star forming regions. Of particular interest is the NGC 6334 region.
- A survey of the Large Magellanic Cloud for massive star formation at L–band (3.5μm). Spending 1 hour per position, a 2° × 2° region can be fully-sampled in a month of telescope time. Several thousand stars should be visible down to L=15.5 magnitudes. While this only allows us to sample the top end of the mass distribution, it will now be possible to survey the LMC in a systematic manner for the regions where these stars form. This will be the first in-depth study of extra-galactic star formation, which is of interest in its own right. However studying star formation in the LMC, occuring in a lower metal abundance environment than our Galaxy, will also yield invaluable insights into how primordial star formation proceeds, as it presumably takes place in near-metal-free environment.
- The star formation history of the Universe is being probed through deep, pencil-beam surveys, of which the Hubble Deep Field (HDF) is the most notable example. A southern field (the HDF–S) is shortly to be obtained. SPIREX/Abu will be capable of complementary deep K– and L–band imaging to the optical bands of the HDF–S. The 11' FOV of the instrument

Table 1. Calculated Sensitivity of SPIREX/Abu at the South Pole

Wavelength (μm)	Bandpass (μm)	Sky Flux Jy/arcsec^2	Continuum Mags/beam	Continuum Jy/beam	Line W/m^2/beam	Filter
2.425	0.034	1 (-4)	17.1	7 (-5)	1 (-18)	H$_2$ Q-branch
3.299	0.074	8 (-2)	14.7	4 (-4)	8 (-18)	PAHs
		0.2	14.0	7 (-4)	2 (-17)	
3.51	0.62	0.1	15.5	2 (-4)		Broad Band L
		0.4	14.7	4 (-4)		
3.82	0.60	0.3	14.8	3 (-4)		Broad Band L'
		0.6	14.1	5 (-4)		
4.051	0.054	0.4	12.9	2 (-3)	2 (-17)	H Br α
		0.7	12.2	3 (-3)	3 (-17)	
4.67	0.16	2	11.7	3 (-3)		Broad Band M
		3	11.0	6 (-3)		

Sensitivities are 3σ in 1 hour of on-source integration, assuming the signals of 5×5 0.66" pixels are summed. For each filter the first line is the estimated winter-time sensitivity (T=213K) and the second-line in summer (T=243K) (except for 2.4μm where daytime measurements would not be sensible). Measured (extrapolated at M) sky fluxes at the Pole are used, and the 60-cm aperture SPIREX telescope assumed to have a 10% emissivity, a system throughput of ~ 0.3, and a detector with 1 e/s dark current, 100 e read noise and $\sim 80\%$ qe. A maximum single integration time period of 600s is assumed, which means the H$_2$ filter is still detector noise limited; in principle another 0.8 mags. of sensitivity gain is possible through it.

includes the three individual fields of the WFPC-2, STIS and NICMOS instruments. With 24 hours of on-source integration a detection threshold of 19.2 magnitudes through the H$_2$ 2.4μm filter and 17.3 magnitudes through the L-band 3.5μm filter would be achieved. Based on the number counts in the HDF this would yield ~ 40 galaxies at K and ~ 4 at L. These galaxies will typically have $<V> = 22.5$, V-K=3.5 and a redshift z=0.6. SPIREX/Abu would thus sample the upper end of the galaxy luminosity function, but with minimal uncertainties from extinction. However, it would also be sensitive to particularly red galaxies at high redshift; for instance an E/S0 galaxy at z=1.4 has an (unreddened) V-L~ 10 and so would be detected by Abu if $V < 27$. Moreover, statistics on the distribution of galaxies in the adjacent fields to the HDF-S would be simultaneously obtained.

The project serves as a demonstrator of the scientific potential of the Antarctic Plateau. While wide-field, deep, thermal infrared surveys are a particular niche this experiment can exploit, the work that a 60 cm telescope will do is limited. SPIREX/Abu is thus designed as a forerunner for an intermediate size infrared telescope, a project we call 'SPIRIT'.

4.1 SPIRIT

The 'South Pole Infrared Imaging Telescope' is envisaged as a 2 m-class infrared optimised telescope capable of yielding near-diffraction limited images at 2.4μm. It will be a wide-field telescope, instrumented with large format focal plane arrays, and operate primarily from 2–5μm and 10–30μm. It will use a tip-tilt mirror, which will recover most of the diffraction limit given that $r_0 \sim$ 36 cm at 2.4μm. With a 0.25″ pixel scale and 4′ field of view, scientific projects would emphasise deep surveys. The first programs would involve surveys of star forming regions at 3.8μm and 11.5μm, to make a complete census of the embedded population of several such regions. Table 2 compares the sensitivity of SPIRIT to both the AAT and the new-generation 8-m telescopes like Gemini and the VLT. A 2.5 m telescope in Antarctica has comparable or better sensitivity than an 8-m telescope for both line and continuum imaging projects across the thermal infrared, with the additional advantage of having a wide field of view. It is clear that SPIRIT could yield unique new science. It would also provide a testbed facility for more ambitious telescopes in the future, capable of exploiting the full potential of the Antarctic Plateau for astronomy. SPIRIT seeks international partners to proceed, as part of the International Antarctic Observatory.

Acknowledgements

SPIREX/Abu has come about due to the dedicated efforts of a great many people working in extenuating and heroic circumstances. We especially wish to thank our colleagues at UNSW; Max Boccas, Mick Edgar, Andre Phillips, Antony Schinckel and Rodney Marks, who is wintering with the instrument. SPIREX/Abu is also the result of a happy and successful collaboration with several US institutions, and we are particularly grateful for the efforts of Bill Ball, Sean Casey, Darren Depoy, Al Fowler, Al Harper, Bob Loewenstein, Paul Martini, Fred Mrozak, Bob Pernic and Nigel Sharp.

References

Ashley, M.C.B., Burton, M.G., Storey, J.W.V., Lloyd, J.P., Bally, J., Briggs J.W., & Harper, D.A. (1996): PASP, **108**, 721

Burton, M.G. et al. (1994): Proc. Astron. Soc. Aust., **11**, 127

Harper, D.A. (1989): 'Astrophysics in Antarctica', *Am. Inst. Phys. Conf. ser.*, **198**, 123. Eds. D.J. Mullan, M.A. Pomerantz & T. Stoner

Hereld, M. (1994): 'Infrared astronomy with arrays: the next generation', *Ast. Ap. Sci. Lib.*, **190**, 248. Ed I.S. McLean (Kluwer).

Table 2. Sensitivity comparison between different Telescopes

		SPIRIT	AAT	Gemini / VLT	
Diameter		2.5 m	3.9 m	8.0 m	
Temperature		213 K	288 K	273 K	
Emissivity		0.05	0.20	0.05	
Location		South Pole	Siding Spring	Mauna Kea / Atacama	
Wavelength	Resolution				Filter
2.15 μm	R=7		20.8 mags 3 (-6) Jy	21.7 mags 1 (-6) Jy	Broad Band, K–narrow
2.37	13	21.9 mags 1 (-6) Jy			Broad Band, K–dark
2.12	100		2 (-19) W/m² 1 (-5) Jy	7 (-20) W/m² 5 (-6) Jy	H$_2$ 1–0 S(1)
2.43	40	7 (-20) W/m² 2 (-6) Jy			H$_2$ 1–0 Q–branch ($\sim$ 3× 1–0 S(1) intensity)
3.30	45	2 (-18) W/m² 7 (-5) Jy	1 (-17) W/m² 6 (-4) Jy	3 (-18) W/m² 1 (-4) Jy	PAHs
3.60	46		14.8 mags 4 (-4) Jy	16.1 mags 1 (-4) Jy	Narrow Band L
3.65	12	17.3 mags 3 (-5) Jy			Broad Band L
11.5	100		9.0 mags 9 (-3) Jy	10.2 mags 3 (-3) Jy	Narrow Band N
11.5	12	11.5 mags 1 (-3) Jy			Broad Band N
20.1	100	8.3 mags 5 (-3) Jy	5.8 mags 5 (-2) Jy	8.2 mags 5 (-3) Jy	Narrow Band Q

Sensitivities are 3σ per square arcsecond for one hour of on-source integration. The comparison is made for the optimum spectral bandpasses for Antarctic and mid-latitude sites, as specified above. Aside from site conditions, aperture size and telescope emissivity, other parameters are held constant between the systems, so that a direct comparison of performance is possible. A system throughput of 0.2, dark current 1 e/s, read noise 10 e and detector qe of 90% is assumed. In the thermal IR both the stability of the background level and the atmospheric transmission will determine whether these sensitivities can be reached in practice; the lower water vapour content at the Pole makes this site superior in this respect.

Marks, R.D., Vernin, J., Azouit, M., Briggs, J.W., Burton, M.G., Ashley, M.C.B. & Manigault, J.F. (1996): A&AS, **118**, 385

Marks, R.D., Vernin, J., Azouit, M., Manigault, J.F. & Clevelin, C. (1998): A&A, in press.

Nguyen, H.T., Rauscher, B.J., Severson, S.A., Hereld, M., Harper, D.A., Loewenstein, R.F., Mrozek, F., & Pernic, R.J. (1996): PASP, **108**, 718

Smith, C.H. & Harper, D.A. (1998): PASP, in press.

Storey, J.W.V. (1998): 'Astrophysics from Antarctica', PASPC, ASP meeting, Chicago, 30 June–2 July 1997, Eds G. Novak & R.H. Landsberg, in press.

Strom, K.M., Strom, S.E., Edwards, S., Cabrit, S. & Strutskie, M.F. (1989): AJ, **97**, 1451

Discussion

Smith: You state that typical seeing at your site is 1.5–2.0″ – for what wavelengths are you quoting this?

Burton: Measured at both V and K (5500 Å, 2.2 μm).

Wagner: How many clear hours do you have at the South Pole per winter and per summer?

Burton: There are 6 months of darkness with about 4 months without any noticeable twilight. We expect that our survey of the LMC requiring about 250 hours of integration time can be completed in about 1 month with allowances for weather and operational matters.

Jauncey: What is the "seeing" like at the pole? I'm puzzled that you need tip-tilt correction to get to 1.4″ diffraction limit at 3.5 microns.

Burton: It's a combination of taking out telescope tracking problems/wind-shake, as well as ensuring that we get the best from the seeing. The image motion will, in general, be at slow frequency and translational motions, in view of the nature of the seeing contribution from the lowest 100m of the inversion layer. For telescope diameters up to 2m tip-tilt will be all that is needed to essentially recover the diffraction limit.

Part 4

THE HIGH-REDSHIFT UNIVERSE

Surveys for High Redshift QSOs

I.M. Hook[1], R.G. McMahon[2], P.A. Shaver[1]

[1] E.S.O., Karl-Schwarzschild-Strasse 2, D–85748 Garching bei München, Germany
[2] Institute of Astronomy, Madingley Road, Cambridge, CB3 OHA, U.K.

Abstract. We present recent results from surveys for radio-loud QSOs. These surveys cover most of the sky to a limiting radio flux density of S = 0.2 − 0.25Jy. In addition we have used the GB6 and VLA FIRST catalogue to survey a smaller area (1600sq deg) to a fainter limit of S = 25mJy. Digitised plate material was used to identify the majority of the radio sources and CCD images to identify the optically fainter sources. The Northern surveys have produced more than 25 $z > 3$ QSOs, one of which, at $z = 4.72$, is the most distant radio source known. In the South redshift information has been obtained for a complete sample of 442 flat-spectrum Parkes QSOs covering the range $z \sim 0$ to $z \sim 4.5$. These QSO samples show a clear drop-off in space density at $z > 3$. The form of this decline is remarkably similar to that seen in optically-selected samples. Since radio emission is unaffected by dust this implies that dust has a minimal effect on the observed drop-off seen in optical samples. Our radio-selected samples show evidence for a decline in space density at high redshift which is more pronounced for less powerful radio sources. We have started a new survey that will be sensitive to bright QSOs with redshifts up to 6.

1 Survey Methods

Whilst radio-loud QSOs are only a small subset of the QSO population, the construction of radio-loud QSO samples is less prone to selection effects than are optical samples, since radio emission is unaffected by absorption due to dust. The rarity of high-redshift QSOs means that surveys to reach $z > 3$ must cover a large area. Our radio samples cover most of the sky to a limit of 0.25Jy or fainter, and are defined at high frequency so they contain a high fraction of flat-spectrum sources, usually identified with QSOs. All the sources have accurate positions ($\sim 1''$) necessary for unambiguous optical identification. In the north we use the $S_{5GHz} \geq 0.2$Jy flat-spectrum sample of Patnaik et al (1992) and the $S_{5GHz} \geq 0.1$Jy MG-VLA sample (Lawrence et al 1986). We have also defined a fainter flat-spectrum sample in a 1600sq deg sub-area, based on $S \geq 25$mJy sources from the 5GHz GB catalogue (Gregory & Condon 1991) and the 1.4GHz VLA-FIRST catalogue (Becker, White & Helfand 1995). The southern survey uses the revised Parkes catalogue (Wright & Otrupcek, 1990), defined at 2.7GHz. The optical identification procedure is described below.

0.1Jy and 0.2Jy Northern Samples: APM scans of POSS-I plates were used to identify flat-spectrum ($\alpha \geq -0.5$, $S \propto \nu^\alpha$) sources to a limit of E($\sim$ R) = 20mag over an area of 12 000 sq deg. Since high-redshift QSOs have red optical colours due to absorption by intervening HI, only red (O − E $\geq$ 1.0), stellar

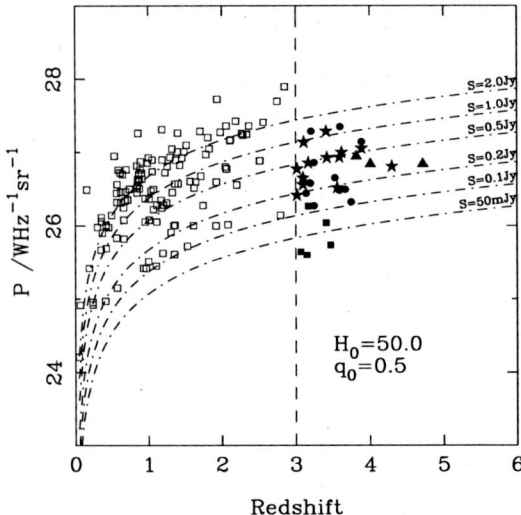

Fig. 1. Radio power vs redshift for $z > 3$ QSOs from the northern samples (solid symbols), and flat-spectrum QSOs with $z < 3$ from Dunlop & Peacock 1990 (open squares). Circles = 0.1Jy MG-VLA sample; stars = APM/POSS identifications of the 0.2Jy sample; triangles = CCD identifications from the 0.2Jy sample; Filled squares = GB/FIRST sample.

identifications were followed up spectroscopically. This resulted in the discovery of 25 $z > 3$ QSOs with $S_{5GHz} \geq 0.1$Jy (Hook 1994, Hook et al 1996, see Figure 1). B and R-band CCD identifications are being obtained for the radio sources with $S \geq 0.2$Jy that remained unidentified ($\sim 16\%$). So far ~ 40 have been observed and an effective area of 2400 sq deg is now 97% identified. Two new $z > 4$ QSOs were found during spectroscopy of the 10 reddest CCD identifications. One, at $z = 4.72$, is the highest redshift radio source known (Hook & McMahon 1998).

GB/FIRST sample: Optical identifications were again made using APM scans of POSS-I plates, this time to a limit of E ≤ 19.5 mag. A complete sample of 73 red (O−E$\geq$1.2), stellar sources was defined (see Hook et al 1998). Spectroscopy of this sample is 70% complete, covering an effective area of 1100sq deg. Six $z > 3$ QSOs have been found, of which two were previously known.

The Parkes QSO Sample Initial identifications of flat-spectrum sources ($\alpha \geq -0.4$) from the Parkes catalogue were made by Shaver et al (1996a) using COSMOS scans of UKST plates to a limit of $B_J \sim 22$mag. CCD identifications were obtained for the optically-fainter sources. Since all the stellar sources have been detected in the B band, there can be no $z > 5$ QSOs in the sample and hence there must be a drop-off in the space density of radio-loud QSOs at $z > 3$ (Shaver et al 1996a). Redshifts for a complete sub-sample of 442 stellar identifications (QSOs) with $S_{2.7GHz} \geq 0.25$Jy have recently been obtained. The most distant QSO found in this sample has $z = 4.46$ (Shaver, Wall & Kellermann 1996).

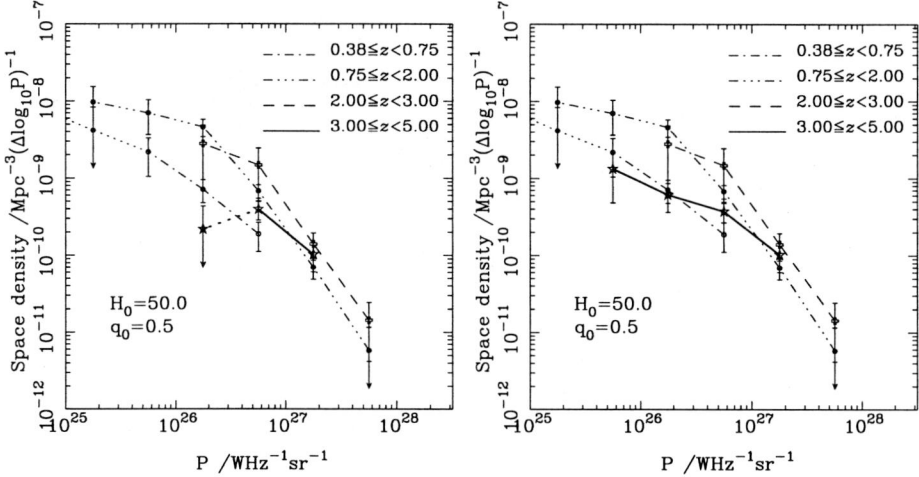

Fig. 2. The differential luminosity function of flat-spectrum QSOs, determined from the data shown in Figure 1. Left: using only sources with S > 0.2Jy at $z > 3$ (CCD identifications have been obtained for sources to this flux limit). Right: using data from all the Northern surveys (the faintest of which reaches of 25mJy). Note that the space density at $z > 3$ is lower than at $z \sim 2$, particularly for weaker sources. Arrows indicate an error bar extending to zero since these bins contain only one point.

2 QSO Evolution & A New Large Survey

The luminosity function for the Northern samples is shown in Figure 2 whereas Figure 3 shows a preliminary analysis of space density as a function of redshift for the Northern and Parkes samples. The space density at $z > 3$ is lower than that at $z \sim 2$, particularly for sources of lower radio luminosity. The decline in space density at $z > 3$ has remarkably similar form to that seen in optically-selected samples of QSOs (e.g. Warren et al 1994). Since the radio sample is unaffected by dust this suggests that dust has a minimal effect on the observed drop-off of optically-selected QSOs. Figure 3 also suggests that the peak in QSO space density occurs at higher redshifts for stronger sources. Thus a survey to reach high redshifts should concentrate on relatively bright objects and cover a large area.

We have started a large survey for $z > 4$ QSOs covering 6000 sq deg and based on sources with S $\geq$ 50mJy from the 5GHz PMN sample. The $S_{1.4\mathrm{GHz}} \geq$ 1mJy NVSS (Condon et al 1994) will provide spectral index information and accurate positions. Optical identifications will be made using APM scans of UKST B, R and I-band plates, and high-redshift QSOs selected based on positional coincidence, red optical colour and stellar optical counterpart. By using I plates this survey is sensitive to bright QSOs to $z \sim 6$. Between 8 and 24 radio-loud QSOs with $z > 4$ should be found (2 – 6 times the number currently known) plus up to 2 with $z > 5$.

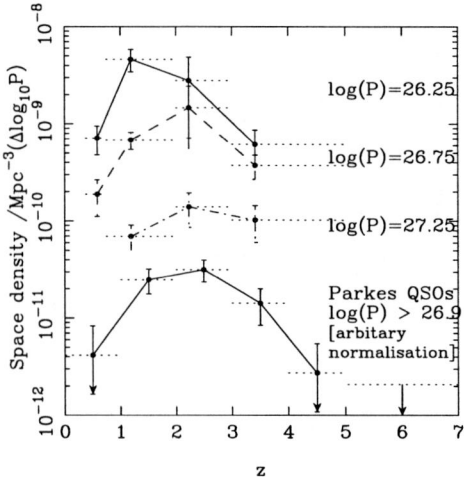

Fig. 3. QSO space density as a function of redshift for various radio luminosities P. The top three curves are from the northern sample combined with data from Dunlop & Peacock (1990) -see Figure 1. The dotted lines show the size of each redshift bin.

3 Conclusions

(1) Radio surveys are efficient for finding high-redshift QSOs. (2) The space density of radio-selected QSOs shows a turnover at $z \gtrsim 3$. Since radio emission is unaffected by dust, this is not an effect of obscuration. (3) The decline in space density seen at high redshift has a similar form to that of optically-selected QSOs, therefore the effect of dust on the observed QSO drop-off is minimal (4) The peak of space density occurs at higher redshift for more luminous sources.

References

Becker, White, Helfand 1995, ApJ, 450, 559
Condon, J. J. et al., 1994 *ADASS III*, ASP, ed. Crabtree et al., 155
Dunlop J.S., Peacock J.A., 1990, MNRAS, 247, 19
Gregory P.C., Condon J.J., 1991, ApJS, 75, 1011
Hook I. M, McMahon, R. G., 1998, MNRAS, 294, L7
Hook I. M., 1994, PhD thesis, Cambridge, U.K.
Hook I.M, McMahon R.G., Irwin, M. J., Hazard, C., 1996, MNRAS, 282, 1274
Hook I. M. et al 1998, MNRAS, submitted.
Lawrence C. R. et al. 1986, ApJS, 61, 105
Patnaik A.R., Browne I.W.A., Wilkinson P.N., Wrobel J.M.,1992, MNRAS, 254, 655
Shaver P.A., Wall J.V., Kellermann K.I., Jackson C.A., Hawkins M.R.S., 1996, Nat, 384, 439
Shaver P.A., Wall J.V., Kellermann K.I., 1996, MNRAS, 278, L11
Warren S.J., Hewett P.C., Osmer P.S., 1994, ApJ, 421, 412
Wright A. E., Otrupcek, R. E., 1990 (Australia Telescope National Facility, CSIRO).

Discussion

van Breugel: Radio quasars have curved radio continuum spectra. How can you be sure that the low and high redshift quasars with flat spectra are of the same populations?

Hook: Shaver et al. have looked into that (see their Nature paper 1996) by looking at the radio spectra of nearby sources and shifting them to high redshift. They came to the conclusion that sources would not appear steeper than the spectral index cut $\alpha > -0.4$ until redshifts well above $z > 10$, those being considered here.

Ekers: How can you be so sure that this flat radio spectrum criteria is not selecting against higher z. We now know that the cores of high z radio galaxies can have steep spectra.

Hook: It is possible that there is an evolutionary effect, i.e. higher redshift cores can have steep spectra, but we are not attempting to address the question of what the flat spectrum sources become at high redshift. We are considering the evolution in space density of flat spectrum objects.

Tsarevsky: The low redshift QSOs show a prominent gap in the radio luminosity histogram between the radio-louds and radio-quiets, the so called radio-intermediate (RI) sub-class. With a few newly found high-redshift QSO, is there an evidence of a similar gap for all of them?

Hook: We are not able to answer that question with our sample since all these new QSOs are strongly radio loud. Radio-intermediates and radio-quiet QSOs at high redshift ($z > 3$) would fall below the radio flux limits of our samples.

Jauncey: I suggest that, to avoid the effects of possible radio spectrum selection effects in your sample, is to use a radio "stellar" selection criterion, in the same way as you are using an optical stellar selection for quasars. The VLA data should be adequate to put a few arcsecond radio size cut-off, rather than use a spectral index cut-off.

Hook: The radio "stellar" size cut-off is very similar to the POSS and UKSTU optical size cut-off.

La Franca: Comment: It is interesting to notice that the redshift evolution of the radio luminosity function is similar to the evolution of the optical luminosity function.

The Evolution of the Clustering of QSOs

F. La Franca[1], P. Andreani[2] and S. Cristiani[2]

[1] Dipartimento di Fisica, Università degli studi "Roma Tre",
Via della Vasca Navale 84, Roma, I-00146
[2] Dipartimento di Astronomia, Università degli studi di Padova,
Vicolo dell'Osservatorio 5, Padova, I-35122

Abstract. The evolution of QSO clustering is investigated with a new sample of 388 QSOs with $0.3 < z \leq 2.2$, $B \leq 20.5$ and $M_B < -23$. Evidence is found for an increase of the clustering amplitude with increasing redshift. These measurements allow us to further distinguish between the various physical scenarios proposed to interpret the QSO phenomenon. A single population model is inconsistent with the observations. The general properties of the QSO population would arise naturally if quasars are short-lived events connected to a characteristic halo mass of $\sim 5 \times 10^{12}$ M$_\odot$.

1 Introduction

The first detections of quasar clustering date back more than one decade (Shaver 1984). Up to now, however, detailed studies of the dependence of clustering on physical parameters like absolute magnitude and redshift was hampered by the small number of quasars in statistically well-defined samples. In recent times, complete samples totalling about 2000 QSOs have been used, resulting in a $4-5\sigma$ detection of the clustering on scales of the order of $6h^{-1}$ comoving Mpc (Andreani & Cristiani 1992, Mo & Fang 1993, Croom & Shanks 1996). The evolution of this clustering is not clear. An amplitude constant in comoving coordinates or marginally decreasing with increasing redshift has been suggested – a value which appears to be consistent or slightly larger than what is observed for present-day galaxies and definitely less than the clustering of clusters.

2 Methods and Results

In an attempt to improve the situation, while waiting for the 2dF QSO redshift survey, we have carried out a survey in the South Galactic Pole (SGP) over a *connected* area of 25 square degrees down to $B_J = 20.5$ (La Franca et al. 1998). Stacked UKSTU plates were used to select UVX candidates and the multi-fiber spectrograph MEFOS at ESO to take spectra of them. The final sample is made up of 388 QSOs with $0.3 < z < 2.2$. The data set was divided into several luminosity, redshift and spatial sub-samples in order to study the autocorrelation function $\xi(r)$ and the integral autocorrelation function $\bar{\xi}(r)$ as a function of the comoving distance, assuming a fixed value of $\gamma = 1.8$. The two point correlation function (TPCF) analysis gives an amplitude $r_o = (6.2 \pm 1.6)\, h^{-1}$ Mpc at an average redshift 1.34. A value of $\bar{\xi}(25) = 0.21 \pm 0.16$ is found, in agreement with

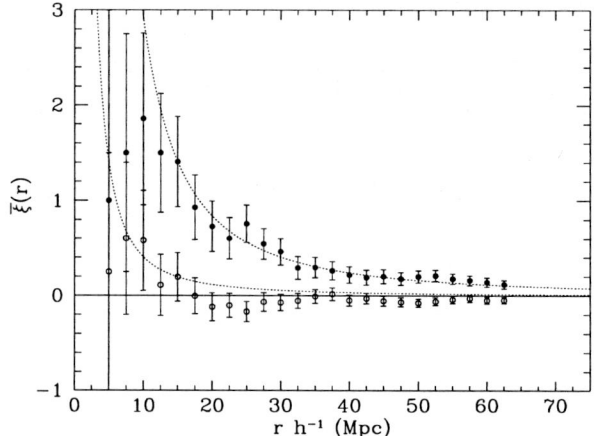

Fig. 1. The integral correlation function $\bar{\xi}(r)$ (defined as $\bar{\xi}(r) = \frac{3}{r^3}\int_0^r x^2 \xi(x)dx$) for the quasars in the SGP sample in two redshift ranges $0.3 < z \leq 1.4$ (open circles) and $1.4 < z \leq 2.2$ (filled circles).

the estimate of Croom and Shanks (1996) of $\bar{\xi}(25) = 0.16 \pm 0.08$. However, when the evolution of the clustering with redshift is analyzed, evidence is found for an *increase* in the clustering with increasing redshift (La Franca, Andreani & Cristiani 1998). The sample was split into the two redshift ranges: $0.3 < z \leq 1.4$ and $1.4 < z \leq 2.2$ (Fig. 1). These were fitted by $\gamma = 1.8$ power laws with r_0 as a free parameter. At low redshift ($z = 0.97$), a value of $r_0 = 4.2\ h^{-1}$ Mpc was found, corresponding to $\bar{\xi}(15) = 0.26 \pm 0.27$, while at high redshift ($z = 1.82$), the measured value was $r_0 = 9.1\ h^{-1}$ Mpc, which corresponds to $\bar{\xi}(15) = 1.03 \pm 0.36$. The effect is small – a 2σ significant discrepancy – but it is interestingly corroborated by other results (at lower and higher redshift) in the literature.

At low redshift, Boyle and Mo (1993) measured the clustering of low-z QSOs in the EMSS, while Georgantopoulos and Shanks (1994) used the IRAS point-source catalog to measure the clustering of Seyferts. Altogether, a low value of the TPCF at 15 Mpc and $z = 0.05$ was obtained: $\bar{\xi} = 0.24 \pm 0.25$. Furthermore, the data of the Palomar Transit Grism Survey (Kundić 1997, Stephens et al. 1997) allow the amplitude of the TPCF to be measured at redshifts higher than 2.7 and the result, $r_o = (18 \pm 8)h^{-1}$ Mpc, suggests that the trend of increasing clustering persists. It may be argued that these surveys tend to select objects with different luminosities and the comparison with the SGP data may not be entirely significant, but an analysis on restricted absolute magnitude slices of the SGP sample shows no correlation of the clustering with the QSO absolute luminosity. If we describe the evolving correlation function in a standard way: $\xi(r,z) = (r/r_0)^{-\gamma}(1+z)^{-(3-\gamma+\epsilon)}$, where ϵ is an arbitrary (and not very physical; see Matarrese et al. 1997) fitting parameter, we obtain $\epsilon = -2.5 \pm 1.0$ (Fig. 2).

In spite of the statistical uncertainties, the measured QSO clustering is able

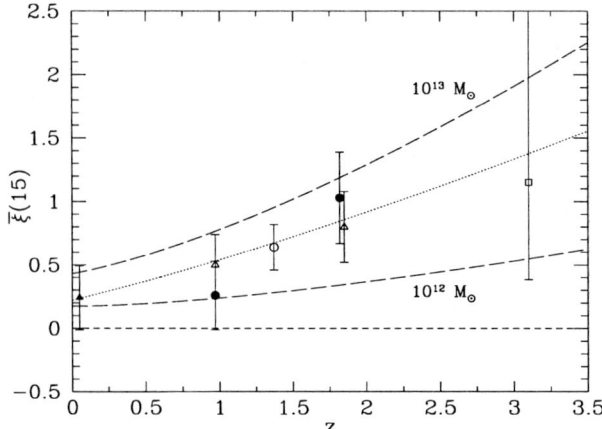

Fig. 2. The amplitude of $\bar{\xi}(15\ h^{-1}$ Mpc) as a function of z. Filled circles: the low- and high-z SGP subsamples; open circle: the SGP sample plus the Boyle et al. (1990), La Franca, Cristiani and Barbieri (1992), and Zitelli et al. (1992) samples; open triangles: same as open circle but divided into two redshift slices; filled triangle: low-z AGNs from Boyle and Mo (1993) and Georgantopoulos and Shanks (1994); open square: the high-z sample from Kundić (1997). The dotted line is the $\epsilon = -2.5$ clustering evolution prediction fitted to the open triangles and the filled triangle data. The dashed lines are the 10^{12} and $10^{13}\ h^{-1}\ M_\odot$ minimum halo mass clustering evolution predictions according to the transient model of Matarrese et al. (1997).

to put interesting constraints on the allowed evolution, being inconsistent with values $\epsilon > 0.0$, such as $\epsilon \simeq 0.8$ observed for faint galaxies at lower redshifts (Le Fèvre et al. 1996, Carlberg et al. 1997, Villumsen et al. 1997). Great care should be exercised, however, when carrying out this comparison. In particular, are the faint lower-redshift galaxies representative of the same population of galaxies for which recent observations by Steidel et al. (1998) show substantial clustering at $z \simeq 3.1$? Are the Lyman-break galaxies progenitors of massive galaxies at the present epoch or precursors of present day cluster galaxies (Governato et al. 1998)?

We already know from energetic arguments that QSOs cannot shine continuously from high redshifts to the present epoch (Cavaliere and Padovani 1989). However, the existing models still do not exclude the possibility that a single population exists, which after having been formed at a certain epoch, has undergone recurrent activity with a sequence of active and quiescent periods. According to Matarrese et al. (1997) and Moscardini et al. (1998) this scenario would correspond to an object-conserving model in which a decrease of the clustering amplitude with redshift is expected. *Thus we can come to the conclusion that the observed increase of the clustering amplitude with redshift is able to rule out a single population model for QSOs.*

If we go back to the model in which quasars are associated with interactions, then we may think in terms of clustering of transient objects, which is definitely different from the case of galaxies which, depending on the physical scenario, can be assimilated into the merging model or the object-conserving paradigm of long-lived objects. In this way the observed clustering is the result of the convolution of the true clustering of the mass with the bias and redshift distribution of the objects. If we think of QSOs as objects sparsely sampling halos with $M > M_{\min}$, we may ask what are the typical masses which are able to reproduce the observed clustering? In this perspective, an increase of the QSO clustering is expected because they are sampling rarer and rarer overdensities with increasing redshift. As we can see from Fig. 2, a value of $M_{\min} = 10^{12} - 10^{13}$ $M_\odot$ would provide the desired amount of clustering and evolution. Similar theoretical results have also been obtained by Bagla (1997).

References

Andreani, P., Cristiani, S. 1992, ApJ, 398, L13
Bagla, J.S., 1997, MNRAS, submitted, astro-ph/9711081
Boyle, B. J., Fong, R., Shanks, T., Peterson, B.A. 1990, MNRAS, 243, 1
Boyle, B.J., Mo, H.J. 1993, MNRAS, 260, 925
Carlberg, R.G., Cowie, L.L., Songaila, A., Hu, E.M. 1997, ApJ, 484, 538
Cavaliere, A., Padovani, P., 1989, ApJ, 340, L5
Croom, S.M., Shanks, T. 1996, MNRAS, 281, 893
Georgantopoulos, I., Shanks, T. 1994, MNRAS, 271, 773
Governato, F., Baugh, C.M., Frenk, C.S., Cole, S., Lacey, C.G., Quinn, T., Stadel, J. 1998, Nature in press
Kundić, T., 1997, ApJ, 482, 631
La Franca, F., Cristiani S., Barbieri, C. 1992, AJ, 103, 1062
La Franca, F., Andreani, P., Cristiani, S. 1998, ApJ, in press, astro-ph/9711048
La Franca, F., Lissandrini, C., Cristiani, S., Miller, L., Hawkins, M.R.S., McGillivray, H.T., 1998, in prep
Le Fèvre, O., Hudon, D., Lilly, S.J., Crampton, D., Hammer, F., and Tresse, L. 1996, ApJ, 461, 534
Matarrese, S., Coles, P., Lucchin, F., and Moscardini, L. 1997, MNRAS, 286, 115
Moscardini, L., Coles, P., Lucchin, F., Matarrese, S., 1998, MNRAS submitted, astro-ph/9712184
Mo, H.J., Fang, L.Z. 1993, ApJ, 410, 493
Shaver, P.A. 1984, A&A, 136, L9
Steidel, C.C., Adelberger, K.L., Dickinson, M., Giavalisco, M., Pettini, M., Kellogg, M 1998 ApJ, 492, 428.
Stephens, A.W., Schneider, D.P., Schmidt, M., Gunn, J.E., Weinberg, D.H., 1997, AJ, 114, 41
Villumsen, J.V., Freudling, W., da Costa, L.N 1997, ApJ, 481, 578
Zitelli, V., Mignoli, M., Zamorani, G., Marano, B., Boyle, B.J. 1992, MNRAS, 256, 349

Discussion

Boyle: The increase in QSO correlation length with redshift, coupled with the observed decrease in the galaxy correlation length with redshift, implies that, even at moderate redshifts ($z \sim 0.5 - 1.0$), QSOs should be strongly biased with respect to galaxies, and therefore be found in rich environments. Have you calculated what your results predict for the environments of intermediate redshift QSOs, and are current observations consistent with this prediction?

La Franca: No, we have no predictions on this effect. Current observations do not have enough data.

Local Population and Evolution of Optically Bright QSOs

Lutz Wisotzki

Hamburger Sternwarte, Gojenbergsweg 112, D-21029 Hamburg, Germany
email: lwisotzki@hs.uni-hamburg.de

Abstract. I present initial results of analysing a new sample of bright quasars, selected from the Hamburg/ESO survey. The sample provides substantial improvement in two regions of the Hubble diagram: The local domain where host galaxy contributions are relevant, and the bright end of the quasar luminosity function for redshifts $z \lesssim 3$.

1 A New Large Sample of Bright QSOs

The wide range of quasar luminosities, altogether spanning over four decades from low-activity Seyfert galaxies up to the most luminous objects in the universe found at high z, together with the steep number-flux relation, restricts individual surveys to cover only small portions of the quasar luminosity function (QLF). While at intermediate flux levels there are now many surveys available, large uncertainties still exist at the bright end of the QLF. In the optical waveband, the Palomar-Green Bright Quasar Sample (BQS; Schmidt & Green 1983) has been the only significant contributor now for 15 years.

In this paper I present a first analysis of a new quasar sample drawn from the Hamburg/ESO survey (HES). The survey design, scientific aims and some initial results were described by Wisotzki et al. (1996), and an update was given recently by Reimers & Wisotzki (1997). Basically, quasar candidates are selected by automated techniques from digitised objective-prism plates taken with the ESO Schmidt telescope. The magnitude limit is typically $B_J \lesssim 17.5$, while there is essentially no bright limit. A well-defined flux-limited sample has been constructed from complete follow-up spectroscopy of 207 ESO Schmidt fields, corresponding to an effective survey area of 3700 deg². The distribution of redshifts and inferred absolute blue magnitudes (for $H_0 = 50\,\mathrm{km\,s^{-1}\,Mpc^{-1}}$ and $q_0 = 0.5$) of the 415 QSOs in the sample is shown in Fig. 1. The sample contains many of the most luminous quasars known and is well suited to studying the evolutionary properties at the bright end of the QLF over a wide range of redshifts. It is also the first quasar sample large enough to effectively *replace* the BQS, with much higher completeness, photometric accuracy, and less affected by redshift-dependent selection biases. As the BQS has influenced most previous studies of quasar evolution, notably the work of Boyle et al. (1991) and their model of 'pure luminosity evolution' (PLE), the new HES sample will provide a powerful testing ground for such models.

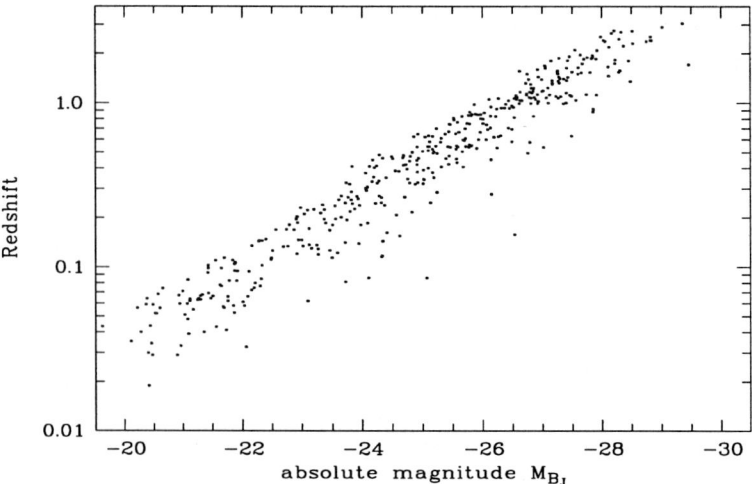

Fig. 1. Distribution of the HES sample QSOs over absolute magnitudes and redshifts.

2 The Local QSO Population

The local universe is the only domain where the full range of luminosities is technically accessible to a single flux-limited survey. The HES is the first optical quasar survey capable of fully exploiting this option, being specifically designed to reduce selection biases due to the presence of extended galaxy envelopes. This has already yielded a first estimate of the local QLF (Köhler et al. 1997). The new sample expands the number of 'local QSOs' by almost an order of magnitude, allowing a much more accurate assessment of QLF shape and slope. Note also that the method used to measure the QSO magnitudes depends less on host galaxy properties than conventional isophotal or total estimates.

The constructed local luminosity function of quasars is shown as the top-left panel of Fig. 2. The most important result is that the local QLF does not display evidence for a break, or significant change of slope, anywhere above magnitudes brighter than $M_{B_J} \simeq -20$, in strong contrast to the PLE prediction. At luminosities fainter than that, there may be a break, but this is statistically insignificant and could also be caused by an onset of incompleteness for very low-luminosity Seyferts with bright host galaxies. The absence of a break-like feature leads to an intersection of the observed local QLF with the PLE prediction around $M_{B_J} \simeq -24$; at higher luminosities, the actual space densities are measured to be much higher, while for $M_{B_J} > -24$, thus for the entire Seyfert galaxy domain, PLE overpredicts the numbers. It is important to realise, however, that because of selection biases against low-redshift quasars common to most other quasar surveys, analyses of the QLF including that of Boyle et al. (1991) have been truncated at low z, typically $z > 0.3$. The construction of a local QLF from such a model, although formally trivial, therefore means an extrapolation outside the validated range.

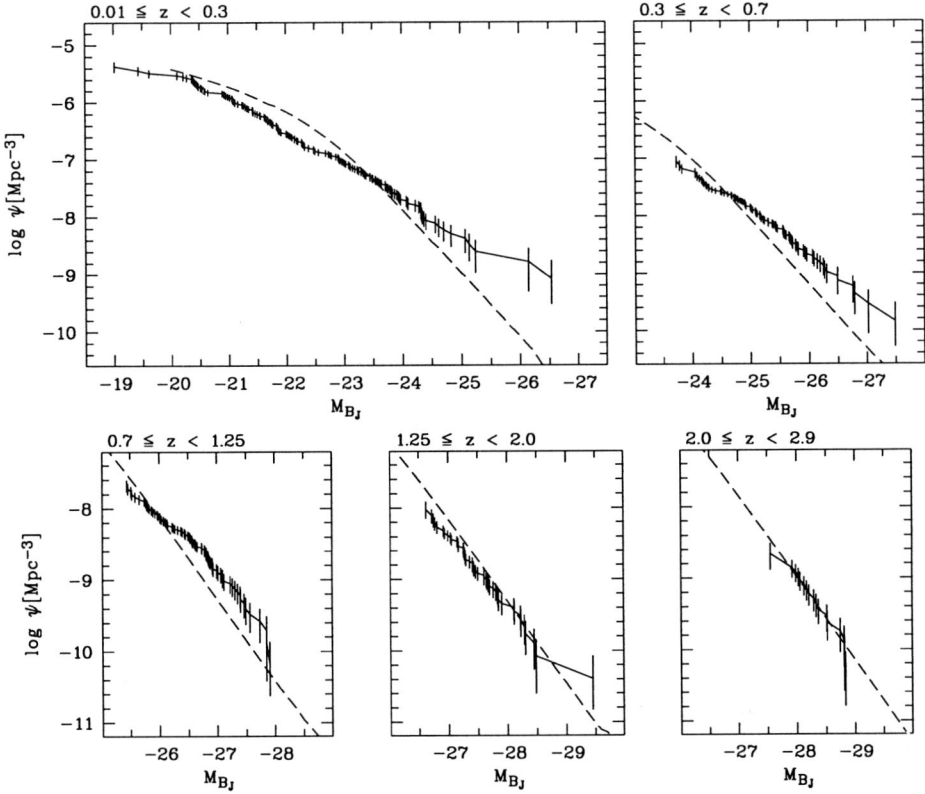

Fig. 2. The cumulative quasar luminosity function determined from the HES sample for five adjacent redshift shells. Each panel shows the result for one redshift domain, including error bars estimated from Poisson statistics. The dashed lines show the predicted relations, integrated over each shell, from the Boyle et al. (1991) PLE model.

3 Evolution of the Most Luminous QSOs

At all redshifts $z > 0.3$, the HES samples the bright end of the quasar luminosity function which is only poorly constrained to date. Figure 2 documents the HES contributions in the form of direct nonparametric estimates of QLF segments in disjoint redshift shells. The main trend is that of a bright-end slope getting considerably steeper from low to high redshifts. This produces a strong excess of luminous low-z QSOs relative to the PLE prediction which disappears only for $z \gtrsim 1$. Similar trends have been detected already by other surveys (Hewett et al. 1993; La Franca & Cristiani 1997), but the large survey area of the HES allows this now to be firmly established on a secure statistical basis. Using a Kolmogorov-Smirnov test, the standard global PLE evolution model is rejected with a confidence level of $P < 0.001\%$.

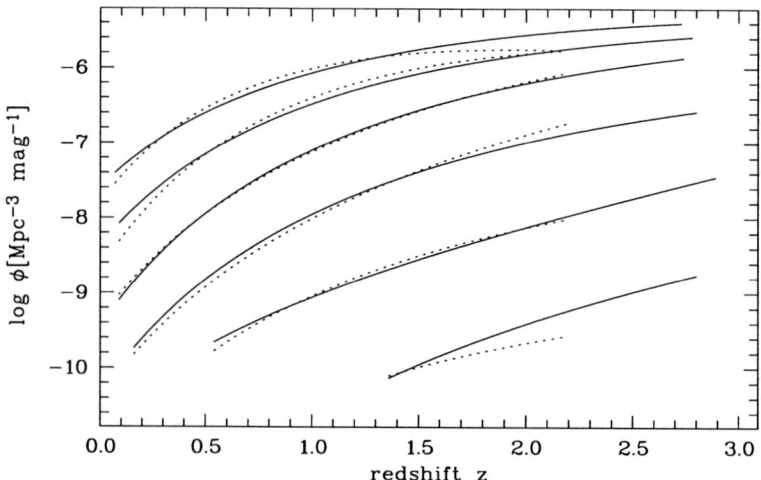

Fig. 3. Density evolution tracks for six different luminosity regimes, stepped by $\Delta M = 1$ mag from $M_{B_J} = -23.5$ (top) to $M_{B_J} = -28.5$ (bottom), computed for the composite sample described in the text (dashed line: including, solid line: excluding AAT sample).

A more detailed assessment of the evolutionary properties can be made by fitting analytic expressions to the observed QSO distribution. The set of 415 HES sources was combined with four other published samples (LBQS, AAT, $(ZM)^2B$, CFRS) to produce a merged sample of almost 2000 QSOs. No currently published evolutionary description, and in fact no global 'pure' evolution picture (neither density nor luminosity evolution), can fit the data acceptably – quasar evolution rates are therefore necessarily luminosity-dependent. To illustrate this, I have determined parametric 'density evolution tracks' for various absolute magnitudes by fitting high-order polynomials to the differential space density as a function of z (see Fig. 3). At low z, the evolution is stronger for the less luminous objects, while at high z the evolution rates are larger for the more luminous sources. There no evidence for a halt or slow-down around $z \simeq 2$ except for the low-luminosity bin; to the contrary, it appears that the most luminous QSOs show the strongest positive evolution up to almost $z = 3$.

References

Boyle B.J., et al., 1991, ASP Conf. Ser. 21, 191
La Franca F., Cristiani S., 1997, AJ 113, 1517
Hewett P.C., Foltz C.B., Chaffee F.H., 1993, ApJ 406, L43
Köhler T., Groote D., Reimers D., Wisotzki L., 1997, A&A 325, 502
Reimers D., Wisotzki L., 1997, The Messenger 88, 14
Schmidt M., Green R.F., 1983, ApJ 269, 352
Wisotzki L., Köhler T., Groote D., Reimers D., 1996, A&AS 115, 227

Discussion

La Franca: I wish just to comment that the bright low redshift ($0.3 < z < 0.6$) QSO excess was already discovered by the HBQS and ROE bright QSO surveys. What I think is important in your survey is that it is the only one able to probe the QSO if at $z < 0.3$.

Wisotzki: I agree. However, both the ROE survey and the HBQS fields are included in our survey, and the HES sample considerably improves the statistics at the bright end of the QSO luminosity function at *all* redshifts $z < 3-4$.

Boyle: Is there any difference between the aperture magnitudes that you have computed for the low luminosity QSOs in your sample and their 'true' nuclear magnitudes; i.e. with the underlying galaxy magnitude subtracted?

Wisotzki: We have done a proper host galaxy subtraction for a subsample of the low-luminosity QSOs. A significant difference is measurable only at the very low-luminosity end, for $M_{B_J} \gtrsim -20$.

Gas and Dust at High Redshift

Richard W. Hunstead

School of Physics, University of Sydney, NSW 2006, Australia

Abstract. Much of our knowledge of the metal enrichment of galaxies at high redshift has come from studying the damped Lyman α absorption systems seen against background QSOs. By combining these results with estimates of the global star formation rate and the rate of consumption of neutral gas, we are moving closer to understanding galaxy formation and evolution.

1 Introduction

The detection and study of high-redshift galaxies has been a major goal of observational cosmology over the past decade. Until recently, searches for the high-z counterparts of galaxies like our own have proved elusive. In the last two years, however, there have been major advances in our ability to detect normal galaxies at $z \sim 3$ and study their stellar populations and star-formation rates (Steidel et al. 1996, Madau et al. 1996, Pettini et al. 1998). To a large extent these studies, and those that have followed, are built on work over the past two decades directed at understanding in detail the physical and chemical properties of the absorbers seen against bright background QSOs.

The rare damped Lyman α (DLA) absorption systems assume special significance in tracing the build-up of metal abundances as a function of redshift. With neutral hydrogen column densities $N(\text{H I}) > 2 \times 10^{20}$ cm^{-2}, the DLA systems encompass an integrated mass of H I at redshifts $z \sim 2\text{--}3$ which is directly comparable with that seen in stars at $z \sim 0$. For this reason, they are believed to be the progenitors of present-day luminous galaxies – or at least the neutral cores of such galaxies – observed at a time prior to the bulk of star formation, when the gas fraction was high. While the circumstantial evidence favours this interpretation, the detailed connection between DLA absorbers and normal galaxies remains a priority research area for the future.

In charting the formation and evolution of galaxies as a function of cosmic time we now have three complementary approaches at our disposal. We can monitor (i) the fuel supply (neutral hydrogen), (ii) the power output (star formation rate), or (iii) the exhaust products (gas-phase metals). Each method, with its strengths and weaknesses, is examined in turn and the review concludes with a look to the future.

2 The Fuel Supply

The fuel for star formation is principally hydrogen and at high redshifts most of the hydrogen is locked up in DLA systems. Observations of these systems

(Lanzetta et al. 1995, Storrie-Lombardi et al. 1996) over the redshift interval $z > 2$ imply a low rate of neutral gas consumption. At lower redshifts $z < 1.5$ the DLA statistics are poor, due to a combination of reduced pathlengths, intrinsic evolution (i.e., gas consumption in star formation) and possible dust bias (Pei and Fall 1995). This means that the DLAs do not provide very tight constraints on the available H I in this important redshift range.

An alternative approach to estimating the mass density of neutral hydrogen at lower redshifts has been used by Natarajan and Pettini (1997). It is based on the local luminosity function of galaxies and uses the relationship between B-band luminosity and H I mass developed by Rao and Briggs (1993) for nearby galaxies. By assuming that the basic relationship does not change, Natarajan and Pettini extended the calculation of $\Omega_{\rm gas}$ from $z = 0$ to 1, using the results of the Canada-France redshift survey (Lilly et al. 1995, 1996), the ESO Slice Project (Zucca et al. 1997) and the Las Campanas Redshift Survey (Shectman et al. 1996). They find that $\Omega_{\rm gas}$ is indeed higher at $z \simeq 0.5$–1 than at present, and lower than the DLA estimates at $z \simeq 2$–3, in each case by factors $\simeq 2$–3. Moreover, both the closed-box and infall models of Pei and Fall (1995), which take account of the biasing effects of dust, fit the data well.

3 The Power Output

Until recently, the search for primeval galaxies – galaxies in the process of assembling a significant fraction of their mass and forming their first generations of stars – has been frustratingly unproductive, relying mostly on serendipitous discoveries. Over the past two years the situation has improved dramatically, following the realisation that colour selection targeted at the Lyman discontinuity at 912 Å was a particularly effective discriminant (Steidel et al. 1995). Objects which match the widely-held picture of a primeval galaxy are now being discovered routinely and in large numbers (Steidel et al. 1996, Pettini et al. 1998).

The spectra of these high redshift ($z \sim 3$) galaxies show many similarities to those of nearby starburst galaxies or H II regions. Moreover, the interstellar absorption lines, seen against the redshifted ultraviolet continuum, bear a striking resemblance to those seen in DLA systems, both in ionisation mix and strengths, reinforcing the connection between normal galaxies and DLA absorbers. On the other hand, the inferred 1500Å luminosities at $z = 3$ are 1–2 orders of magnitude greater than the most luminous nearby starburst galaxies.

Pettini et al. (1998) claim that the integrated UV continuum light from O and B stars can, in principle, give a more direct estimate of star-formation rate (SFR) at $z \sim 3$ than the Balmer recombination lines. The estimates, which depend on the particular star-formation model, the initial mass function and the age of the starburst, are typically SFR $\simeq 4h^{-2}$ M$_\odot$ yr^{-1}, where $h = H_0/100$. The similarity of the SFR per unit area between the high-z galaxies and local starbursts (~ 13 M$_\odot$ yr^{-1} kpc^{-2}) points to the same limiting mechanisms (Meurer et al. 1997) operating at these two very different epochs.

On the other hand, the spectral slopes in the restframe UV are substantially reddened, suggesting that interstellar dust is already present in these young galaxies. Estimates of the amount of dust present depend on the reddening law and the parameters of the starburst model. Pettini et al. (1998) conclude that the likely dust correction to L_{1500} is a factor ~ 3, with a possible range ~ 2–6; measurements of redshifted Hβ emission confirm these relatively low values of UV extinction. Their observations also show that the weakness of Lyα emission in these high-z galaxies can be explained by resonant scattering in an outflowing ISM, and this probably accounts for the lack of success of previous searches based on this emission line.

4 The Exhaust Products

Since 1990, we (M. Pettini, L. Smith, D. King and RWH) have been carrying out a survey of metallicity and dust in DLA absorbers seen against high-z QSOs, using as diagnostics the weak multiplets of Zn II and Cr II (Pettini et al. 1990, 1994, 1997a, 1997b). While [Zn/H][1] is a straightforward measure of the degree of metal enrichment of the absorbing interstellar medium (ISM), analogous to [Fe/H] in stars, [Cr/Zn] measures the extent of element depletion onto grains, and therefore gives an indication of the ratio of dust to metals. In carrying out comparisons between DLA and Milky Way sightlines, it is important to acknowledge that in young galaxies there may be intrinsic departures from solar relative abundances and Galactic dust composition. Nevertheless, it has proven extremely fruitful to compare the DLA abundances with those measured in Galactic stars and the local ISM.

The first results of the survey were reported in Pettini et al. (1994), with an extension to higher redshifts and a general reappraisal of the whole dataset in Pettini et al. (1997b). Two examples of the Zn II and Cr II regions, at redshifts near the limits for ground-based abundance studies, are shown in Fig. 1. It is worth noting that in cases where DLAs from the Pettini et al. sample have been reobserved at higher signal-to-noise and spectral resolution with HIRES on the Keck 10-m telescope (e.g., Wolfe 1995, Prochaska & Wolfe 1997), [Zn/H] has been found to be in good agreement with the earlier values based on 4-m telescope data.

4.1 Metal Abundances

The full dataset in Pettini et al. (1997b) consists of measurements (or upper limits) of [Zn/H] in 34 DLAs over the redshift range $z_{\rm abs} = 0.69$–3.39. Figure 2 shows the abundance of Zn as a function of redshift, using the most recent f-values of the Zn II doublet (Bergeson & Lawler 1993) and the meteoritic solar abundance of Zn (Anders & Grevesse 1989). The main conclusions to be drawn from Fig. 2 are that DLA systems are metal poor at *all* redshifts probed and

---

[1] The conventional notation is used, where [Zn/H] = log(Zn/H) − log(Zn/H)$_\odot$

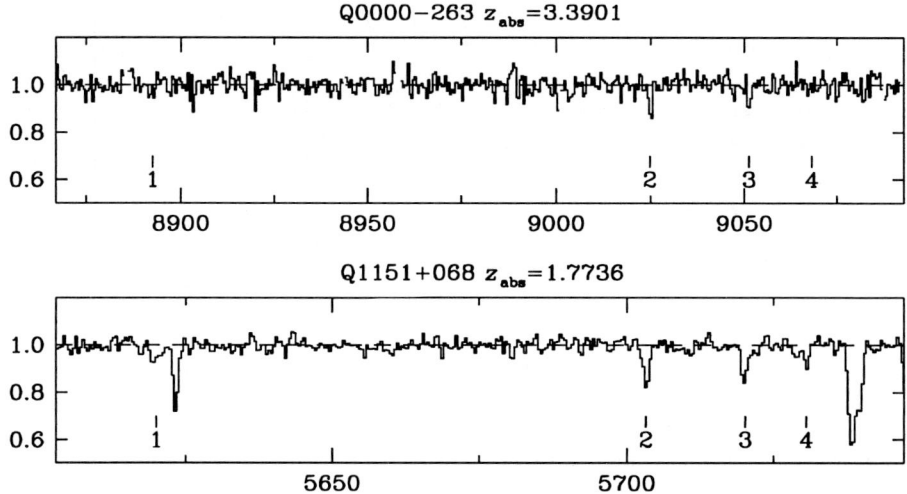

Fig. 1. Portions of two QSO spectra from the study of Pettini et al. (1997b) in the region of the Zn II $\lambda\lambda 2025.48, 2062.01$ and the Cr II $\lambda\lambda 2055.60, 2061.58, 2065.50$ multiplets. Residual intensity (expanded scale) is shown as a function of wavelength in Å. The vertical tick marks indicate the expected positions of the absorption lines of Zn II (1,3) and Cr II (2,3,4).

presumably arise in galaxies at an early stage of chemical evolution. Moreover, there is a large range in metallicity among galaxies at the same redshift, pointing to a protracted epoch of galaxy formation and/or different rates of enrichment in different DLA galaxies. The mean column density-weighted abundance for the full sample, $[\langle Zn/H \rangle]$, is -1.13 ± 0.38, or 1/13 solar. For redshifts between 1.5 and 3 there is little change from this mean value. However, for redshifts above 3 the Pettini et al. upper limit of -1.39 provides tentative evidence that rapid element build-up began at this epoch, and is supported by the data of Lu et al. (1996) who found $[Fe/H] \leq -2$ in three DLAs at $z > 3$. Moreover, the increase in Z_{DLA} over the time interval from $z \sim 4$–2 is broadly consistent with the metal ejection models developed independently by Madau et al. (1996) for the Hubble Deep Field galaxies.

The low DLA abundances at $z_{abs} < 1.5$ are more difficult to explain. While the number of systems (4, see Fig. 2) is too small for definitive conclusions, it is possible that selection effects favouring low-metallicity/low-dust galaxies begin to operate in this redshift range. The identification of a much larger sample of intermediate redshift DLAs remains an urgent priority.

4.2 Dust

The dust-to-gas ratio in DLA systems can be obtained from the relative gas-phase abundances of Zn and Cr because of their very different dust depletion

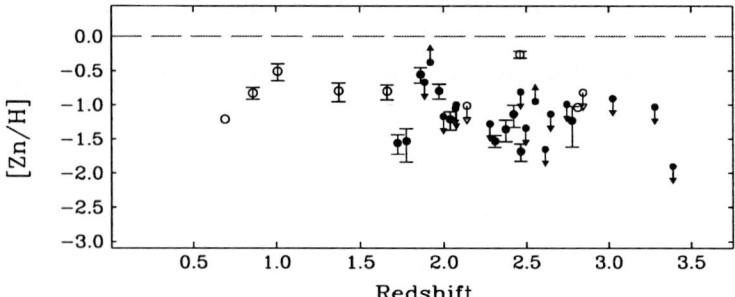

Fig. 2. Zn abundance versus redshift for the 34 DLAs in the Pettini et al. (1997b) survey. Open symbols indicate a measurement from the literature. Upper limits are indicated by downward pointing arrows, and lower limits are marked in two cases where the Zn II lines may be saturated.

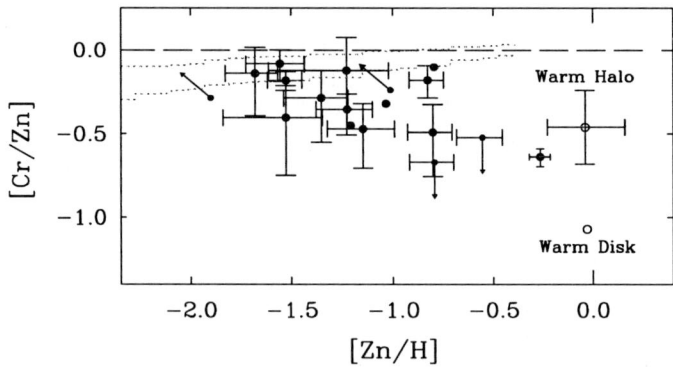

Fig. 3. Abundance of Cr relative to Zn in 18 DLA systems (filled symbols) plotted as a function of metallicity, [Zn/H]. The corresponding values in warm interstellar clouds in the disk and halo (open circles) are from Savage and Sembach (1996). The dotted lines indicate the upper and lower quartiles of [Cr/Fe] for a large sample of Galactic stars (Ryan et al. 1996) in this metallicity regime (from Pettini et al. 1997a).

properties in the local ISM. While Zn is at most only lightly depleted in warm interstellar clouds (Savage and Sembach 1996), Cr is amongst the most depleted elements, with 90–99% in solid form. On the other hand, in stars both [Zn/Fe] and [Cr/Fe] are close to zero over a wide range of metallicities, from [Fe/H] $\simeq 0$ to -2. Therefore, based on what is known of the relative abundances of Zn and Cr in stars and the ISM of our Galaxy, the [Cr/Zn] ratio can be used to infer the presence of dust in DLA galaxies.

Figure 3 shows [Cr/Zn] plotted against metallicity [Zn/H] for 18 DLA systems, together with corresponding data for Galactic stars and for warm interstellar clouds in the Galactic disk and halo. There is a spread in [Cr/Zn] in DLA systems ranging from values consistent with stars to values comparable with local halo clouds. The mean $\langle[\text{Cr}/\text{Zn}]\rangle$ of $-0.3^{+0.15}_{-0.2}$ is interpreted as indicating that about half the Cr is locked up in grains or, putting it another way, that the dust-to-metals ratio in the DLAs is approximately half that in the Galaxy. The lower overall level of heavy element depletion in DLAs can be understood in terms of the low abundances of the grain constituents themselves, more effective grain destruction at the higher temperatures accompanying the lower metallicities, and more frequent grain processing as a result of higher supernova rates.

Combining the mean metallicity (-1.13) and mean dust-to-metals ratio (-0.3), Pettini et al. (1997a,b) conclude that the 'typical' dust-to-gas ratio in DLA systems at $z \sim 2$–3 is $\approx 1/25$ of that found in the Galactic ISM. Although this ratio is lower than previously reported (due to revised f-values) it is still consistent with the estimates of Pei and Fall (1995) based on the reddening of QSO continua. However, given the spread in [Cr/Zn] in Fig. 3 and the scatter in [Zn/H] in Fig. 2 the concept of a 'typical' dust-to-gas ratio may be of limited use.

5 Summary and Future Work

It is encouraging that, to a first approximation, three different approaches to exploring the epoch of galaxy formation – the consumption of neutral gas, the UV luminosity of high-redshift galaxies, and the metal abundance of DLA absorbers – are giving a broadly consistent picture of galaxy evolution at large lookback times. The wide spread in DLA metal abundances points to a protracted epoch of galaxy formation and, most likely, to a mix of different morphological types. Near infrared observations – with 8–10 m telescopes – of the familiar optical emission lines from star-forming regions in the absorbing galaxies will help to link kinematics and abundances and, at the same time, clarify whether thick disk rotation is a viable explanation for the absorption-line profiles, as claimed by Wolfe & Prochaska (1998).

At redshifts $z \lesssim 1.5$ Z_{DLA} does not increase as expected from simple models of cosmic chemical evolution. While explanations have been sought in terms of dust and other selection effects, the real problem is the paucity of DLA systems at low to intermediate redshifts. The detection of DLAs at $z < 1$ is proceeding slowly, and STIS on the *HST* will allow measurements of [Zn/H] below the ground-based limit of $z_{\text{abs}} \simeq 0.65$.

With the 2dF QSO survey now underway (Boyle et al. 1998), the number of DLAs is expected to increase by an order of magnitude. Being substantially fainter – and potentially more reddened – than existing QSOs, follow-up observations of Zn and Cr on 8–10 m telescopes will lead to a better assessment of the dust bias in existing DLA samples.

It is a pleasure to acknowledge the input of my collaborators, Linda Smith, David King, and especially Max Pettini whose insight and persistence ensured the success of the DLA project.

References

Anders, E., Grevesse, N. (1989), Geochim. Cosmoschim. Acta, 53, 197.
Bergeson, S.D., Lawler, J.E. (1993), ApJ, 408, 382.
Boyle, B.J., Smith, R.J., Shanks, T., Croom, S.M., Miller, L. (1998), IAU Symp. 183, in press (astro-ph/9710202)
Lanzetta, K.M., Wolfe, A.M., Turnshek, D.A. (1995), ApJ, 440, 435.
Lilly, S.J., Tresse, L., Hammer, F., Crampton, D., Le Fevre, O. (1995), ApJ, 455, 108.
Lilly, S.J., Le Fevre, O., Hammer, F., Crampton, D. (1996), ApJ, 460, L1.
Lu, L., Sargent, W.L.W., Barlow, T.A., Churchill, C.W., Vogt, S.S. (1996), ApJS, 107, 475.
Madau, P., Ferguson, H.C., Dickinson, M., Giavalisco, M., Steidel, C.C., Fruchter, A. (1996), MNRAS, 238, 1388.
Meurer, G.R., Heckman, T.M., Lehnert, M.D., Leitherer, C., Lowenthal, J. (1997), AJ, 114, 54.
Natarajan, P., Pettini, M. (1997), MNRAS, 291, L28.
Pei, Y.C., Fall, S.M. (1995), ApJ, 454, 69.
Pettini. M., Boksenberg, A., Hunstead, R.W. (1990), ApJ, 348, 48.
Pettini, M,. Smith, L.J., Hunstead, R.W., King, D.L. (1994), ApJ, 426, 79.
Pettini, M., King, D.L., Smith, L.J., Hunstead, R.W. (1997a), ApJ, 478, 536.
Pettini, M., Smith, L.J., King, D.L., Hunstead, R.W. (1997), ApJ, 486, 665.
Pettini, M., Steidel, C.C., Adelberger, K.L., Kellogg, M., Dickinson, M., Giavalisco, M. (1998), in *Origins*, eds. C.E. Woodward, J.M. Shull & H.A. Thronson, Jr, ASP Conf Ser, 148, in press.
Prochaska, J.X., Wolfe, A.M. (1997), ApJ, 474, 140.
Rao, S., Briggs, F. (1993), ApJ, 419, 515.
Ryan, S.G., Norris, J.E., Beers, T.C. (1996), ApJ, 471, 254.
Savage, B.D., Sembach, K.R. (1996), ARA&A, 34, 279.
Shectman, S.A., Landy, S.D., Oemler, A., Tucker, D.L., Lin, H., Kirschner, R.P., Schechter, P.L. (1996), ApJ, 470, 172.
Steidel, C.C., Pettini, M., Hamilton, D. (1995), AJ, 110, 2519.
Steidel, C.C., Giavalisco, M., Pettini, M., Dickinson, M., Adelberger, K.L. (1996), ApJ, 462, L17.
Storrie-Lombardi, L., McMahon, R.G., Irwin, M.J. (1996), MNRAS, 283, L79.
Wolfe, A.M. (1995), in *QSO Absorption Lines*, ed. G. Meylan, Berlin:Springer, 13.
Wolfe, A.M., Prochaska, J.X. (1998), ApJ, 494, L15.
Zucca, E. et al. (1997), A&A, 326, 477.

Discussion

Illingworth: What reddening [$E(B-V)$] does your dust/gas ratio infer? Even a small $E(B-V)$ can be important for high-z galaxies observed in the UV.

Hunstead: At the typical DLA dust/gas ratios we expect $E(B-V)$ to be around 0.01–0.02. Note that the UV spectra of the Lyman break galaxies are *more* reddened than this, probably because the abundances and H I column densities are greater.

Rocca-Volmerange: What about the nucleosynthesis of Cr and Zn? If these elements are produced by different stellar masses, the variation of the Cr/Zn ratio would indicate an IMF variation and not a dust variation.

Hunstead: This is unlikely because Zn and Cr have a similar nucleosynthetic origin. In the Galaxy, [Zn/Cr] = 0 (i.e., solar relative abundances) in stars with [Fe/H] ranging from 0 to -2.5.

Evidence of Structure in the Lyman-α Forest

Jochen Liske and John K. Webb

Department of Astrophysics & Optics, University of New South Wales, Sydney 2052, Australia

Abstract. The spatial distribution of Ly-α forest absorption systems toward a group of eight closely spaced QSOs has been analysed and evidence for large scale structure has been found at $\langle z \rangle = 2.8$. Our technique is based on the first and second moments of the transmission probability density function which is capable of identifying and assessing the significance of regions of over- or underdense Ly-α absorption. The data has revealed at least two interesting features. 1. An overdense structure at $z = 2.27$ which extends at least over $\sim 8\ h^{-1}$ comoving Mpc ($q_0 = 0.5$) in the plane of the sky. Metal absorption lines have been found at the same redshift and thus a cluster or proto-cluster of galaxies seems to have been discovered. 2. A void at $z = 2.97$, extending over $\sim 20\ h^{-1}$ comoving Mpc in the plane of the sky, possibly caused by a locally increased UV ionising flux due to a foreground QSO.

1 Introduction

Recent work (Lanzetta et al. 1995) has shown that at least some fraction of the Ly-α absorption lines seen in the spectra of low redshift QSOs arises in the extended haloes of galaxies. At high redshift, Ly-α absorbers are found to be strongly clustered (Fernández-Soto et al. 1996) and many contain ionised carbon (Cowie et al. 1995). This suggests that some fraction of high redshift Ly-α absorbers may also be identified with the haloes of galaxies. Thus it seems likely that the Ly-α forest may exhibit large scale structure.

However, it has been shown (Fernández-Soto et al. 1996) that significant clustering may be missed when using the classical tool of cluster analysis, the two-point correlation function (Peebles 1993, Sargent et al. 1980). Here, we present a new technique to search for non-randomness in the spatial distribution of the Ly-α forest based on the first and second moments of the transmission probability density function. This method is able to identify the strength, position and scale of individual structures since it retains spatial information. It is fairly insensitive to noise and resolution characteristics and is easy to apply in practice. The new technique has been tested with the help of synthetic spectra and it was found to be substantially more sensitive than a two-point correlation function analysis.

2 Results

The method was applied to the spectra of a close group of eight QSOs with a mean redshift of 2.97. The data (Williger et al. 1996) was kindly made available to us by Gerry Williger.

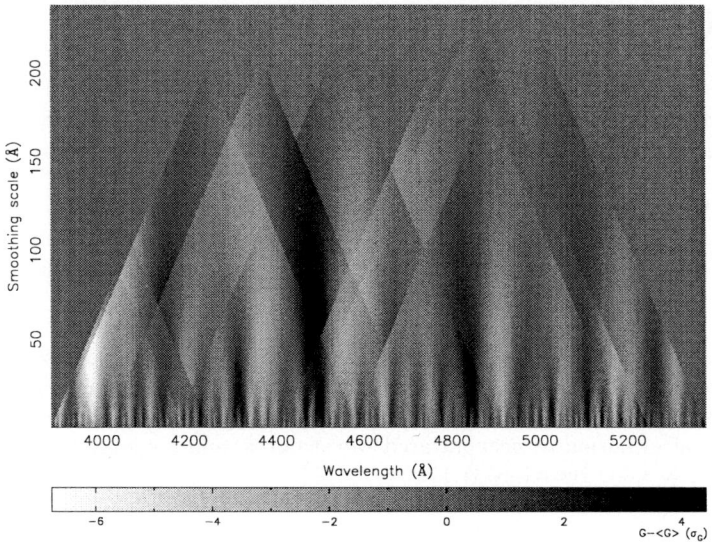

Fig. 1. Transmission in the spectra of a group of eight closely spaced QSOs as a function of wavelength and smoothing scale. The transmission is measured relative to a theoretical expectation value and in units of the theoretical standard deviation by convolving the spectra with a Gaussian of varying size. Each "triangle" corresponds to one spectrum, where the base is the original spectrum itself and the tip is a value comparable to $1-D_A$, D_A being the flux deficit parameter (Oke and Korycansky 1982).

Figure 1 shows the result of the analysis. The most prominent feature is a 6.7σ overdensity of absorption at 3976 Å ($z = 2.271$). It is due to the spectra of Q0041-2707 and Q0041-2658. The two lines of sight are separated by 2.4 h^{-1} proper Mpc ($q_0 = 0.5$) and the feature covers ~ 2700 km s^{-1} in velocity space. Williger et al. (1996) find metal absorption at redshift 2.2722 in the spectrum of Q0041-2658, which is remarkably consistent with the redshift of the overdense structure.

There are also two noticeable voids at ~ 4490 Å and at 4842 Å. The second void is possibly due to a foreground QSO which lies within 500 km s^{-1} of the underdensity.

References

Cowie L.L., Songaila A., Kim T., Hu E.M., 1995, AJ 109, 1522
Fernández-Soto A., et al., 1996, ApJ 460, L85
Lanzetta K.M., Bowen D.B., Tytler D., Webb J.K., 1995, ApJ 442, 538
Oke J.B., Korycansky D.G., 1982, ApJ 255, 11
Peebles P.J.E., 1993, *Principles of Physical Cosmology*, Princeton Univ. Press
Sargent W.L.W., Young P. J., Boksenberg A., Tytler D., 1980, ApJS 42, 41
Williger G.M., et al., 1996, ApJS 104, 145

Very High Redshift Radio Galaxies

Wil van Breugel[1], Carlos De Breuck[1,2], Huub Röttgering[2], George Miley[2], and Adam Stanford[1]

[1] University of California Inst. of Geophysics and Planetary Physics,
   LLNL L-413, P.O. Box 808, Livermore, CA 94550, USA
[2] Leiden Observatory, P.O. Box 9513, 2300 RA, Leiden, The Netherlands

Abstract. High redshift radio galaxies (HzRGs) provide unique targets for the study of the formation and evolution of massive galaxies and galaxy clusters at very high redshifts. We discuss how efficient HzRG samples are selected, the evidence for strong morphological evolution at near-infrared wavelengths, and for jet-induced star formation in the $z = 3.800$ HzRG 4C41.17.

1 Introduction

Radio sources are surprisingly effective beacons for identifying galaxies at extremely high redshifts. Optical/near–IR campaigns during the past few years by several groups have resulted in the discovery of more than 120 radio galaxies at $z > 2$, including 17 with $z > 3$, and 3 with $z > 4$. At low redshifts powerful radio galaxies are uniquely identified with massive ellipticals. If this is true also at high redshift, as seems reasonable given the surprisingly good Hubble $K - z$ relation for radio galaxies at $0 < z < 4.4$, then we should be able to systematically study the evolution of massive elliptical galaxies over large lookback times using samples selected by their radio emission. While recently developed techniques of finding very distant star–forming galaxies (*e.g.* Steidel *et al.* 1996) are yielding substantial galaxy populations at $z \sim 3$, radio galaxy samples remain the best means of finding the most massive galaxies at $z \sim 3$, and even higher.

Hierarchical galaxy formation scenarios suggest that these massive galaxies are assembled from smaller structures at relatively late cosmic epochs (*e.g.* Baron and White 1987). Observations of HzRGs may thus provide a unique opportunity to study the beginning of this process. Furthermore, when they have been found, HzRGs may also be used for 'color-dropout' searches of galaxy clusters around them, and begin cluster evolution studies at very high redshift.

2 How to Find High Redshift Radio Galaxies

2.1 Ultra Steep Spectrum Sources

It has been known for many years that radio sources with very steep spectra are mostly associated with faint, distant galaxies (*e.g.* Tielens *et al.* 1979; Blumenthal & Miley 1979). An example of a steep radio spectrum, for the powerful nearby radio galaxy Cygnus A, is shown in figure 1. One can also see that the radio spectrum steepens at higher frequency. If all powerful radio galaxies have such curving spectra, due to synchrotron and/or inverse Compton losses, then with increasing redshift we would observe the steeper parts of their spectra. Analysis of the 3CR radio sample indeed showed such an expected $\alpha - z$ correlation (van Breugel and McCarthy 1990; Fig 2 shows an updated version). A similar result has recently been reported for a sample of ultra-luminous radio galaxies at high redshifts (Carilli et al. 1998).

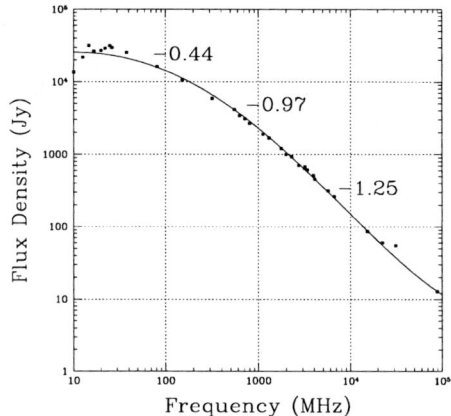

Fig. 1. The radio spectral energy distribution of Cygnus A, using single dish measurements from the literature (*e.g.* Kellermann, Pauliny-Toth and Williams 1969, and others). Note the spectral steepening with frequency.

Thus this ultra-steep spectrum (USS) 'red radio colors' selection technique may be used as a very effective tool for identifying very high redshift radio galaxies. The first comprehensive USS search for HzRGs was begun by Chambers and collaborators and resulted in the discovery of the $z = 3.800$ radio galaxy 4C41.17 (Chambers, Miley and van Breugel 1990). For many years this was the most distant galaxy known. This relatively bright object, with its extended Ly-α halo, radio-aligned UV continuum, and evidence for star formation (see below) in many ways stimulated the continued search for more USS HzRGs.

With the advent of several new, deep radio surveys it is now possible to define much better USS samples which reach 10 – 100 times fainter flux density levels than previous samples, and which allow much more accurate spectral index determinations. We have used several such surveys to define USS samples for study in the northern and southern hemispheres. Our primary northern hemisphere sample uses the WENSS 325 MHz survey (Rengelink et al. 1997) together with the NVSS 1.4 GHz survey (Condon et al. 1998), to define the spectral indices, and the FIRST 1.4 GHz survey (Becker et al. 1995), to obtain radio maps.

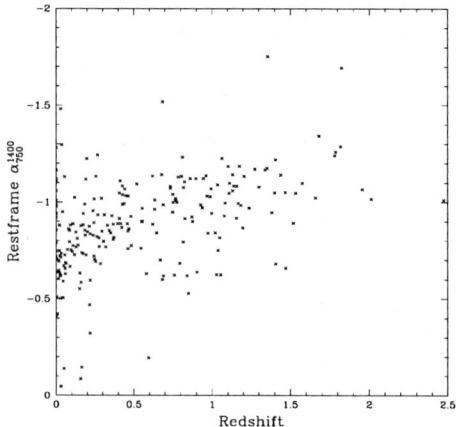

Fig. 2. Spectral steepening with redshift for all 3CR radio galaxies.

In the southern hemisphere we use the low frequency Texas 365 MHz (Douglas et al. 1996; Dec $> -35°$) and MRC 408 MHz (Large et al. 1981; Dec $< -35°$) surveys, in combination with the NVSS and PMN 5 GHz (Griffith & Wright 1993; Dec $< -35°$) surveys.

The sources in our southern hemisphere sample all have spectral indices $\alpha < -1.2$. We have obtained radio images of all our southern USS targets with the Australia Telescope and VLA so that we can make accurate identifications. A deep near-infrared identification program of this sample was begun with the CTIO 4m telescope. Spectroscopic observations of some objects from this sample with the ESO 3.6 m have shown that USS sample is indeed extremely efficient at finding HzRGs, and have already resulted in the discovery of the most distant radio galaxy in the Southern Hemisphere known to date (TN J1338−1942 at $z = 4.13$; De Breuck et al., these proceedings). A summary of the current status of HzRG identification programs is given in Table 1.

Sample	Definition	Known Redshift			Unknown
		$z < 2$	$2 < z < 3$	$z > 3$	
3CR	$S_{178} > 10$ Jy	99.5 %	0.5 %	0 %	1%
MRC/1Jy	$S_{408} > 0.95$ Jy	93.4 %	5.9 %	0.7 %	41 %
4C USS	$\alpha_{178}^{1414} < -1.0$	53 %	35 %	12 %	50 %
New USS	$\alpha_{325}^{1400} < -1.3$	35 %	35 %	30 %	45% (faint)

Table 1. The HzRG content for four samples of radio sources as a function of z: 3CR (Spinrad, private communication), MRC/1Jy (McCarthy et al. 1996), 4C (Chambers et al. 1996), and our new USS sample.

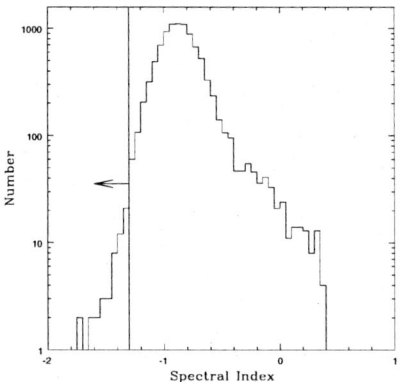

Fig. 3. Hystogram of southern USS sources at DEC < -35° from the MRC and PMN surveys. In total 76 out of 2915 sources have $\alpha_{408MHz}^{5GHz} < -1.20$.

2.2 The $K - z$ Diagram

Around the time that 4C41.17 was discovered another systematic search for HzRGs also proved to be successful. This method uses the magnitude – redshift relationship at infrared wavelengths, the $K - z$ 'Hubble' diagram, for powerful radio galaxies galaxies to find promising targets. This resulted in the discovery of the $z = 3.395$ radio galaxy B2 0902+34 (Lilly 1988). Indeed, the $K - z$ diagram has since proved to be a very reliable, if little understood, tool. Despite strong morphological evolution seen at near-infrared wavelengths, discussed below, the most powerful radio sources continue to follow the $K-z$ relationship even to z ~ 4.4 (Fig 4). It is our experience that, by combining the USS and $K-z$ techniques and using near-infrared identifications of USS sources which are unidentified on UKST/POSS2 plates, one is virtually guaranteed to be successful in HzRG hunts. Without near-IR identifications and photometry approximately 2/3 of the galaxies are at $z > 2$. With near-IR identifications and photometry one can select the redshift range one wishes to study and to date all galaxies observed at Keck were found to be at the redshift predicted from the $K - z$ diagram, at least for $z \lesssim 4.4$.

3 Morphological Evolution of HzRGs

Near-infrared images obtained with the W. M. Keck I Telescope of HzRGs with $1.9 < z < 4.4$ show strong morphological evolution at *rest–frame optical* ($\lambda_{\rm rest} > 4000$ Å) wavelengths (van Breugel *et al.* 1998; Fig 5). At the highest redshifts, $z > 3$, the rest–frame visual morphologies exhibit structure on at least two different scales: relatively bright, compact components with typical sizes of ~10 kpc surrounded by large–scale (~ 50 − 100 kpc) diffuse emission.

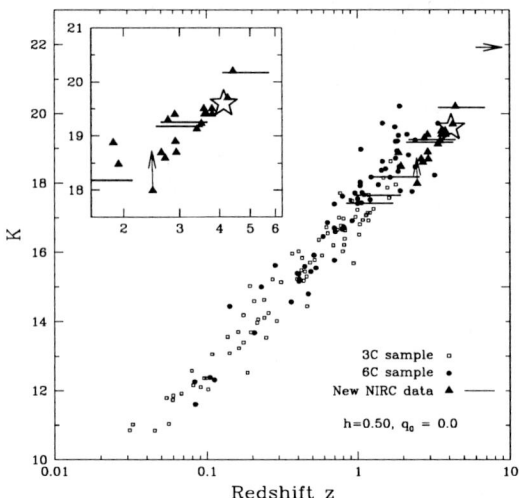

Fig. 4. Hubble K-z diagram for the 3C and 6C surveys (Eales et al. 1997), and NIRC data for HzRGs from van Breugel et al. (1998). The new USS K-band detections and very high redshift candidates are shown by horizontal bars at the predicted redshift. The horizontal arrow indicates the range of predicted redshift for our faintest K-band object. The open star represents TN J1338-1942 (De Breuck et al. , this volume). All magnitudes are corrected to a 64 kpc metric diameter, assuming $H_0 = 50$, $q_0 = 0$. The inset shows a blowup of the NIRC data in the $1.5 < z < 5.5$ and $17.5 < K < 21$ region.

The brightest components are often aligned with the radio sources, and their *individual* luminosities are $M_B \sim -20$ to -22. For comparison, present–epoch $L_\star$ galaxies and, perhaps more appropriately, ultra-luminous infrared starburst galaxies, have, on average, $M_B \sim -21.0$. The *total, integrated* rest–frame B-band luminosities are $3 - 5$ magnitudes more luminous than present epoch $L_\star$ galaxies.

At lower redshifts, $z < 3$, the rest–frame optical morphologies become smaller, more centrally concentrated, and less aligned with the radio structure. Galaxy surface brightness profiles for the $z < 3$ HzRGs are much steeper than those of at $z > 3$. We attempted to fit the $z < 3$ surface brightness profiles with a de Vaucouleurs $r^{1/4}$ law and with an exponential law, the forms commonly used to fit elliptical and spiral galaxy profiles, respectively. We demonstrate the fitting for our best resolved object at $z < 3$, 3C 257 at $z = 2.474$ (Fig 5). Within the limited dynamical range of the data, both functional forms fit the observed profiles—neither is preferred. Interestingly, despite this strong morphological evolution the $K - z$ 'Hubble' diagram for the most luminous radio galaxies remains valid even at the highest redshifts, where a large fraction of the K-band continuum is due to a radio–aligned component.

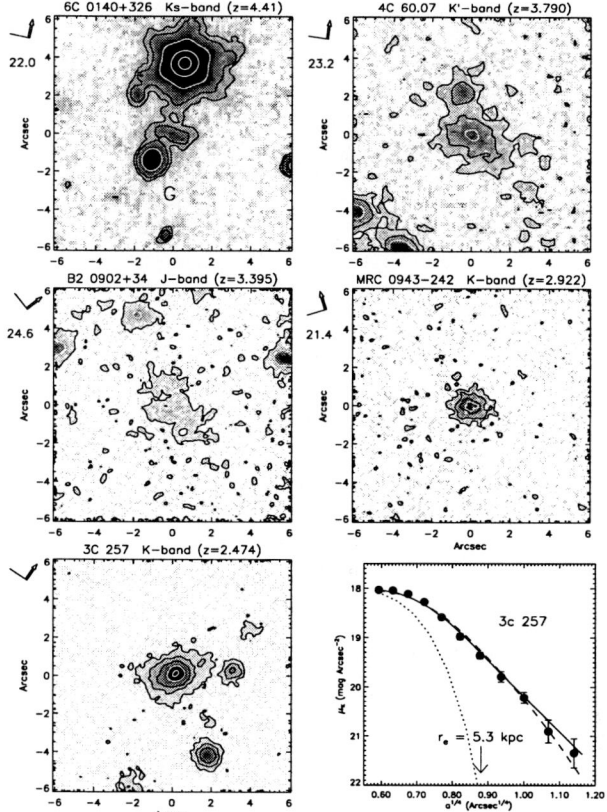

Fig. 5. Selected near–IR images of HzRGs, presented in order of decreasing redshift, and the surface brightness profile of 3C257.

4 Jet-Induced Star Formation in 4C41.17: HST Imaging

Deep HST images of 4C41.17 at 'R'-band (rest-frame UV) and in Ly-α show that the 4C41.17 system consists of two components: 4C41.17-North with a bright string of UV knots and Ly-α emission along the radio axis, and 4C41.17-South with several much fainter UV knots, distributed in random fashion throughout a low surface brightness halo. The brightest radio knot in 4C41.17-North is associated with the brightest UV knot and arc-like Ly-α emission. One of the field objects in the HST images was also seen at near–IR (Graham et al. 1994; object # 16) and radio wavelengths (Carilli et al. 1994). This was used to align the HST and radio frames with an estimated relative accuracy of $\sim$ 0.1″. The central, radio–aligned UV and Ly-α emission is shown in Figure 6 with the 0.21″ resolution radio X-band image from Carilli et al. overlaid. Figure 7 shows the HST continuum image of the entire 4C41.17 system smoothed to 0.3″ resolution to enhance low surface brightness features.

Downstream from the bright radio/UV/Ly-α knot the radio source curves

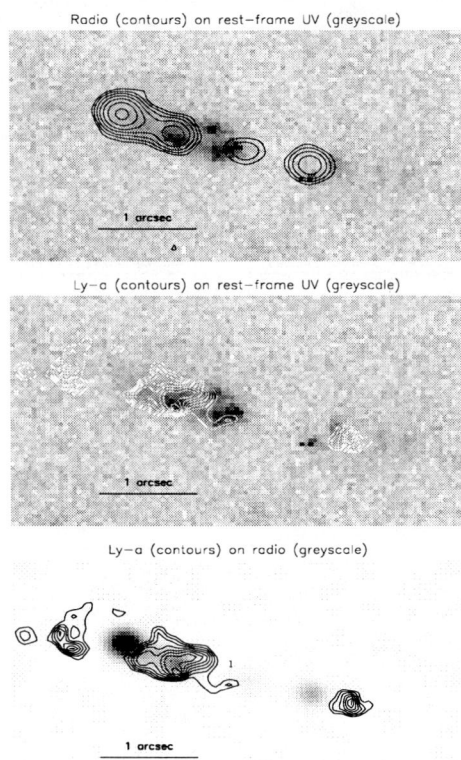

Fig. 6. HST WFPC2 images of 4C41.17-North with the X-band map of Carilli *et al.* 1994 superimposed.

towards a faint, very steep spectrum NE lobe (see Carilli *et al.* 1994), while upstream from this knot, towards the central radio core, the UV continuum appears edge–brightened. These morphological features suggest a strong interaction between the radio jet and dense ambient gas and, in fact, are as expected in jet–induced star formation models where sideways shocks induce star formation in the dense medium of forming galaxies (*e.g.* De Young 1989; Bicknell *et al.* 1998).

The star formation rates (SFR) of the various components as deduced from the rest–frame UV HST photometry range from 5 – 40 $M_\odot$/yr for the knots in 4C41.17-North, to 5 – 10 $M_\odot$/yr in 4C41.17-South. Here we have assumed $L_{1500\text{Å}} \sim 10^{40.1}$ erg s^{-1}Å^{-1} for a $SFR = 1$ $M_\odot$/yr (Conti *et al.* 1996) and no dust reddening. The derived values are surely lower limits, given the detection of dust at sub—mm wavelengths in 4C41.17 by Dunlop *et al.* (1994).

The entire 4C41.17 system is embedded in a common halo of diffuse, low surface brightness emission which extends over a very large area of $54h_{50}^{-1}$ kpc × $76h_{50}^{-1}$ kpc (5″ × 7″). This includes a faint region, 4C41.17-South, with half a dozen compact knots distributed in random fashion. Spectroscopic observations have shown that 4C41.17-South is indeed at the same redshift as 4C41.17-North (Dey et al. 1999). The range of UV luminosities and SFR rates for the individual knots in 4C41.17-South is lower than in 4C41.17-North, and very similar to the 'normal' (radio-quiet) Lyman-break galaxies discovered by Steidel et al. (1996). The random distribution and on average lower SFR in the 4C41.17-South knots suggests that star formation here is unaided by bowshocks from the radio jet. The total star formation rate, integrated over the entire 4C41.17 system and including the low surface brightness emission, is $\sim$ 660 $M_\odot$/yr. Of this perhaps as much as 2/3 of the star formation may be occurring in the inter-knot regions. If the total star formation would continue at this rate for $2 \times 10^8 - 2 \times 10^9$ yrs an entire massive elliptical galaxy of $10^{11}\,M_\odot - 10^{12}\,M_\odot$ might be assembled between $z \sim 4$ and $z \sim 2.5$, consistent with the morphological evolution for HzRGs seen in the near-IR Keck observations.

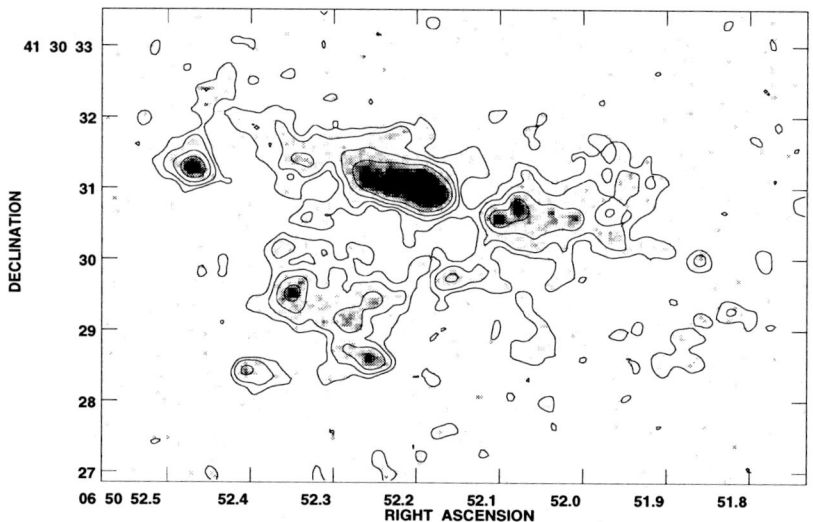

Fig. 7. Smoothed version of the HST WFPC2/F702W image showing the entire 4C41.17 system, including the clumpy companion system 4C41.17-South.

5 Jet-Induced Star Formation in 4C41.17: Keck Spectroscopy

Deep spectropolarimetric observations with the Keck II telescope by Dey et al. (1997) have provided strong evidence in support of the jet–induced star formation model for 4C41.17-North suggested above on the basis of the HST and radio morphologies. These observations showed that the bright, radio–aligned rest–frame UV continuum is unpolarized ($P_{UV}(2\sigma) < 4\%$). This implies that scattered AGN light, which is generally the dominant contributor to the rest-frame UV emission in $z \sim 1$ radio galaxies, is unlikely to be a major component of the UV flux from 4C 41.17. Instead, the total light spectrum shows absorption lines and P–Cygni–like features that are similar to those detected in the spectra of the recently discovered population of star forming galaxies at slightly lower ($z \sim 2-3$) redshifts (Fig 8). The detection of the S Vλ1502 stellar photospheric absorption line, the shape of the blue wing of the Si IV profile, the unpolarized continuum emission, the inability of other AGN–related processes to account for the UV continuum flux, and the overall similarity of the UV continuum spectra of 4C 41.17 and the nearby star forming region NGC 1741B strongly suggest that the light from 4C 41.17 is dominated by young, hot stars. The presence of radio–aligned features in many of the $z > 3$ HzRGs suggests, by analogy to 4C41.17, that jet–induced star formation may be a common phenomenon at these very high redshifts.

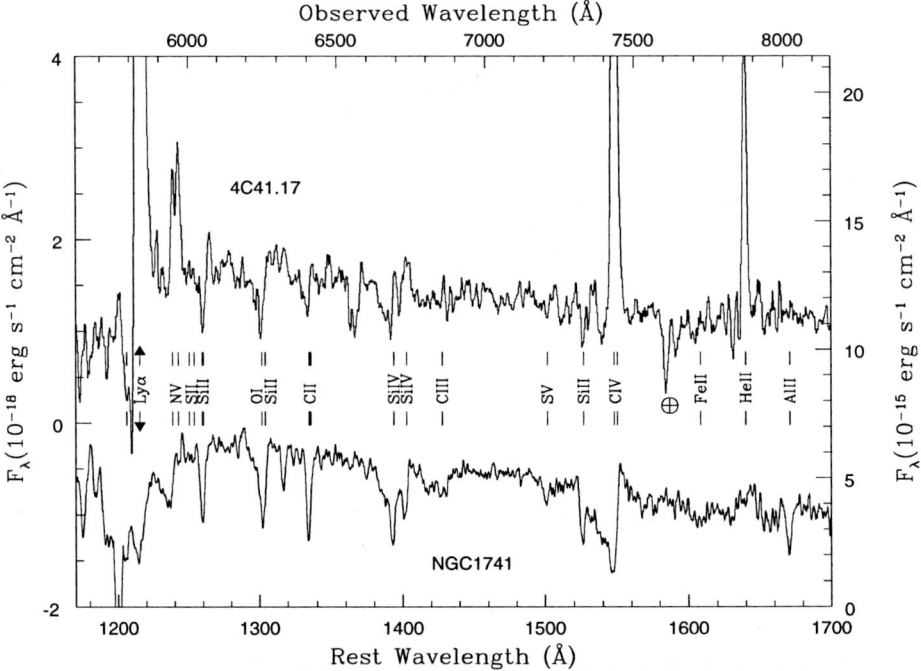

Fig. 8. Keck spectrum (Dey *et al.* 1997) of the radio-aligned component 4C41.17-North, compared with a UV spectrum of the Wolf-Rayet starburst galaxy NGC 1741 (Conti *et al.* 1996).

Acknowledgments

Part of the work described here was performed while WvB was on sabbatical leave during January 1997 – April 1997 at the Anglo-Australian Observatory, the Australian National Telescope Facility, and the Mount Stromlo and Sidings Springs Observatories. He appreciates the support provided by these institutes. He thanks the Kookaburra's for their wake–up calls, and his Australian colleagues for their warm hospitality and invigorating discussions, with special thanks to Drs. J. Bland–Hawthorn, G. Bicknell, M. Dopita, and R. Sutherland. We gratefully acknowledge S. Rawlings for advance information regarding 6C 0140+326 and 8C 1435+635 and his work on the $K - z$ diagram, and C. Carilli for providing high quality radio images of 4C41.17 which allowed to improve on the relative radio/optical astrometry for this source. The research by WvB, CDB and AS at IGPP/LLNL is performed under the auspices of the US Department of Energy under contract W–7405–ENG–48.

References

Baron, E., & White, S. D. M., ApJ, 322, 585
Becker, R. H., White, R. L., & Helfand, D. J. 1995, ApJ, 45, 559
Bicknell, G.V. et al. 1998, ApJ, (in preparation)
Blumentahl, G., & Miley, G. 1979, A&A, 80, 13
Carilli, C., et al. 1998, in Proc. "The Most Distant Radio Galaxies", Amsterdam, in press, astro-ph/9801128
Carilli, C.L., Owen, F.N. & Harris, D.E. 1994, AJ, 107, 480
Chambers, K.C., Miley, G.K. & van Breugel, W. 1990, ApJ, 363, 21
Chambers, K., Miley, G., van Breugel, W., & Huang, J., 1996, ApJS 106, 215.
Condon, J. et al. 1998, AJ, 115, 1693
Conti, P.S., Leitherer, C. & Vacca, W.D. 1996, ApJ, 461, 87
De Breuck, C., van Breugel, W., Röttgering, H., & Miley, G. 1998, in Proc. Radio Surveys Workshop, Tenerife Spain, ed. M. Bremer (Dordrecht: Kluwer), p 185
De Young, D. S. 1989, ApJ, 342, 59
Dey, A., van Breugel, W., Vacca, W., & Antonucci, R. 1997, ApJ, 490, 698
Dey, A. et al. 1999, (in preparation)
Douglas, J., Bask, F., Bozyan, F., Torrence, G., & Wolfe, C., 1996, AJ 111, 1945
Dunlop, J.S. et al. 1994, Nature, 370, 347
Eales, S., et al. 1997, MNRAS, 291, 593
Graham, J.R. et al. 1994, ApJ, 420, 5
Griffith, M. & Wright, A. E. 1993, AJ, 105, 1666
Kellerman, K. I., Pauliny-Toth, I. I. K, & Williams, P. J. S. 1969, ApJ, 157, 1
Large, M. et al. 1981, MNRAS, 194, 693
Lilly, S. J. 1988, ApJ, 333, 161
McCarthy, P., et al. 1996, ApJS 107, 19
Rengelink, R., et al. 1997, A&A, 124, 259
Steidel, C.C. et al. 1996, AJ, 112, 352
Tielens, A., Miley, G., & Willis, A. 1979, A&AS, 35, 153
van Breugel, W., & McCarthy, P. 1990, in ASP Conference Series, Vol 10, Evolution of the Universe of Galaxies, ed. R. G. Kron (San Francisco: ASP), 359
van Breugel, W. et al. 1998, ApJ, 502, astro-ph/9803019

The Highest Redshift Radio Galaxy Known in the Southern Hemisphere

Carlos De Breuck[1,2], Wil van Breugel[1], Huub Röttgering[2], George Miley[2] and Chris Carilli[3]

[1] Institute for Geophysics and Planetary Physics, Lawrence Livermore National Laboratory, L−413, P.O. Box 808, Livermore, CA 94550, U.S.A.
[2] Leiden Observatory, P.O. Box 9513, 2300 RA Leiden, The Netherlands
[3] National Radio Astronomy Observatory, Socorro, NM ,U.S.A.

Abstract. We present the discovery of a $z = 4.13$ galaxy TN J1338-1942, the most distant radio galaxy in the southern hemisphere known to date. The source was selected from a sample of Ultra Steep Spectrum (USS; $\alpha < -1.3$; $S \propto \nu^\alpha$) radio sources using the Texas and NVSS catalogs. The discovery spectrum, obtained with the ESO 3.6m telescope, shows bright extended Ly-α emission. The radio source has a very asymmetric morphology, suggesting a strong interaction with an inhomogeneous surrounding medium.

1 Southern High Redshift Radio Galaxy Searches

High Redshift Radio Galaxies (HzRGs) may be used to study the formation and evolution of massive elliptical galaxies (see, *e.g.* van Breugel, this volume). However, the sample of HzRGs at the highest redshifts ($z > 3$) is extremely small, despite vigorous searches by several groups. This is especially true in the southern hemisphere: of the 20 $z > 3$ radio galaxies known, only 3 are in the South; below declination $-40°$, only one $z > 2$ radio galaxy is known!

To provide samples of HzRGs to study with the soon to be operational 6−8m class telescopes in the southern hemisphere (VLT, Gemini−South, Magellan), we have constructed a sample of USS sources from the TEXAS 365 MHz (Douglas *et al.* 1996) and NVSS 1.4 GHz (Condon *et al.* 1998) surveys ($\delta > -35°$), and from the MRC 408 MHz (Large *et al.* 1981) and PMN 4.85 GHz (Griffith & Wright 1993) surveys ($\delta < -35°$). This USS selection makes our sample $\sim 65\%$ efficient in selecting $z > 2$ radio galaxies (see van Breugel, this volume, and De Breuck *et al.* 1998).

2 The First Southern $z > 4$ Radio Galaxy

The highest redshift USS object we have found thus far is the radio galaxy TN J1338-1942. The source has an integrated spectral index $\alpha = -1.33$, and a straight power-law spectrum between 365 MHz, 1.4 GHz and 4.8 GHz. High resolution VLA imaging at 4.8 GHz and 8.3 GHz shows that the source has a core with $\alpha^{4.8}_{8.3} = -0.6$, a northwestern lobe with $\alpha^{4.8}_{8.3} = -1.6$ at $1''$ from the core, and a southeastern lobe with $\alpha_{8.3}^{4.8} = -2.5$ at $3\rlap.{''}6$ from the core (Fig. 1). The coincidence of extended Ly-α emission with the NW lobe suggests that this asymmetry may be due to an inhomogeneous ambient medium.

The R−band identification and spectroscopic observations were obtained with the EFOSC1 imaging spectrograph on the ESO 3.6m telescope. We obtained two spectra, one covering 3725Å to 6940Å (not shown here), and another covering 6000Å to 9200Å (Fig. 2). The 2″ slit used to obtain the blue spectrum was offset 2″ from the radio core, but nevertheless showed bright Ly-α, proving the large extent of the Ly-α emission.

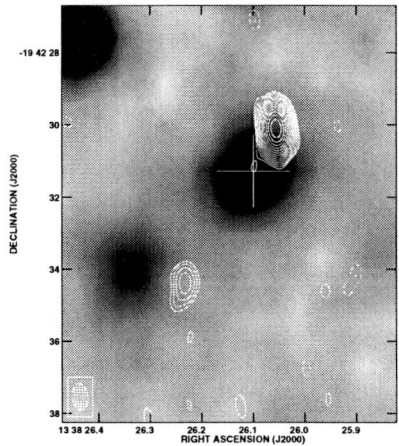

Figure 1. *Greyscales:* R−band image ($t_{int} = 10$ min, $1\rlap.{''}3$ seeing), dominated by Ly-α emission in the passband. *Contours:* VLA 4.71 GHz map showing the asymmetric lobes. The flat-spectrum core is indicated by a cross.

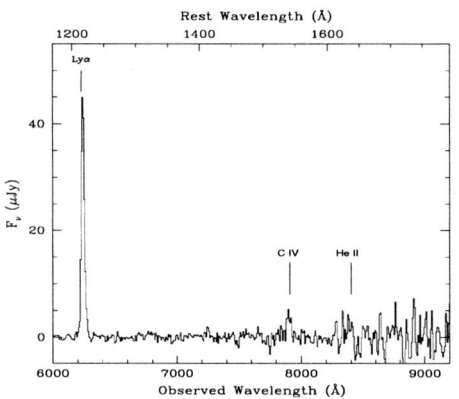

Figure 2. ESO 3.6m spectrum of TN J1338−1942 ($t_{int} = 105$ min). The bright (1×10^{-15} erg s^{-1}cm^{-2}) Ly-α is extended (6″ ; $\sim$ 50 kpc) and has spatially extended absorption on the blue side, as seen in many $z > 3$ radio galaxies (*e.g.* Dey 1998). Weak C IV and He II lines confirm the redshift $z = 4.13$.

Acknowledgments: The research at IGPP/LLNL is performed under the auspices of the US Department of Energy under contract W–7405–ENG–48.

References

Condon, J. J., *et al.* 1998, AJ, 115, 1693
De Breuck, C., *et al.* 1998, in Proc. Radio Surveys Workshop, Tenerife Spain, ed. M. Bremer (Dordrecht: Kluwer), p 185
Dey 1998, in Proc. *The Most Distant Radio Galaxies*, Amsterdam, The Netherlands, in press, astro-ph/9802163
Douglas, J., *et al.* 1996, AJ 111, 1945
Griffith, M., & Wright, A. E. 1993, AJ, 105, 1666
Large, M. I., *et al.* 1981, MNRAS, 194, 693

Radio Continuum and Emission Line Morphologies of Southern Seyfert Galaxies

Z. Tsvetanov[1], R. Morganti[2,3], R.A.E. Fosbury[4], M.G. Allen[5], and J. Gallimore[6]

[1] Johns Hopkins University, Baltimore, MD 21218, USA
[2] Australia Telescope National Facility, PO Box 76, Epping NSW 2121, Australia
[3] Instituto di Radioastronomia, CNR, I-40129 Bologna, Italy
[4] ST-ECF, D-85748 Garching bei München, Germany
[5] Mount Stromlo and Siding Spring Observatories, ACT 2611, Australia
[6] Max Planck Institut für Extraterrestische Physik, Garching bei München, Germany

1 Background

Modern active galactic nuclei (AGN) research is greatly concerned with geometry. The radiation and absorption anisotropies, and the orientation to the line of sight, are fundamentally important for both the object classification and for understanding the physical nature of the copious energy release.

There are two types of observed anisotropies – highly collimated radio emission, i.e., beams, jets (opening angle $\theta \sim 10°$), and ionization cones ($\theta \sim 100°$). The axial symmetry is determined on a subparsec scale but the two types of collimated radiation are traced out to kiloparsec scales. It is particularly important to understand the connection between the AGN axis and any global symmetry properties of the host galaxy. With no exception, the radio and ionization cone axis are aligned to within the measurement errors (e.g., Wilson & Tsvetanov 1994), but there is little relation to other galaxian scale axes.

To address the important questions of the AGN – host galaxy relationship we have collected extensive optical emission line and radio continuum imaging data for a volume limited sample of southern Seyfert galaxies. Our sample consists of 50 well classified galaxies with $cz \leq 3600$ km s^{-1} and $\delta \leq 0°$.

2 Observations

Optical [O III] $\lambda 5007$ and Hα+[N II] emission line and their adjacent continua images of all the galaxies in the sample were obtained at ESO using the 2.2 m, NTT and 3.6 m telescopes. The typical resolution of the emission line maps is $1''$, with a noise level of $\sim 1 \times 10^{-16}$ ergs cm^{-2} s^{-1} arcsec^{-2}. In addition to the emission line maps, for each galaxy in the sample we have formed an excitation/reddening map, the ratio [O III] $\lambda 5007$ / (Hα+[N II]), and a continuum color map. The later is affected by a combination of extinction and color effects.

New radio continuum observations were obtained for 29 of the galaxies in the sample. Objects with $-30° < \delta < 0°$, 8 in total, were observed with the Very Large Array (VLA) at 4.9 GHz (6 cm) and objects with $\delta < -30°$ were observed with the Australia Telescope Compact Array (ATCA) at 8.4 GHz (3 cm). Both the VLA and ATCA radio maps have a resolution of $\sim 1''$ matching that of the optical images. All but two of the observed sources were detected above the noise limit of ∼ 0.15 mJy. Our radio observations were combined with data available from the literature to achieve almost 85% coverage of the sample.

3 Highlights of Results

In the radio, 30% of the sources show linear structure, 25% are only slightly resolved or diffused, and 45% remain unresolved at the ∼ 1″ resolution. As in previous work, a correlation is found between the size of the radio structure and the radio power. The radio sources in Seyfert 2 galaxies have, on average, larger linear size than their type 1 counterparts (see Fig. 1), but there is no significant difference in radio power between types 1 and 2, although all the most powerful objects appear to be Seyfert 2's. No significant difference is found in the spectral indices of the two Seyfert types.

Extended emission is common in Seyfert galaxies – essentially all objects observed show extended Hα+[N II] and nearly 50% show extended [O III] emission. At least 40% (18 out of 50) of the galaxies show high excitation extended emission well outlined in the excitation map. The morphology of the high excitation extended emission line region vary from linear to conical to S-shaped and even X-shaped. Almost exclusively all elongated EELR are in Seyfert 2 galaxies. The orientation of the EELR appears to be random relative to either the major (or minor) axis or relative to the non-axisymmetric structures, such as bars or ovals, when present, and there is a hint of relation to the morphological type (Fig. 2).

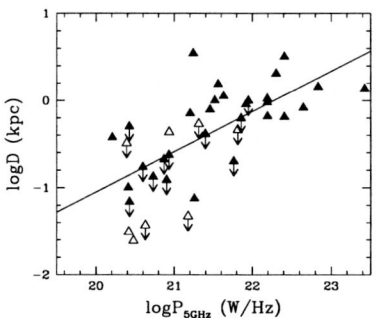

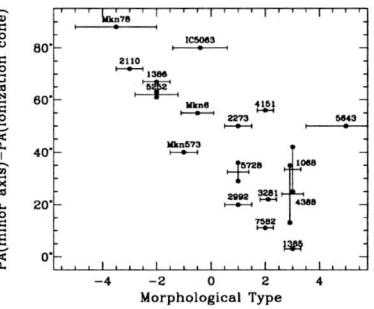

Fig.1. Radio power versus linear size of the radio structure. Open triangles are Seyfert 1's, filled symbols are Seyfert 2's. The K-S test suggests that the two distributions are significantly different. The median size of the radio structure is 0.32 kpc in Seyfert 1's and 0.66 kpc in Seyfert 2's (56% and 44% of measurements being upper limits, respectively).

Fig.2. Orientation of the galactic axis with respect to the ionization cone axis plotted as a function of the morphological type of the host galaxy. Objects are indicated by their NGC or Markarian number. Vertical lines connect the values on opposite sides of the nucleus. There seems to exist a clear tendency, but selection effects may play significant role.

References

Wilson, A.S., & Tsvetanov, Z.I. 1994, AJ, 107, 1227

Surveying High z Galaxies with HST and Keck

Garth Illingworth

UCO/Lick Observatory/Department of Astronomy and Astrophysics,
University of California, Santa Cruz, CA, 95064, USA

Abstract. Imaging and spectroscopic surveys of distant galaxies are entering a new era. It has been recognized for some time that characterizing and understanding galaxy evolution is ultimately dependent on establishing mass scales for galaxies as a function of redshift and environment. This is now possible by utilizing high-resolution imaging from HST and combining it with spatially resolved spectroscopy from large ground-based telescopes. Such an approach has been adopted for the field by the Santa Cruz DEEP group and for clusters by a Netherlands-Santa Cruz collaboration. For the field project, spectra of a large sample of galaxies to I $\sim$ 25 mag, derived from the HST WFPC2 HDF and "Groth Strip" images, have been obtained by the Santa Cruz DEEP group with the Keck LRIS multislit spectrograph and combined with detailed structural measurements from the WFPC2 images to (i) derive evolution for a sample of disk galaxies at $z \sim 0.5$ using the Tully-Fisher relation, (ii) characterize, from size and velocity width data, the global properties to $z \sim 1$ of a sample of "compact" galaxies, (iii) determine star formation rates in that "compact" sample, and (iv) derive structure and redshifts for a sample of U and B-band "dropout" objects. For the cluster project, the goals are to characterize the evolution of the galaxy population using relations such as the fundamental plane for the early-type galaxies, as well as through the derivation of color-magnitude relations as a function of position across the cluster from very accurate photometry. Work on the first cluster, Cl1358+62 at $z = 0.33$, has progressed to where velocity dispersions have been obtained for 53 early-type galaxies, and used to derive the fundamental plane relation. This is one of the largest FP samples, *at any redshift*. Interestingly, this cluster program has also resulted in the discovery of some of the highest redshift galaxies reported to date, both of which are $z = 4.92$. One of these two objects is strongly gravitationally-lensed and so provides the clearest view yet of the structure of star-forming regions at early times. These results, from both programs, demonstrate that length scales, surface brightnesses and morphological data from HST imaging, when combined with Keck redshifts, velocity width and rotation curve data, provide an unprecedented opportunity to characterize the nature of distant galaxies, particularly with relations such as the fundamental plane and Tully-Fisher.

1   HST and Keck – The Quantitative Survey Era

The combination of HST imaging and Keck spectroscopy (where "Keck" represents the new generation of 8-10 m class telescopes) has proven to be a remarkably powerful for characterizing the nature of galaxies at redshifts $z \sim 1-3$, and even beyond. The reasons why these telescopes together have revolutionized this field is that, quite simply, distant galaxies are (i) small, i.e., $<< 1''$ in size, and (ii) faint, i.e., $I > 23 - 24$ mag.

HST can image with resolution that overcomes the problem that the small size (i) poses for ground-based telescopes, and can now allow us characterize the structural features and components ("morphology" and Hubble types), as well as providing quantitative measurements such as scale lengths, and surface brightness and color distributions. Keck, and other 8-10 m ground-based telescopes, on the other hand are the spectroscopy engines that allow us to (partly) overcome the faintness (ii) of distant galaxies, and provide measurements, not only of redshifts, but also kinematics (velocity widths and rotational velocities), and emission and absorption line strengths. In particular, what these HST and Keck data enable us to do is step beyond luminosities to the measurement of masses. The ability to think about "mass evolution" as well as "luminosity evolution" adds a new dimension to the scientific insight that can be derived from a comparison of models with the measured universe. The recent results from the Hubble Deep Field, the HDF (Williams *et al.* 1996), are a striking example of what can be achieved by combining these capabilities.

1.1 Field and Cluster Surveys

Many groups are now using these new capabilities to work on HST-imaged fields, such as the HDF, to delineate the properties of distant galaxies. Two major projects that exemplify this new direction are the Keck Deep project, developed in collaboration with Sandra Faber and David Koo at Santa Cruz, and the HST intermediate redshift cluster environment project, a joint Leiden–Santa Cruz collaboration, initiated by Marijn Franx.

A central guiding philosophy in both these programs is that establishing mass scales, as noted above, is key to characterizing and understanding the evolution of galaxies from intermediate to high redshifts. The very large variance in luminosities at a given mass, due to widely varying star formation rates, dust content and dust distribution, greatly complicates the interpretation of galaxy fluxes and colors. High spatial resolution images from HST for length scales and structural characteristics, combined with spectroscopy for kinematics from Keck are key to putting galaxy evolution onto a much more quantitative footing.

1.2 The DEEP Project

The DEEP Project (Deep Extragalactic Evolutionary Probe) was developed initially with the Center for Particle Astrophysics at Berkeley with the goal of elucidating the physical processes and time scales of the formation and evolution of galaxies. The Keck telescopes and their multi-object spectrographs (LRIS – the Low Resolution Imaging Spectrograph, Oke *et al.* 1995, and, when it becomes available in 1999, DEIMOS, the Deep Imaging Multi-Object Spectrograph) are central to this program, providing redshifts and velocity widths as well as absorption and emission line characteristics. Of equal importance is the high spatial resolution imaging capability provided by HST and its WFPC2 camera (and NICMOS in 1998, and later the Advanced Camera, the ACS, in 2000). These

HST instruments provide the structural and morphological data that complements the 8-10 m telescope spectroscopic data. As has been demonstrated with the HDF (see, e.g., papers by Vogt et al. 1997, Phillips et al. 1997, and Guzmán et al. 1997), such HST-imaged regions are crucial for progress in studying galaxies at intermediate to high redshifts. Ground-based telescopes, even with adaptive optical systems, will not provide comparable capability for the foreseeable future. Broadly, the goal of DEEP is to obtain kinematical data with DEIMOS on Keck on $\sim 10,000$ galaxies from several HST-imaged fields spread at ~ 6 hour intervals around the sky.

1.3 The Intermediate Redshift Cluster Program

The second of these new survey programs is the intermediate redshift cluster program. This is complementary to the DEEP project in that it is focused on exploring and establishing the kinematic, structural and morphological properties of galaxies in rich clusters and their immediate neighborhood. It differs from other cluster programs at similar redshifts $z \sim 0.3 - 1$ in that the observations are made over a wide field (from many pointings with HST), and so provides results across a wide range of galaxy densities and environments. We are studying three clusters, Cl1358+62 at $z = 0.33$, MS2053-04 at $z = 0.58$, and MS1054-03 at $z = 0.83$. These are all strong emitters in X-rays (they are all EMSS clusters). This program is focused primarily on early-type galaxies which preferentially lie in such clusters, and much of the attention of the program is being given to establishing the fundamental plane for the early-type galaxies in these clusters (see Kelson et al. 1997 and van Dokkum et al. 1998). The fundamental plane for Cl1358+62 (Kelson et al. 1998a, 1998b, 1998c) is now one of the best established *at any redshift*, with data for 53 early-type galaxies at $z = 0.33$.

2 Galaxies at $z \sim 5$

One of the remarkable aspects of acquiring new data with HST or large telescopes like Keck are the serendipitous discoveries that arise. This was certainly the case for the Cl1358+62 HST image. While checking the image soon after the HST data were taken, Franx noted that the cluster contained a strongly-lensed arc. However, the arc was unusually red in (F606W-F814W). This provoked us into including it in our Keck observational program of the fundamental plane galaxies. The first results from the spectroscopic observations were tantalizing, since it was clear that this arc was potentially a very high redshift object if the single emission line seen at 7204 Å was Lyα. The redshift would then be $z = 4.92$. Subsequent higher S/N data allowed us to confirm this result, as reported in detail in Franx et al. (1997) By a remarkable coincidence, the serendipitous discovery of the arc in the HST image was matched by another serendipitous discovery of a second $z = 4.92$ galaxy. This galaxy just happened to fall on one of the slitlets in the multi-slit mask that was being used to measure velocity dispersions for the cluster members. This second object is $150 h_{65}^{-1}$ kpc from the first, and differs in redshift by only 450 km s^{-1}! The spectrum of one of the two high redshift galaxies is shown in Figure 1.

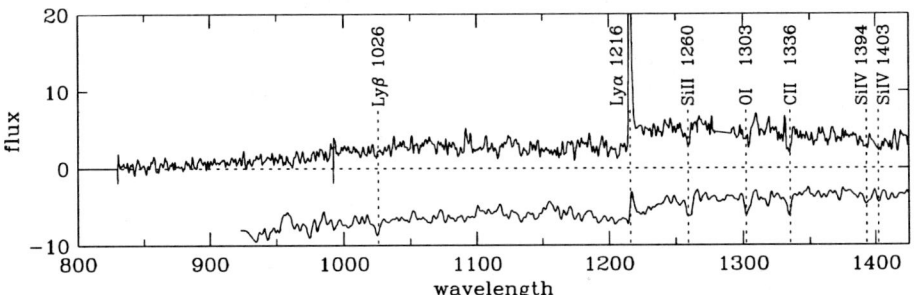

Fig. 1. The spectrum of Cl1358-G1, averaged along the slit. Comparison is made with the average spectrum of 12 $z = 3$ galaxies from Lowenthal et al. (1997 – plotted beneath the G1 spectrum). The Lyα line, the sharp break at Lyα due to intervening absorption by the Lyα forest, the drop to zero flux (within the noise) at the Lyman continuum, and a number of interstellar absorption features in the galaxy are all consistent with a redshift $z = 4.92$ object. The smaller-than-expected decrement at Lyα is due to a weak contribution from an elliptical galaxy in the $z = 0.33$ cluster that is near to the arc.

Steidel and his collaborators (e.g., Steidel et al. 1996a, 1996b; Giavalisco et al. 1996) have made the detection and acquisition of redshifts at $z \sim 3$ a remarkably routine activity – so much so that it is hard to recall the dark ages of just a few years ago when such galaxies were very few, and unusual in their properties, being mostly radio galaxies. While the $z = 4.92$ galaxies, Cl1358-G1 and Cl1358-G2, were distinguished, momentarily, by being the highest redshift galaxies, and the first galaxies which exceeded the redshift of all know QSOs since the discovery of QSOs, what is most valuable about the strongly-lensed object is that the spatial resolution on the reconstructed source galaxy of G1 is better than 20 mas in some parts of the image. This corresponds to about $100 h_{65}^{-1}$ pc for $q_0 = 0.5$. The reconstructed images are shown in Franx et al. (1997). The high magnification in this object allows one to study the distribution of star formation in a way that is not possible for virtually all other high redshift galaxies (the number of strongly-lensed sources at $z \sim 3$ or greater is still very small).

For example, it is striking that a very substantial fraction ($\sim$50%) of the star formation in G2 is taking place within a single "knot" or "complex" that is only $200 h_{65}^{-1}$ pc in size (FWHM). The optical data suggest a star formation rate of 36 $M_\odot yr^{-1}$ for G1. As Franx et al. note, even this modest rate allows for the formation of 20-40% of the Galactic bulge in 10^8 yr, a period that is about 10-20% of a Hubble time at $z \sim 5$. However, subsequent Keck IR imaging (Soifer et al. 1998) shows that significant dust (E(B-V)$\sim$0.3) is likely to be present, raising the SFR by an order-of-magnitude and changing the likely mass scale of this object.

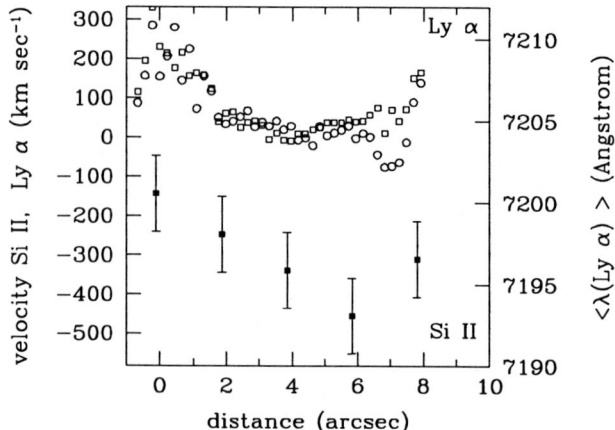

Fig. 2. The velocity structure in the emission and absorption lines in Cl1358+62 G1. This velocity structure was measured along the arc using the long-slit mode of LRIS. The open symbols are measurements of the Lyα emission, while the filled symbols are the interstellar absorption feature Si II 1260 Å. The absorption line is systematically blueshifted with respect to the Lyα emission. Such effects are seen in gas outflows where the blue wing of Lyα is absorbed, resulting in a "blueshift" of the Si II absorption line. The asymmetry of the Lyα profile, with a sharp blue edge and a red "wing" is also indicative of such effects.

The high magnification allows us to also investigate one of the other key issues surrounding these high redshift galaxies. A central question that has been asked a number of times in connection with these objects, as noted above in the discussion of the Lowenthal *et al.* (1997) paper is: "what are their masses?" The high magnification allows one to reconsider this issue since the arc is resolved for ground-based observations (the total length is about 8″). Using the long-slit mode of LRIS, we obtained data that could be useful for such purposes. However, in endeavoring to interpret the results it became clear that the "motions" we were seeing were likely to be due to turbulent motions resulting from the large energy input from supernovae and winds from massive, young stars. The velocity structure was most naturally due to outflows and not gravity. The velocity variations that we found are shown in Figure 2, where the velocities from Lyα are compared to those from the interstellar absorption line of Si II 1260 Å. Such velocity differences are seen locally in starburst galaxies (e.g., Heckman, Armus and Miley 1990), as well as other high redshift galaxies (see e.g., Lowenthal *et al.* 1997; Trager *et al.* 1997). These results thus make it difficult to use the observed velocity structure in high redshift galaxies to determine the masses of these very actively star forming galaxies, as Steidel *et al.* (1996a), as well as others, including us, hoped to do. While it is likely that a variety of techniques will be developed that will place constraints on the masses of high redshift galaxies, it is not going to be as easy as we hoped!

3 Future

There are a number of instrumental developments underway that will provide a huge increase in our ability to explore galaxy formation and evolution at redshifts $z > 4$. The first major development in survey capability comes from the order-of-magnitude improvement in imaging sensitivity and field of view that will be provided in 2000 by the HST ACS, the Advanced Camera. On the same time scale a new generation of multi-slit spectroscopic instruments, such as DEIMOS on Keck, will be operating with comparable gains. The next major development will be using the $0.8 \sim 1.7 \mu$ region for multi-slit observations. The impact of such capability at redshifts $z \sim 5 - 6$ will be marked. All these advances will be overshadowed, however, as NGST turns its gaze upon the distant universe. For most of us currently working on distant galaxies, NGST will provide the high point of our observational careers by offering us insights that no other telescope will provide into the foreseeable future.

Acknowledgments

The work reported here is the result of collaborations with many inspiring colleagues, namely, Pieter van Dokkum, Sandra Faber, Marijn Franx, Jesus Gallego, Rafael Guzmán, Dan Kelson, David Koo, James Lowenthal, Dan Magee, Drew Phillips, Luc Simard, Tom Soifer, Kim-Vy Tran & Nicole Vogt. We are particularly grateful for funding from STScI, NASA and the NSF.

References

Franx, M., *et al.*, ApJ, 486, L75, 1997.
Giavalisco, M. *et al.*, ApJ, 470, 189, 1996.
Guzmán, R., *et al.*, ApJ, 460, L5, 1996.
Heckman, T. M., Armus, L., & Miley, G. K., ApJS, 74, 833, 1990.
Kelson, D. D., *et al.*, ApJ, 478, L13, 1997.
Kelson, D. D., *et al.*, ApJ, in preparation, 1998a, 1998b, 1998c.
Lowenthal, J. D. *et al.*, ApJ, 481, 673, 1997.
Oke, J. B., *et al.*, PASP, 107, 375, 1995.
Phillips, A. C., *et al.*, ApJ, 489, 543, 1997.
Soifer, B. T., *et al.*, ApJ, in press, 1998.
Steidel, C. C., *et al.*, AJ, 112, 352, 1996a.
Steidel, C. C., *et al.*, ApJ, 462, L17, 1996b.
Trager, S. C., *et al.*, ApJ, 485, 92, 1997.
van Dokkum, P. G., *et al.*, ApJ, submitted, 1998.
Vogt, N. P., *et al.*, ApJ, 479, L121, 1997.
Williams, R. E., *et al.*, AJ, 112, 1335, 1996.

Discussion

Mould: Two orders of magnitude range in M/L seems very different from the current epoch. Isn't it inconsistent with a Tully-Fisher relation or a fundamental plane at the epoch that Keck is observing?

Illingworth: The figure showing M/L vs mass actually comprises a local sample as well as the (numerically) smaller more distant sample. The very large M/L spread occurs because we have plotted galaxies of all morphological types from ellipticals to very late, star forming dwarf galaxies. Typically, both with the fundamental plane and for Tully-Fisher, one restricts the sample to small subsets of the local galaxy population for which these correlations exist. However, such a priori selection is difficult at high redshift since the morphological/Hubble type is poorly determined (if at all!) and so a broad comparison with the totality of the current epoch galaxy population is appropriate.

Boyle: You mentioned that there was significant dust associated with your galaxy at $z = 4.92$, yet the spectrum clearly shows strong Lyman alpha. Presumably this requires either critical amounts of dust or a special distribution of dust?

Illingworth: It certainly is interesting that significant extinction and strong Lyman alpha co-exists. A number of high redshift objects, including Cl1358-G1 discussed here, have significant extinctions (E(B-V) $\sim$ 0.3 mag) inferred from the slope of their UV continua, yet also have measurable Lyman alpha. Since such effects are also seen at lower redshift in strongly star-forming galaxies, it presumably results from a combination of geometry and velocity structure.

AUSTRALIS: A Multi-Fibre Near-IR Spectrograph for the VLT

Keith Taylor

Anglo-Australian Observatory, PO Box 296, Epping, NSW 2121, Australia

Abstract. An Australian consortium of astronomers and engineers (based at AAO, MSSSO and UNSW) were contracted by the European Southern Observatory to carry out a one-year concept design study for a near-infrared multi-object spectrograph for the VLT. The underlying instrumental philosophy was to supply a significant object multiplex at a high enough spectral resolution to resolve the internal kinematics of galaxies. A full contiguous wavelength coverage from 0.9μm to 1.8μm is achieved through the use of multiple HgCdTe-based spectrograph cameras. A preliminary optical design for the spectrographs has been achieved as has a detailed concept design for the 400-fibre positioner and deployable integral field units.

1 Introduction

The AUSTRALIS Concept Design Study (Taylor and Colless 1996) was initiated in response to an ESO *Announcement of Opportunity* for the study of optimal approaches to near infra-red multi-object spectroscopy on the VLT in parallel with a similar study by a consortium of European astronomers which later became know as the VIRMOS group. While accepting the VIRMOS proposal for its imaging and low dispersion multi-slit capabilities, the arguments for a fibre-based system were regarded as sufficiently strong to encourage ESO to explore ways of incorporating such a facility for the VLT in their longer-term plans. This eventually led to the Wide-Field Fibre Facility proposal (FLAMES) for one of the Nasmyth platforms of the Unit 2 telescope (Avila et al. 1998) which is planned to be commissioned in mid-2001.

2 Instrument Concept Design

2.1 Optimum Aperture Size

Determining an optimal fibre aperture size for multi-object work is a key design issue which drives much of the instrumental constraints. For this reason we took great care to quantify the effects of aperture size on predicted S/N performance as a function of target image size, seeing and instrumental performance.

In order to quantify this for typical faint galaxies the predicted S/N was obtained as a function of aperture size and seeing for a sample of suitably faint galaxies from the Hubble Deep Field (HDF) convolved with gaussians to simulate a range of seeing conditions. The I-band HDF magnitudes were converted to J using the mean colour I-J$\sim$2 expected for galaxies at z$\leq$1. The S/N was then computed using this J magnitude along with the suitable values for J-band sky continuum levels, dark current and readout noise.

While an optimum aperture of $\sim 1.5''$ is obtained it is clear from the S/N curves that some significant leeway is available should ancillary arguments, such as spectral resolution, push the design to smaller values ($\rightarrow 1''$).

2.2 Requirement for 1-to-7 Hex Relay

The camera f-ratio required to image a $1.5''$ aperture onto 2.35 ($18.5\mu m$) pixels of a 2K-by-2K Rockwell array on the VLT is f/0.75. Furthermore, this camera has to have a linear field of 53.6mm equivalent to an angular field (at f/0.75) of $\sim \pm 10°$ assuming a beam-size of 200mm. This is clearly impractical; the most tractable solution is to adopt some form of fibre image slicing. Inevitably, by re-formatting aperture information along the input slit of the spectrograph, such a solution automatically compromises object multiplex, however, this is seen to be an acceptable compromise.

Instead of segmenting the in-coming f/15 Nasmyth image, we place a $1.5''$ aperture stop at the focal-plane and image the telescope pupil onto a single fibre thus avoiding the geometrical losses inherent in such a non-telecentric design. When fed at f/3, the resulting single fibre has a diameter of $175\mu m$ whose output is then segmented down-stream in order to supply a narrower slit for the spectrograph. The segmentation is provided by relay optics (located in the fibre optic switchyard - as detailed in 2.6) which take as their input the f/3 output of the single fibre, re-imaging this onto the a close-packed hexagonal lenslet array. These lenslets then form individual f/5 pupil images onto the input faces of 7 smaller, $130\mu m$, fibres which are themselves re-formatted to form individual object slitlets at the spectrograph input slit.

2.3 Focal Plane Changer

The demand to facilitate the positioning of a large number of fibres, both single object and IFUs, is a familiar problem which has seen a great deal of attention through the years that astronomical fibre systems have evolved. AUSTRALIS uses magnetic buttons on a magnetic field-plate with a double plate changer permitting field configuration on the second plate while observing with the first. This exactly follows the 2dF (Taylor et al. 1997, Smith and Lankshear 1998) design philosophy; we see no reason to make any fundamental change for the VLT.

A corrected VLT Nasmyth focus which provides a mildly curved but telecentric focal surface, does, however, make fibre positioning substantially easier and a design is proposed which allows for the fibre positioning to be performed on a stable, orientation invariant, mount. The basic 2dF fibre button design is retained as is the fibre retractor scheme which keeps each fibre tensioned so that its locus is well determined. In contrast to the 2dF, however, the AUSTRALIS positioner is an $r - \theta$ robot whose radial arm is curved to match the curvature of the corrected focal-plane.

2.4 NIR Spectrograph Concept Design

The basic requirement is to disperse the light from the fibre input slitlets and feed the resulting spectra at an f-ratio which matches the projected slit-width onto ~2 detector pixels. In order to capture all the light from the 130μm, f/5 input fibres we have chosen a faster, f/4.5, collimator to accommodate the expected fibre FRD. Given the final camera f-ratio then needs to be faster than f/1.3, delivering a very wide angle of view, a white pupil design has been employed.

The system is an off-axis Schmidt, near-Littrow, spectrograph (205mm beam-size) giving a curved intermediate spectra. In the case of full 900nm to 1800nm wavelength coverage, the input slit is located in the J-H absorption band to avoid spectral vignetting. The 452 line/mm grating (1.14μm blaze) delivers a ~430mm long spectrum at $\mathcal{R}$ ~4,000 which is split into three sections (the two ends by reflection facets; the central region undeviated) which are individually re-imaged by f/1.23 finite conjugate Schmidt cameras onto their respective 2K-by-2K detectors. Such a configuration implies that each 7-fibre object slitlets project onto ~1.9 pixels in the dispersion direction. The length of each slitlet, defined by close packing fibres with a core/clad ratio of 1:1.4, is ~18 pixels. With 2 pixels between each object spectrum, 100 critically sampled spectra can thus be recorded over the full wavelength range.

Since only half of a large aspheric corrector plate is employed it is natural to build two such spectrographs implying 6 2K-by-2K HgCdTe detectors servicing 200 objects in all. Furthermore, by deploying slits at the ends of the intermediate spectra as well as at its centre, options for recording just half the spectral range (*semi*-spectra: j=900-1350nm or h=1350-1800nm) but with twice the number of fibres become available. Thus a 400-fibre system is proposed.

2.5 Integral Field Units

The IFUs are designed to provide the same number of fibre spatial samples as there are *spectroscopic* fibres (ie: 7-times the object multiplex, hence the *on-sky* IFU fibres are fed, through the switchyard, into the same set of invariant spectrograph fibres thus requiring a 7-to-7 hex fibre relay in the switchyard rather than the 1-to-7 hex relay for the multi-object mode.

Large-IFU Here we propose a contiguous field of 2800 lenslets feeding 130μm fibres through a variable focal-plane re-imager to effect scale changes at its telecentric input. Scale changes over a wide range can be envisaged but we would suggest sampling options of perhaps 0.2", 0.4" and a maximum of 0.7", corresponding to fields of 11", 21" and 36" on a side.

Small-IFU Given that the deployable IFUs are positioned on the field-plate and hence mounted on fibre buttons, to effect image segmentation they have to receive the raw Nasmyth f/15 beam. In this case the $\mathcal{A}\Omega$ product of the telescope is not easily retained at the f/5 input of the fibre and hence the 130μ

fibre cores will subtend a maximum 0.45" implying lenslet sizes up to 0.23mm (centre-to-centre).

The number and size of these deployable IFUs is an astronomical specification which is completely uncritical to the design, however the fibre positioner has two plates to allow for rapid field changes and hence two choices can be made. As an example it is suggested that one field-plate has 8 deployable IFUs each with 350 lenslets in a square array with a 4.7" FoV sampled at 0.25" while the other has 28 deployable IFUs each with 100 lenslets in a 5-by-20 array with a 2.2-by-8.6" FoV sampled at 0.43".

2.6 Fibre Optic Switchyard

AUSTRALIS uses a fibre optic switchyard to connect fibres from the active focal plane to the cooled grating spectrographs housed in the cool-room while enabling back-illumination of the configuring set for accurate positioning. A single fibre runs ~6m from each button on the telescope focal plane to the switchyard mounted on top of the NIR spectrograph cool-room. At the switchyard a microlens collimates the f/3 beam emerging from each *thick* fibre and forms a new pupil ~2mm in diameter onto a group of seven hex-packed microlenses. These hex packed microlenses segment the new pupil and feed a group of seven *thin*, 130μm fibres at f/5. These *thin* fibres are relatively short and lead from the switchyard through the cool-room roof and down to the slits in the spectrographs. The slits and the last 1.5m of the fibres are cooled along with the spectrographs to minimize background signal at longer wavelengths. There are no back illumination lamps inside the cool-room, this function is performed at the switchyard and prevents signal contamination inside the spectrograph.

In addition, by use of the 7-to-7 hex relay all the IFU modes detailed in 2.5 have access to all the spectrograph modes. Most important of all of the switchyard functions, however, is the ability to place a dichroic between input and output feeds of the relay to feed a separate optical spectrograph; all of the above options then become available simultaneously between 450 and 1800nm.

3 Conclusions

The basic concept for the AUSTRALIS positioner has now been incorporated into ESO's Wide-Field Fibre Facility (FLAMES). The double-buffered positioner facility (OzPoz) has a 560 fibre (or deployable IFU) capacity (a 1320 capacity in total) and will initially supply 8 fibres to UVES (the VLT UV Echelle Spectrograph) and 150 fibres to GIRAFFE (a new optical-only intermediate dispersion fibre spectrograph dedicated to FLAMES). The long-term goal is to acquire an AUSTRALIS-like, Near-IR spectrograph facility targeted for higher dispersions than NIRMOS can supply: ie: $\mathcal{R} \geq 5,000$ to give adequate digital OH-suppression while being complementary to NIRMOS.

Acknowledgments

The author would like to thank Ian Parry (IoA), Peter Conroy (MSSSO) and Damien Jones (Prime Optics, Qld) for their assistance during the technical phases of this program. The AUSTRALIS Science Team was headed by Matthew Colless (MSSSO) and compromised Michael Ashley, Warrick Couch and John Storey from the University of New South Wales; Paul Francis from Melbourne University; Heath Jones, Charlene Heisler, Peter McGregor, Jeremy Mould, Bruce Peterson and Peter Wood from Mount Stromlo Observatory; and Joss Bland-Hawthorn, Karl Glazebrook and Fred Watson from the Anglo-Australian Observatory. More recently Peter Gillingham and Stan Miziarski put together the costed proposal for OzPoz.

References

Taylor, K., Colless, M.M. et al. (1996): *AUSTRALIS Concept Study Report* (Internal report document (AAO/MSSSO))

Avila, G. et al., (1998) *Optical astronomical instrumentation*, S. D'Odorico, ed., *Proc. SPIE* **3355**, In press.

Taylor, K., Cannon, R.D. and Parker, Q.A. (1997) *IAU Symposium*, **179**, p135.

Smith, G.A. and Lankshear, A.F. (1998) *Optical astronomical instrumentation*, S. D'Odorico, ed., *Proc. SPIE* **3355**, In press.

Surveys for Galaxies at $z > 2$, and an Introduction to the HDF–South

Mark Dickinson

The Johns Hopkins University, Dept. of Physics and Astronomy,
and the Space Telescope Science Institute, Baltimore MD 21218 USA

Abstract. After years observational progress exploring the evolution of galaxies from $0 < z \lesssim 1$, the universe at $z > 2$ is now becoming similarly accessible to wholesale investigation. I briefly review one survey of galaxies at $2 < z < 4.5$, and then highlight plans for the southern counterpart to the Hubble Deep Field, which will be observed by HST in October 1998.

1 Introduction

The lure of the distant universe has drawn many astronomers into the search for galaxies at high redshifts. Observers have made great progress in mapping the history of galaxy evolution during the last half of cosmic history using deep, magnitude–limited redshift surveys. Thousands of redshifts have been collected for galaxies at cosmologically interesting distances, allowing properties such as luminosity functions, color distributions, emission line activity and clustering to be measured. HST has provided imaging and surface photometry with kpc–scale resolution, offering the opportunity to connect these distant galaxies to their more familiar, well–studied nearby counterparts. These surveys have been "hemispherically unprejudiced," with important advances coming from surveys carried out with both northern and southern 4m–class telescopes.

The most extensive magnitude–limited redshift surveys (e.g. Lilly et al. 1995; Ellis et al. 1996) contain relatively few objects at $z > 1$. In part, this "$z = 1$ barrier" represents a transition between "hard" and "very hard" observational regimes. At $z > 1$ the spectral features commonly used to measure redshifts move to portions of the night sky dominated by strong OH emission, making their identification much more difficult. However, the $z \sim 1$ limit also represents a physical property of the galaxy population: at magnitudes readily accessible to 4m–class telescopes (e.g. $I = 22$ or $B = 24$), most galaxies really *are* at $z < 1$. The median redshift of galaxies increases only slowly from (e.g.) $B = 21$ to 24. In the CFRS it is $\langle z \rangle \approx 0.6$. These surveys have a tail of very luminous galaxies extending to $z \approx 1.2$, but most lie at $z < 1$.

At fainter magnitudes and higher redshifts, the W.M. Keck Telescopes in the north have indeed led to a hemispheric imbalance, soon to be rectified by the advent of VLT, Gemini South, and Magellan. The deeper Keck redshift surveys of Cowie et al. (1996) have a larger $z > 1$ "tail," in part because of their fainter flux limits, and in part because those observers have systematically pushed to redder spectrographic settings in order to track the [OII]λ3727 emission line out to $z \approx 1.6$. Even at these faint magnitudes, which already tax the patience of both observers and time allocation committees faced with the prospect of large redshift surveys, the large majority of galaxies lie at $z < 2$. One exciting prospect for the next few years is the availability of high–performance multi–object infrared spectrographs like VIRMOS on the VLT, which should help to map the "redshift desert" at $1 < z < 2$.

2 Color–Selected Surveys for High Redshift Galaxies

Progress at higher redshifts has required a more specialized approach to target selection than simple magnitude–limited surveys. Faint radio sources offer one means: the strong redshift evolution of the radio source luminosity function ensures that even bright radio samples like the 3CR contain objects out to $z = 2.5$, and fainter radio samples have identified objects at much larger redshifts, including the first galaxies known at $z > 4$. Narrow band imaging has also been employed with some success, searching for redshifted Lyman α at optical wavelengths or nebular and Balmer line emission in the infrared. But it is difficult to probe large co–moving volumes to interestingly deep flux limits.

One technique which has been impressively successful has been the use of broad band colors keyed to detecting the redshifted 912Å Lyman limit. Color selection methods have been used for many years in QSO surveys, but have only recently been pushed to the faint limits needed to find ordinary, star–forming *galaxies*. Due to space limitations I will not attempt to review the method and its application here, but direct the reader to a series of papers describing the incremental results of our own survey using this technique (Steidel & Hamilton 1992; Steidel, Pettini, & Hamilton 1995; Steidel et al. 1996a,b; Giavalisco, Steidel & Macchetto 1996; Dickinson, 1998; Steidel et al. 1998; Giavalisco et al. 1998; Adelberger et al. 1998; Pettini et al. 1998). As of this writing, we have measured redshifts for more than 600 galaxies at $2 < z < 4$. This sample, comparable in size to the CFRS, is large enough to permit systematic study of the luminosity function, spectral and photometric properties, and clustering of galaxies at $z \approx 3$.

What fraction of the faint galaxy population lies at $z > 2$? This was first addressed by Guhathakurta et al. (1990), who pioneered the application of the Lyman break technique to faint galaxies. At their magnitude limits they found few U–band dropout objects and concluded that only a small fraction of faint galaxies lie at $z \approx 3$. However this fraction rises as we go fainter. Figure 1 shows the overall number counts of galaxies in our ground–based survey and in the HDF compared to the number of Lyman break candidates in each. Brighter than $\mathcal{R} = 22.9$, all of the color selected objects in the ground–based sample have turned out to be galactic stars (plus a small number of high redshift QSOs) – we have yet to find a Lyman break galaxy with $\mathcal{R} < 22.9$ in the 0.3 square degrees which we have surveyed. Fainter than this, the number of Lyman break objects rises rapidly, reaching 5% of the galaxy population at $\mathcal{R} = 25.0$. In the HDF the counts of Lyman break galaxies are higher. This is partially due to the larger volume probed by the color selection technique in the HDF because of the

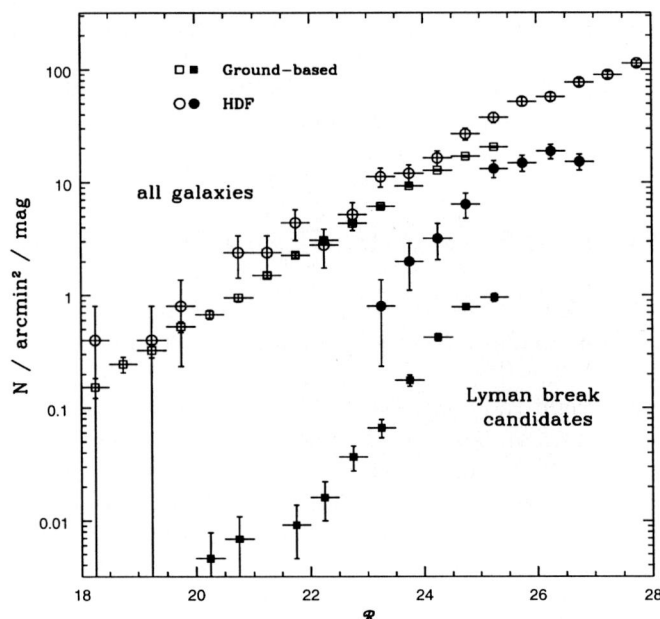

Fig. 1. Number counts of Lyman break galaxies from ground–based and HDF samples compared to the overall counts of faint field galaxies. R magnitudes are approximated for HDF galaxies as $(V_{606} + I_{814})/2$. At $R \approx 26.5$, nearly 1 in 4 galaxies in the HDF is a Lyman break "dropout" with a likely redshift $z > 2$.

bluer F300W filter. It may also reflect genuine redshift evolution in the galaxy population, however, since the HDF F300W–dropout galaxies have, on average, lower redshifts than do the objects in the ground–based sample, and the total UV luminosity density in galaxies is evidently rising from $z \approx 4$ to 2 (Madau et al. 1996). At $R \approx 26.5$, nearly 1 in 4 galaxies in the HDF is probably at $z \gtrsim 2$.

In principle these color techniques may be readily extended to identify galaxies at higher redshifts. The HDF B dropouts, candidate $z \sim 4$ galaxies, are fainter and less numerous than the U–dropouts, and are correspondingly more difficult to observe spectroscopically – only two have spectroscopic redshifts (cf. Dickinson 1998). We have extended our ground–based survey to look for "G–dropouts," taking advantage of the much larger solid angle coverage to find the rare examples bright enough for spectroscopy. Thus far we have confirmed redshifts for ∼20 galaxies at $z \approx 4$ (figure 2). Extending two–color selection to $z > 5$ will probably require very deep near–infrared imaging to extend R and I-band optical photometry, although some progress may be made simply by selecting extremely red $R-I$ objects for spectroscopy. The first confirmed galaxy at $z > 5$ was found serendipitously (Dey et al. 1998), but qualifies photometrically as an "R-dropout."

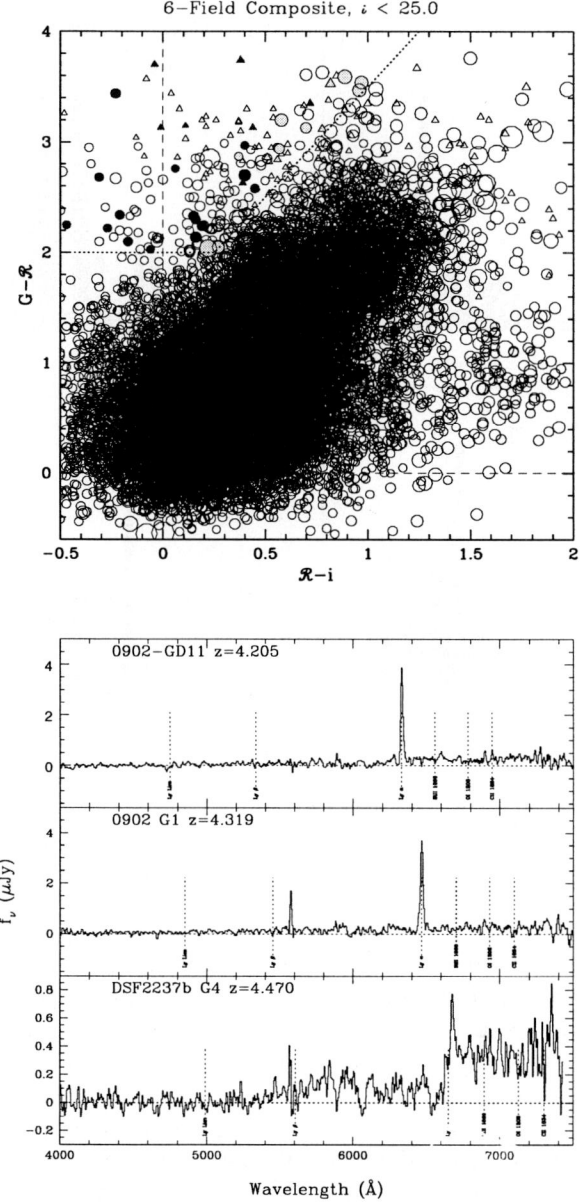

Fig. 2. *Above:* GRi color–color plot showing the selection of candidate $z \approx 4$ galaxies. Our color selection range is indicated, and spectroscopically confirmed galaxies with $z \sim 4$ are shown as dark filled symbols. Separating the $z \sim 4$ galaxies from the locus of lower redshift galaxies is more difficult than for the $UG\mathcal{R}$ technique at $z \sim 3$ – the lightly shaded symbols are "interlopers," mostly red, early-type galaxies at $0.7 < z < 1$. *Below:* Spectra of galaxies at $z \approx 4$ selected as "G–dropouts" using the Lyman break color technique.

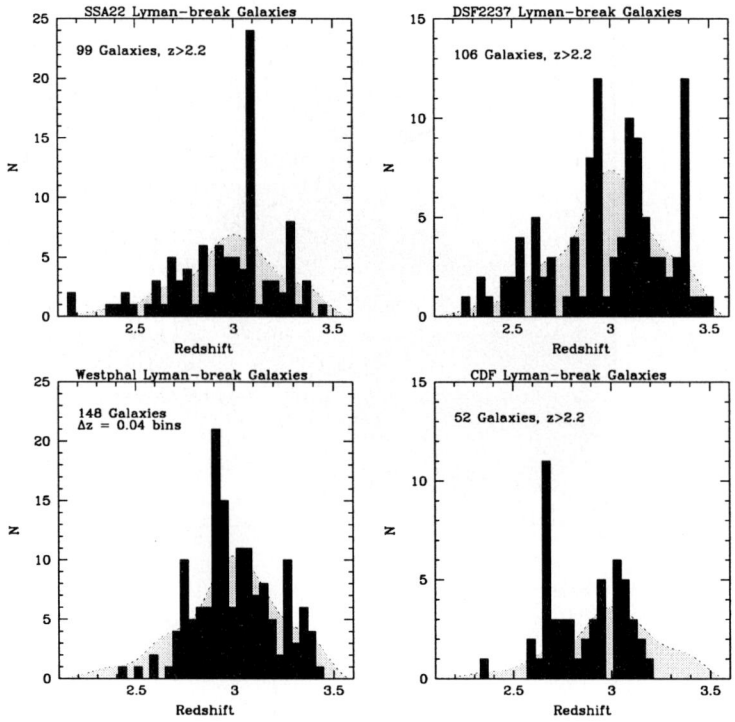

Fig. 3. Redshift distributions of Lyman break galaxies along four lines of sight. The lightly shaded curve shows the redshift distribution of objects averaged over all of our survey fields. Along each line of sight, however, the galaxies exhibit strong clustering.

With hundreds of confirmed objects at $z \approx 3$ we may begin to study the clustering of galaxies in the early universe. The redshift distribution of galaxies within our individual survey fields (see figure 3) shows very strong clustering, with prominent "spikes" seen along virtually every line of sight (Steidel et al. 1998, Adelberger et al. 1998). Similarly, the angular correlation function of the Lyman break galaxies (Giavalisco et al. 1998) provides another robust measurement of clustering strength. Both the angular and redshift–space correlations show that the Lyman break galaxies exhibit clustering with similar strength to that found for bright galaxies in the local ($z \approx 0$) universe. This might be considered surprising given that in almost any structure formation model one would expect the underlying *mass* distribution to have been much more uniform at $z = 3$ than it is today. In the context of hierarchical models for galaxy formation, the observed clustering of Lyman break galaxies suggests that these objects are found in the most massive halos which have collapsed at that era, which in turn are predicted to be a highly biased tracer of the overall mass distribution, clustering much more strongly than does the dark matter in general.

3 A Southern Hubble Deep Field

The Hubble Deep Field has provided considerable stimulus to observers and theorists studying the high redshift universe. It is important to remember, however, what a small cosmological volume the HDF actually probes, given its extremely narrow pencil–beam geometry. Table 1 lists the comoving volume out to various redshifts for two cosmologies (both with zero cosmological constant). The two rightmost columns multiply this volume by ϕ^*, the local normalization of the galaxy luminosity function (here taken to be $\phi^* = 0.0166h^3$ Mpc^{-3}, from the K–band survey of Gardner et al. 1997). This gives an indication of the number of L^* galaxies, or their progenitors at high redshift, which should be found in the HDF volume. These numbers are quite small, particularly for the case of an Einstein–de Sitter cosmology. Statistical uncertainties alone provide reason for caution about any conclusions about the properties of L^* galaxies which one might draw from a small–volume survey like the HDF.[1] Clustering, which might cause number densities and galaxy properties to vary significantly along any one, narrow sightline, can only further aggravate this concern (cf. figure 3).

Table 1. Co–moving volume of the HDF

Redshift	$V(<z)$ (h^{-3} Mpc^{-3})		$V \times \phi^*$	
	$q_0=0.5$	$q_0=0.1$	$q_0=0.5$	$q_0=0.1$
1	765	1202	13	20
2	2300	4798	38	80
3	3808	9728	63	161
4	5145	15348	85	255
5	6312	21305	105	354

For this reason among others, STScI has decided to carry out a second Hubble Deep Field observation, in order to provide an independent faint galaxy sample as a check on the results of the first HDF. The imminent arrival of an impressive fleet of large–aperture telescopes and powerful new instruments in the southern hemisphere encourages us to locate this new field in the southern sky. Finally, the new HST instruments available since the second HST servicing mission, STIS and NICMOS, provide new observational capabilities which ensure that the HDF–South will offer new information and research opportunities not available with the northern HDF.

One new angle is provided by the presence of a $z \approx 2.24$ QSO (see the contribution by Boyle to these proceedings) which will be located in the STIS field, approximately 5 arcmin east of the WFPC2 deep pointing. STIS will obtain high

---

[1] Of course the HDF has thousands of galaxies brighter than its $V \approx 30$ detection limit, but the simplistic volume exercise shown in Table 1 reinforces the fact that most of these are either objects with intrinsically low luminosities, or are high redshift fragments which must merge together later to become the rarer L^* galaxies.

signal–to–noise, high resolution UV spectra of the QSO in order to study the forest of absorption lines along that line of sight, which may then be correlated with faint galaxies identified in the vicinity. STIS will also collect very deep images at optical and UV wavelengths of the $50'' \times 50''$ field around the QSO. The plan for WFPC2 imaging is very similar to that of HDF–N, with the same four filters using similar exposure times. Meanwhile, NICMOS Camera 3 will image yet another parallel field, roughly 5.75 arcmin south of the WFPC2 field, gathering deep infrared data through two filters at at 1.1 and 1.6 microns, as well as shallower 2.2 micron images (the HST thermal background limits the depth which can be achieved in the K–band). Despite the slight defocus of this camera due to the NICMOS dewar distortion, it still provides excellent images of unprecedented depth. Some optical imaging of the NICMOS field will be obtained, probably using STIS, and there will be a "Flanking Field" configuration of shallower WFPC2 images providing high angular resolution HST imaging covering a wider solid angle.

It is also noteworthy that the HDF–S should contain many more faint stars (and, less desirably, bright stars as well) than did HDF–N. The coordinates of the field ($l = 328°.2$, $b = -49°.2$, in Tucana) point through the outskirts of the Galactic bulge; the SMC is also not far away. The observations may therefore provide more interesting fodder for galactic astronomers as well.

The HST observations of HDF–S are planned for October 1998, and the data will be distributed to the community in late November. Planning for the project is still underway. A detailed and continuously updated description of the project can be found at
http://www.stsci.edu/ftp/science/hdf/hdfsouth/hdfs.html.
This URL also provides links to sites maintained by other groups conducting observations on the HDF–S. One unique feature of the HDF–S is that much of the "follow–up" work is being done in advance of the actual HST observations. As we have heard at this workshop, various groups have already carried out optical, radio, and mid–infrared (with ISO) observations of the HDF–S.

I would like to thank the organizers of this meeting for being excellent hosts, and the editors of this volume for their patience. I thank Chuck Steidel, Mauro Giavalisco, Kurt Adelberger, Max Pettini and Mindy Kellogg for allowing me to summarize our work here. Finally, the HDF–South is a collaborative effort involving many people at STScI: my contribution here simply highlights their enormous efforts.

References

Adelberger, K., Steidel, C., Giavalisco, M., Dickinson, M., Pettini, M., & Kellogg, M., 1998, ApJ, in press.
Cowie, L.L., Songaila, A., Hu, E.M., & Cohen, J.G., 1996, AJ, 112, 839.
Dey, A., Spinrad, H., Stern, D., Graham, J.R., & Chaffee, F.H., 1998, ApJ, 498, L93.
Dickinson, M., 1998, in *The Hubble Deep Field,* eds. M. Livio, S.M. Fall and P. Madau, Cambridge Univ. Press.

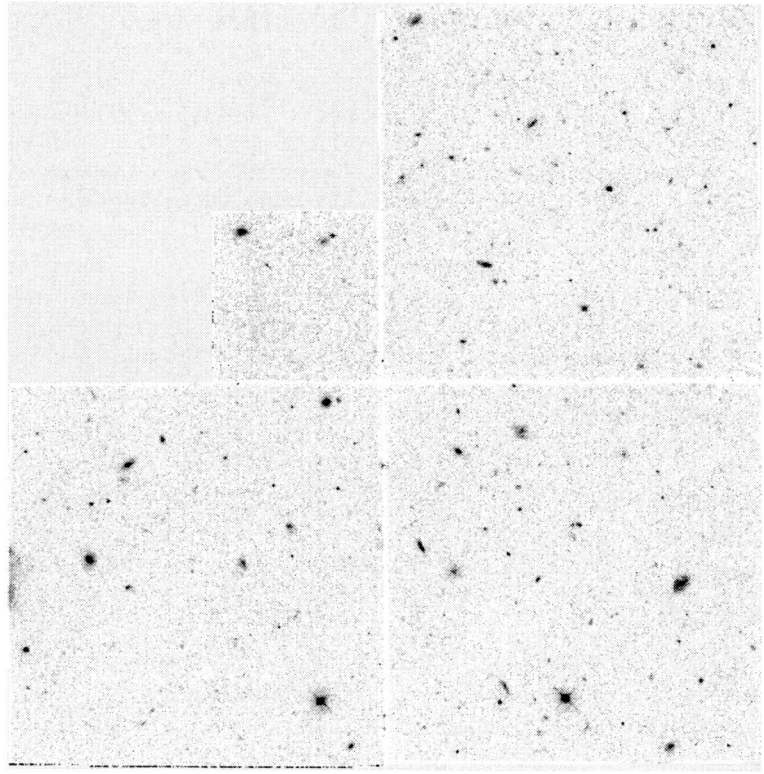

Fig. 4. 3–orbit WFPC2 test exposure of the HDF–S.

Ellis, R.S., Colless, M., Broadhurst, T., Heyl, J., & Glazebrook, K., 1996, MNRAS, 280, 235.
Gardner, J.P., Sharples, R.M., Frenk, C.S., & Carrasco, B.E., 1997, ApJ, 480, L99.
Giavalisco, M., Steidel, C.C., & Macchetto, F.D., 1996, ApJ, 470, 189.
Giavalisco, M., Steidel, C.C., Adelberger, K.L., Dickinson, M., Pettini, M., & Kellogg, M., 1998, ApJ, in press.
Guhathakurta, P., Tyson, J.A., & Majewski, S.R., 1990, ApJ, 357, L9.
Lilly, S.J., LeFèvre, O., Crampton, D., Hammer, F., & Tresse, L., 1995, ApJ, 455, 50.
Madau, P., Ferguson, H.C., Dickinson, M., Giavalisco, M., Steidel, C.C., & Fruchter, A., 1996, MNRAS, 283, 1388.
Pettini, M., Kellogg, M., Steidel, C.C., Dickinson, M., Adelberger, K.L., & Giavalisco, M., 1998, ApJ, in press.
Steidel, C.C., & Hamilton, D., 1992, AJ, 104, 941.
Steidel, C.C., Pettini, M., & Hamilton, D., 1995, AJ, 110, 2519.
Steidel, C.C., Giavalisco, M., Pettini, M., Dickinson, M., & Adelberger, K.L., 1996a, ApJ, 462, L17.
Steidel, C.C., Giavalisco, M., Dickinson, M., & Adelberger, K.L., 1996b, AJ, 112, 352.
Steidel, C.C., Adelberger, K.L., Dickinson, M., Giavalisco, M., Pettini, M., & Kellogg, M., 1998, ApJ, 492, 428.

High-Redshift Galaxies: The HDF and More

Alberto Fernández-Soto[1], Kenneth M. Lanzetta[2], and Amos Yahil[2]

[1] Dept. of Astrophysics and Optics, UNSW, Sydney, NSW2052, Australia
[2] Dept. of Physics and Astronomy, SUNY, Stony Brook, NY11794-3800, USA

Abstract. We review our present knowledge of high-redshift galaxies, emphasizing particularly their physical properties and the ways in which they relate to present-day galaxies. We also present a catalogue of photometric redshifts of galaxies in the Hubble Deep Field and discuss the possibilities that this kind of study offers to complement the standard spectroscopically-based surveys.

1 Introduction

For a long time, models for galaxy formation and evolution advanced unhampered by observations. Nowadays, however, the rapid increase in both observational capabilities and efficiency of the selection methods (see Steidel et al. 1995 [S95]) has converted the task of looking for distant galaxies from one of the most difficult challenges to an almost routine job, and large databases of high-z galaxies are already being compiled (Dickinson 1998, this volume). Observations can now constrain the models, and this obliges us to understand the properties of these objects in order to get a complete image of the processes involved in the formation and evolution of galaxies.

This study of the properties of high-z galaxies is twofold. We need to understand the information provided by the confirmed high-z galaxies. In this way we will learn about the spectral and morphological properties of the bright end of the galaxy population, i.e., the putative progenitors of present-day large ($L > L_*$) galaxies. Second, the use of photometric redshift techniques applied to deep multi-colour images (like the HDF; Williams et al. 1996) opens a wealth of statistical methods to study those faint objects for which we cannot obtain spectroscopic information in the near future. These studies will yield further results on the general distribution and evolution of galaxies. The main problem for both methods resides in the $z \approx 1-2$ range, where spectroscopic identification of galaxies at optical wavelengths is made difficult by the lack of spectral features.

2 Physical Properties of the High-Redshift Galaxies

We start with a brief review of the physical properties of high-z galaxies, most of which have been selected applying colour techniques (S95). The HDF triggered a wave of intense spectroscopic follow-up observation (Steidel et al. 1996 [S96], Lowenthal et al. 1997 [L97], Zepf et al. 1996) that added a large number of galaxies to the sample. Nature also provides us with a telescope capable of amplifying

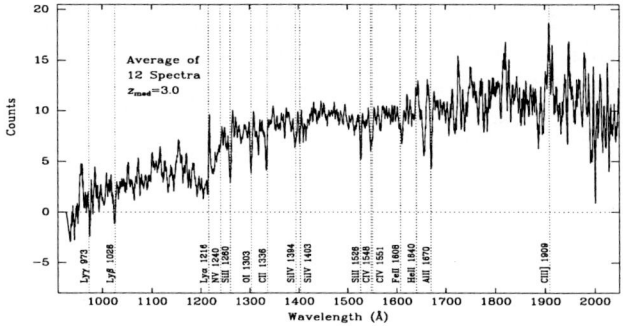

Fig. 1. Average spectrum of 12 high-redshift galaxies with $z_{med} = 3.0$ (from L97).

the light from distant galaxies, although plagued by geometric aberrations: *gravitational lensing* has been used by several groups to discover some of the most distant known galaxies (Trager et al. 1997 [T97], Franx et al. 1997).

2.1 Spectral Features: Dust, Metal and Gas Content

High-z galaxies (see Fig. 1) are characterized by a flat continuum. Their Lyα emission lines vary considerably, from weak or even absent – with superposed damped absorption profiles – to rest equivalent widths (EW) of up to 60 Å. All galaxies show optically thick Lyman limits and a strong discontinuity in the continuum bluewards of Lyα, due to the onset of the Lyα forest. Stellar and interstellar absorption lines are present, showing narrow profiles that are weaker for high-ionization species. A detailed study of the different emission and absorption lines and the slopes of the observed spectra suggests that high-z galaxies are low metallicity systems ($Z \approx 0.1\ Z_\odot$), with high neutral gas content that allows them to imprint damped HI absorption profiles on background objects. The amount of dust seems to be moderate ($E(B-V) \approx 0.10$), but a better measurement is necessary in order to establish firmer values for the extinctions, luminosities and star-formation rates.

2.2 Morphology and Luminosity

Most of the observed galaxies show compact cores (with half-light radii on the order of a few kpc) surrounded by irregular asymmetric halos (Giavalisco et al. 1996 [G96]). Although far from homogeneous, they look more regular than the galaxies observed at $z \approx 1$ – only one of the galaxies at $z > 2$ shows the "chain" morphology reported by Cowie et al. (1995) to be usual in moderate-z galaxies. A joint analysis of our photometric redshift catalogue and a morphological catalogue of galaxies in the HDF is presented by Simon Driver in this same volume (see also Driver et al. 1998 [D98]). It must be remarked that we are observing these objects in the rest-frame UV range, so passband effects are indeed important. Direct comparison of their morphologies with those of their low-z counterparts will have to wait until high-resolution IR imaging is available.

The total B-band luminosities lie in the range $1 - 10L_*$, with a strong concentration in the compact, high surface brightness cores. The SFRs range from 1 to 50 $M_\odot \text{yr}^{-1}$, although dust extinction might increase this by a factor of perhaps 3 or even more (Pettini et al. 1997).

2.3 Number Densities and Clustering

Some measured number densities (expressed as galaxies per arcmin^2) are: 0.6 ± 0.2 ($R < 25.0, 3.0 < z < 3.5$, S95), 3.2 ± 1.9 ($R < 25.3, 2.4 < z < 3.4$, S96), 6.5 ± 2.0 ($R < 25.5, 2.0 < z < 3.5$, L97). For the same ranges our catalogue gives 0.6 ± 0.3, 3.2 ± 0.8 and 7.7 ± 1.2 gals arcmin^{-2}, respectively. We also estimate that approximately 5, 15 and 25 % of all galaxies brighter than $AB(8140) = 24$, 26 and 28, respectively, are at $z > 2$. Evidence of large scale structure in the distribution of high-z galaxies is presented by Mark Dickinson in this volume.

3 Photometric Redshifts in the HDF

The determination of redshift via photometric methods (the "Poor person's z machine", as stated in Koo 1985) is a long-known technique. We present here some results from our catalogue, based on $UBVIJHK$ photometry of the HDF –IR images provided by Mark Dickinson (Dickinson et al. 1998, in prep). Full details are given in Fernández-Soto et al. (1998, in prep). The catalogue is essentially complete down to $AB(8140) = 28$ and contains 1067 objects.

Comparison with a sample of 106 spectroscopically-determined redshifts shows that the results are very good up to $z = 1.4$ ($\Delta z_{rms} = 0.13$). At $z > 2$ there is a 7% error rate (high-z galaxies that are assigned a low redshift in our analysis), while for the rest we obtain $\Delta z_{rms} = 0.45$. Lanzetta et al. (1997) have shown that the number of incorrect redshifts in the spectroscopic measurements (due to misidentification of lines or operator error) are comparable to this rate.

The advantage of this technique is the ability to estimate redshifts for large samples of objects that are too faint to have their redshifts spectroscopically measured ($AB(8140) \approx 28$ vs. $AB(8140) \approx 24$). With our sample we can estimate the $N - m - z$ distribution, the Hubble Diagram for different spectral types (see Fig. 2), the morphological evolution of galaxies (see D98), SFR densities (Lanzetta et al. 1998, in prep), and other characteristics.

4 Interpretation and Conclusions

The available data allow for different interpretations. While S95, S96 and G96 support the hypothesis that the observed high-z galaxies are the progenitors of present-day luminous galaxies at the epoch of formation of the first stars in their spheroidal components, T97 suggests that these objects will evolve to form

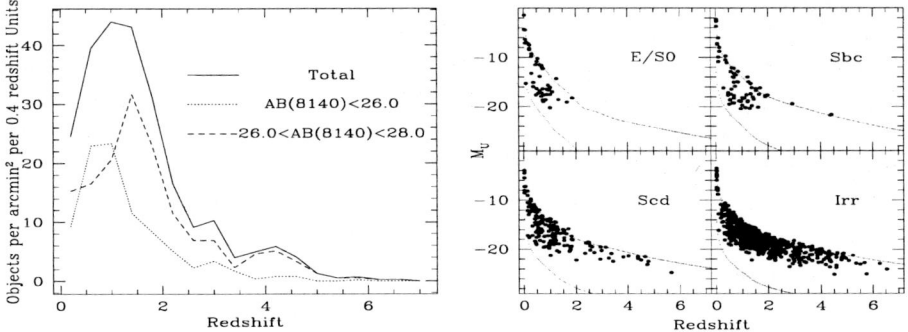

Fig. 2. Redshift distribution and Hubble Diagram of galaxies in the HDF.

the Population II components of early-type spirals. Another interpretation (L97) maintains that these objects represent a range of physical processes and stages of galaxy formation and evolution rather than any particular class of object.

While this third interpretation might be closer to reality, we are still missing an important piece of the puzzle. Detailed IR imaging and spectroscopy is needed in order to: a) shed light on the $z = 1 - 2$ galaxies allowing us to constrain evolutionary models, b) obtain images of the $z > 2$ galaxies at optical rest-frame wavelengths to be compared with their low-z counterparts, and c) perform moderate resolution spectroscopy of the $z > 2$ galaxies to accurately measure their metallicities and the importance of dust corrections.

We expect that these observations, with the support of techniques like cosmological simulations and stellar population evolutionary models, will lead us closer to the long-searched-for understanding of the process by which the Universe came to be as we see it. Perhaps it is not the moment for us to "look deeper in the Southern Sky", but to look at it with different eyes.

References

Cowie, L.L., Hu, E.M. & Songaila, A. 1995, AJ, 110, 1576
Driver, S.P. *et al.* 1998, ApJ, *in press*
Franx, M. *et al.* 1997, ApJ, 486, L75
Giavalisco, M., Steidel, C.C. & Macchetto, F.D. 1996, ApJ, 470, 189
Koo, D.C. 1985, AJ, 90, 418
Lanzetta, K.M., Fernández-Soto, A. & Yahil, A. 1997, in *The Hubble Deep Field*, eds. M. Livio, S.M. Fall & P. Madau (Cambridge, Cambridge University Press)
Lowenthal, J.D. *et al.* 1997, ApJ, 481, 673
Pettini, M. *et al.* 1997, in *The ultraviolet Universe at low and high redshift*, ed. W. Waller (Woodbury, AIP Press)
Steidel, C.C., Giavalisco, M., Dickinson, M. & Adelberger, K.L. 1996, AJ, 112, 352
Steidel, C.C., Pettini, M. & Hamilton, D. 1995, AJ, 110, 2519
Trager, S.C., Faber, S.M., Dressler, A. & Oemler, A. 1997, ApJ, 485, 92
Williams, R.E. *et al.* 1996, AJ, 112, 1335
Zepf, S.E., Moustakas, L.A. & Davis, M. 1996, ApJ, 474, L1

Discussion

Bland-Hawthorn: TAURUS Tunable Filter Surveys are turning up many compact Lyα emitters at $z = 4.5$. It is surprising that previous imaging surveys failed so badly. Our objects are very compact with low luminosity. It may be that lower mass objects have far less dust so that Lyα is quenched only by resonant scattering.

Fernandez-Soto: That may well be the case. There is a large variety of Lyα strengths in high redshift galaxy spectra. The equivalent width of Lyα might be checked against size and perhaps colour, to check whether the hypothesis is right.

Tsvetanov: Time and time again we hear that high-z galaxies look very compact, core radius of order 1–2 kpc. However, as you pointed out, we see them in the rest UV frame. We now know from the available HST and UIT images that local galaxies (say NGC 1068) look very much the same as the high-z ones. So is there a problem and if so, what is it?

Fernandez-Soto: As Mark Dickinson pointed out, the surface brightness at the core of the high-z galaxies are more similar to local star-burst galaxies. Apart from that, there is no real evidence of different geometries, and we will have to wait for high-resolution IR images to come to solve the question.

Illingworth: How did you estimate the dust content/reddening?

Fernandez-Soto: The measurements I've presented did not come from our photometric redshift catalogue, but from the spectroscopically confirmed galaxies. They are estimates based on colours and continuum slopes.

AAO Support Observations for the Hubble Deep Field South

B.J. Boyle

Anglo Australian Observatory, PO Box 296, Epping, NSW 2121

Abstract. We present proposed ground-based support observations at the AAO for the forthcoming Hubble Deep Field South (HDF-S) campaign.

1 Introduction

In October 1998, the Hubble Space Telescope (HDF-S) will once again spend several hundred hours imaging an area of sky to yield a deep image of the distant Universe. The region was chosen to lie in the Southern Continuous Viewing Zone (CVZ) at declination $-61°$ and includes a $z = 2.24$ QSO (Q2233.6-6033), first identified by the UK Schmidt Telescope and confirmed spectroscopically by the 2dF instrument at the Anglo-Australian Telescope (Boyle 1997, see also Sealey et al. these proceedings). In the HST campaign, the STIS observations will be centred on the QSO, with the WFPC2 deep field offset by 9 arcmin west. Positions of the HST fields are given in table 1. Full details of the HST observations may be obtained from the STScI website http://www.stsci.edu/.

Table 1. HDF-S positions

Instrument	RA (2000) Dec
WFPC2	22 32 56.2 −60 33 03
STIS	22 33 37.7 −60 33 29
NICMOS3	22 32 52.4 -60 38 33

2 AAO Observations

The HST observations provide a unique opportunity for Southern Hemisphere facilities to provide essential ground-based support for the HDF-S. Although there is currently no operational 8m-class telescope in the South, the existing 4m-class telescopes can still play a major role in this early stage of the HDF-S campaign. We detail below the proposed observations which the Anglo-Australian Telescope will under in support of the HDF-S. All data obtained will be made available to the community as quickly as possible, hopefully timed to coincide with the release of the HST imaging and spectroscopic data. A WWW page has been set-up at the AAO http://www.aao.guv.au/hdfs/ to distribute information and data relating to these proposed observations.

1. **Echelle observations of Q2233.6-6033.** UCLES observations of the QSO ($B = 17.5$) at the centre of the STIS field will be used to generate a list of absorption lines with $W_\lambda > 24$mÅ (3σ) over the wavelength region 3334Å–5045Å corresponding to Lyβ – CIV at the redshift of the QSO. Observations are planned for July.
2. **Intermediate-depth prime-focus imaging.** Broadband BRI imaging will be used to generate an catalogue of $R < 24$ objects in a 9×9 arcmin region comprising both the STIS and WFPC fields. Observations will be obtained in May and used to provide an input catalogue for the LDSS++ observations planned in July.
3. **Spectroscopy of $R < 24$ galaxies in the HDF-S.** An innovative upgrade to the LDSS instrument at the AAT (LDSS++, Glazebrook et al. 1998) will enable spectra for up to 300 faint ($R < 24$) galaxies to be obtained over a 9×3 arcmin region comprising both the WFPC and STIS deep fields. LDSS++ observations are currently planned for July.
4. **Flanking field QSO absorption line systems.** 2dF observations of colour and/or prism-selected QSO candidates in the 3-deg^2 region centred on the HDF-S have been proposed by Hewett et al. These observations would be used to derive information on the structure of QSO absorption line systems on the largest possible scales.

In addition, the Australia Telescope Compact Array (PI: Norris) will be used to construct a deep radio map (few μJy rms) of the HDF-S at 13cm. Initial ATCA observations (at 6 and 20cm) of the field have already been obtained confirming the lack of any strong (> 100mJy) radio sources near the HDF-S.

References

Boyle, B.J., (1997): AAO Newsletter, **83**, 4
Glazebrook, K., (1998): AAO Newsletter, **84**, 9

Discussion

Jones: Are the Taurus observations of the southern HDF aimed at the redshift of the $z > 2$ QSO?

Boyle: No. The Taurus observations currently planned are without the Tunable Filter – just broadband I. However, when the blue TTF is delivered, we will be able to image in Lyα at the redshift of the QSO.

Renzini: With EIS-DEEP we plan to observe a $15' \times 15'$ area centred on the HDF-S in six filters from U to K'. The data will be immediately public, so we may think to co-ordinate efforts on HDF-S.

Boyle: Yes. Absolutely. The STScI have given astronomers in the South a wonderful opportunity with the HDF-S. We must make every effort to co-ordinate ground-based preparatory observations to maximise the scientific exploitation of the HDF-S.

The Hubble Deep Field-South QSO

Katrina Sealey[1], Michael Drinkwater[1], John Webb[1], Brian Boyle[2]

[1] Physics, University of New South Wales, Sydney 2052, Australia
[2] Anglo-Australian Observatory, PO Box 296, Epping 2121, Australia

Abstract. The Hubble Deep Field-South was chosen to have a QSO (RA 22:33:37.6 Dec -60:33:29 J2000 and B=17.5) in the field to allow for studies of absorption systems intersecting the sight line to the QSO. To assist in the planning of HDF-S observations we present here a ground-based spectrum of the QSO. We measure a redshift of $z = 2.24$ for the quasar and find associated absorption in the spectrum at $z = 2.204$ as well as additional absorption features.

1 Introduction

Unlike the original Hubble Deep Field, the Hubble Deep Field South (HDF-S) was chosen specifically to contain a $z > 2$ QSO suitable for studying the relationship between the high redshift galaxies identified in the HDF-S and the absorption lines in the spectrum of the HDF-S QSO. The QSO was found on a UK Schmidt Telescope objective prism plate scanned by Mike Irwin using the Automated Plate Measuring facility in Cambridge analysed by Paul Hewett and then confirmed by observations at the Anglo-Australian Telescope (Boyle 1997). To aid future observations of the HDF-S we present here a low-resolution spectrum of the QSO; the data are available at http://bat.phys.unsw.edu.au/~kms/hdfs/.

2 The QSO Spectrum

We observed the HDF-S QSO on 1997 October 21 with the Australian National University 2.3m Telescope at Siding Spring with ≈ 2 arcsec seeing. The Double Beam Spectrograph was used with a 300 l/mm grating in the blue arm (binned by 2 in dispersion), and a 158 l/mm grating in the red giving a resolution of 9 Å in each arm. A dichroic was used to split the light at ≈ 5400–5500 Å (there is some uncertainty in the spectrum in this region). The spectrum shown in Fig. 1 is the weighted average of 2 exposures (1200s and 900s).

2.1 Emission Lines and QSO Continuum

The spectrum has emission lines due to Lyman-α + NV, SiIV + OIV] , CIV, CIII] and MgII. Broad FeII emission is seen around 7200–8700Å. The redshift of the QSO derived by simultaneously fitting to CIV, CIII] and MgII is $z = 2.24$. The individual redshifts for each line are z=2.23 (CIII] and CIV) and z=2.25 (MgII); the redshift difference is not unusual (Espey et al. 1989). A power law fit to the QSO continuum, avoiding obvious emission features (both broad and narrow) gives a slope of $\alpha = -0.8 \pm 0.1$ (where $f_\nu \propto \nu^\alpha$) and is shown in Fig. 1.

2.2 Absorption Systems

An initial investigation of the QSO shows absorption present from intervening systems. There appears to be "associated" absorption at $z_{abs} = 2.204$, suggested by the strong, narrow absorption lines in the blue wing of the CIV and NV emission lines (the latter falls directly between the Lyman-α and NV emission peaks). The blended Lyman-α + NV emission profile drops very sharply from Lyα towards NV (compared to the blue wing of Lyman-α, which must be cut into by multiple Lyman-α systems), probably due to absorption, rather than the 2 emission lines being resolved. The absorption line is most likely to be NV because of the presence of strong CIV absorption at the same redshift. If the above interpretation is correct, there are at least 3 additional strong absorption lines (the 2 conspicuous lines straddling the SiIV emission line and one mid-way between the NV and SiIV emission lines) which we can not yet identify reliably. If the lines are interpreted to be CIV we are seeing a rich complex of CIV absorption. These lines nevertheless reveal additional structure in an extensive system of absorbers, which may arise in a cluster of galaxies around the QSO.

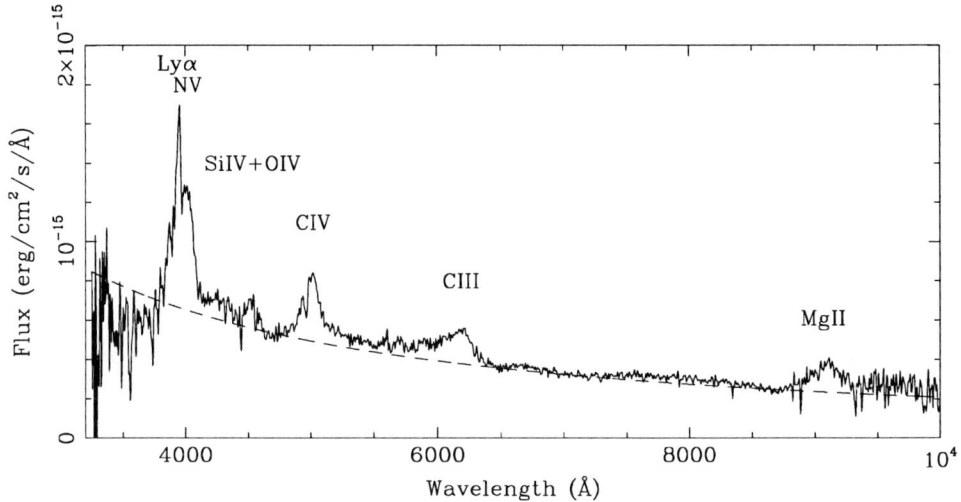

Fig. 1. Spectrum of the HDF-South QSO.

References

Boyle, B.J. (1997): AAO Newsletter, 83, 4
Espey, B. R., Carswell, R. F., Bailey, J. A., Smith, M. G., Ward, M. J., 1989, ApJ, 342, 666

Note: Katrina Sealey's new address is Macquarie University, NSW 2109, Australia.

Hubble Deep Fever: A Faint Galaxy Diagnosis

Simon P. Driver

School of Physics, University of New South Wales, Sydney, NSW 2052, Australia

Abstract. The longstanding faint blue galaxy problem is gradually subsiding as a result of technological advancement, most notably from high-resolution Hubble Space Telescope imaging. In particular two categorical facts have recently been established, these are:
1) The excess faint blue galaxies are of irregular morphologies,
   and,
2) the majority of these irregulars occur at redshifts $1 < z < 2$.

These conclusions are based on the powerful combination of morphological and photometric redshift data for all galaxies in the Hubble Deep Field to $I < 26$. Our interpretation is that the faint blue galaxy excess, which incidentally coincides with the peak in the observed mean galaxy star formation rate, represents the final formation epoch of the familiar spiral galaxy population. This conclusion is corroborated by the low abundance of normal spirals at $z > 2$. Taking these facts together we favour a scenario where the faint blue excess is primarily due to the formation epoch of spiral systems via merging at redshifts $1 < z < 2$. The final interpretation now awaits refinements in our understanding of the *local* galaxy population.

1 Introduction

The faint blue galaxy problem has been with us for several decades and is comprehensively reviewed in Koo & Kron (1992) and Ellis (1997). The problem is surmised as: *an observed excess of faint galaxies over the zero or passive evolution model predictions*. This excess first arises at $b_J = 22$ mags and extends to the faintest magnitudes probed ($b_J = 28.5$ mags, c.f. Metcalfe *et al.* 1995). The problem is compounded when one also considers the redshift distributions of galaxies at $b_J = 22 - 24$ mags, e.g. Glazebrook *et al.* (1995a), which in shape agree well with the zero-to-passive model predictions, but of course not in amplitude (*i.e.* a reiteration of the original faint blue galaxy problem). Spectroscopic surveys to fainter magnitudes are currently limited by aperture and signal-to-noise considerations. The unfortunate situation then, is that the models require a continuous renormalisation to match the observations and such a solution often results in contrived and implausible physical implications. In this overview I summarise the recent substantial developments in the observational data from Hubble Space Telescope imaging and in particular the Hubble Deep Field (HDF), the current interpretation and finally the future observations required. The expectation is that through these refinements, as opposed to speculative retro-fitting, comes concordance.

2 Hubble Deep Field Imaging

The past three years have seen two large strides forward into new parameter space within this field, these are: the ability to discern morphologies (c.f. Odewahn et al. 1995) and the ability to reliably estimate redshifts/distances (c.f. Fernández-Soto, these proceedings; Hogg et al. 1998). Even more powerful however, is the combination of these two approaches and the generation of morphological number-count N(m,T), AND morphological redshift distributions N(z,T), to $I < 26$ (c.f. Driver et al. 1998). Figures 1 and 2 are adapted from Driver et al. and show the latest comparison between faint galaxy observations and models. Before comparing the observations with the models it is first worth highlighting a number of purely observational points:

1. At bright magnitudes the total counts are dominated by the classical Hubble types (*i.e.* ellipticals and spirals).
2. The elliptical galaxy counts are almost totally flat at faint magnitudes and eventually contribute negligibly to the total galaxy counts.
3. The galaxy counts of disk and irregular systems are steep and almost linear over a broad magnitude range, with no sign of flattening and with gradients of ~ 0.3 and ~ 0.4 respectively.
4. At faint magnitudes the counts are divided almost equally between disks and irregulars over a broad magnitude range.
5. All redshift distributions become broad towards fainter magnitudes and the Euclidean correlation between faintness and distance breaks down entirely.
6. Few spiral systems are seen at $z > 2$.
7. The late-type/irregular distributions are the broadest but typically have a *higher* mean redshift then either of the so called "giant" classes.

At this point it is important to note that the data presented in these Figures is only feasible from a space borne imaging system, such as that on board the Hubble Space Telescope. Plate 1 (Figure 3) shows a qualitative true colour representation of this same dataset subdivided into redshift intervals and arranged according to apparent magnitude within these redshift intervals (and therefore crudely absolute magnitude). Note that no correction can/has been made for the K-correction in this plate. The plate essentially provides a good sanity check on the data quality and believability of the quantitative results presented in Figures 1 and 2. The trend toward higher irregularity at $z > 1.5$ is irrefutable and the degree of irregularity is far higher than that seen in the limited UV observations of local normal galaxies - *i.e.* the first indication is that the increase in irregularity towards higher-z is intrinsic rather than a manifestation of the shifting bandpass.

3 Faint Galaxy Modeling

The models shown on Figures 1 and 2 are described in detail in Driver et al. (1998) and are based on the following: Local morphological luminosity functions

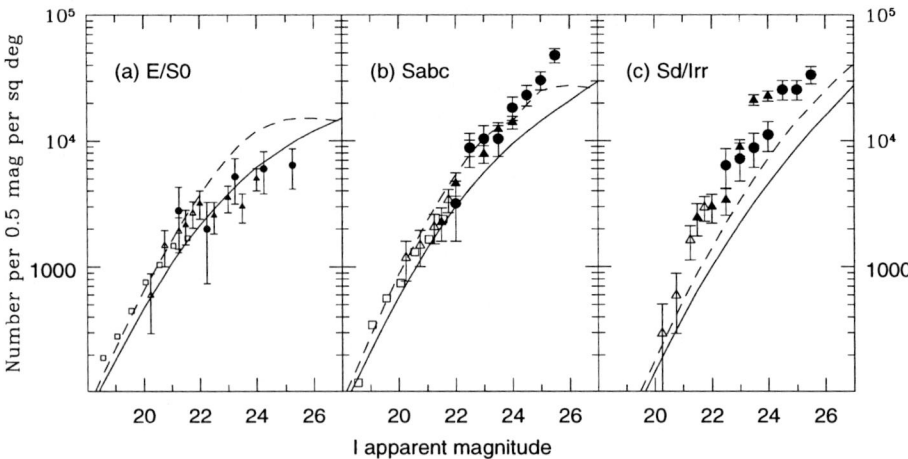

Fig. 1. Morphological galaxy number counts for: (a) ellipticals, (b) spirals, and (c) irregulars. The models lines show the zero- (solid) and maximal passive- (broken) evolution models based on a global renormalisation at $b_J = 18$.

from the CfA survey (c.f. Marzke et al. 1994); a standard flat cosmology; zero or passive E- and K- corrections (c.f. Poggianti); and a global renormalisation at $b_J = 18$. By comparison with Figure 1 we see that the ellipticals are consistent with zero-evolution, spirals with passive-evolution and irregulars require strong-evolution. These form the basic conclusions derived in a number of previous papers based on morphological galaxy counts alone (e.g. Driver et al. 1995; Glazebrook et al. 1995b) and independently from the Canada France Redshift Survey (c.f. Lilly et al. 1995). This interpretation has the obvious result that the level of evolution correlates with the mean local colour of the respective galaxy sub-class, i.e. the systems with the oldest stellar populations require the least evolution and vice-versa. However with the inclusion of photometric redshift data (described in Fernández-Soto, these proceedings) we can now take the comparison between observations and models one step further and, immediately, we see from Figure 2 that the description above is overly simplistic. Considering each class independently:

The ellipticals' N(z,T=E/S0)s are well embraced by the zero-to-passive evolution N(z) models (bear in mind the strong clustering nature of ellipticals and the limiting statistics in this single sight-line). However, there is a tendency towards an underdensity of ellipticals in the faintest magnitude intervals suggesting some moderate obscuration, disassembly or transmorphing. Although taking into consideration the fact that the observed N(z,T=E/S0) distribution is broad and not obviously truncated argues against a homogeneous evolutionary path (i.e. no single epoch of formation !) and leads us to ask whether this population is formed via a continuous ongoing crystallisation out of a non-elliptical population. Given the limited statistics for this population more observations are

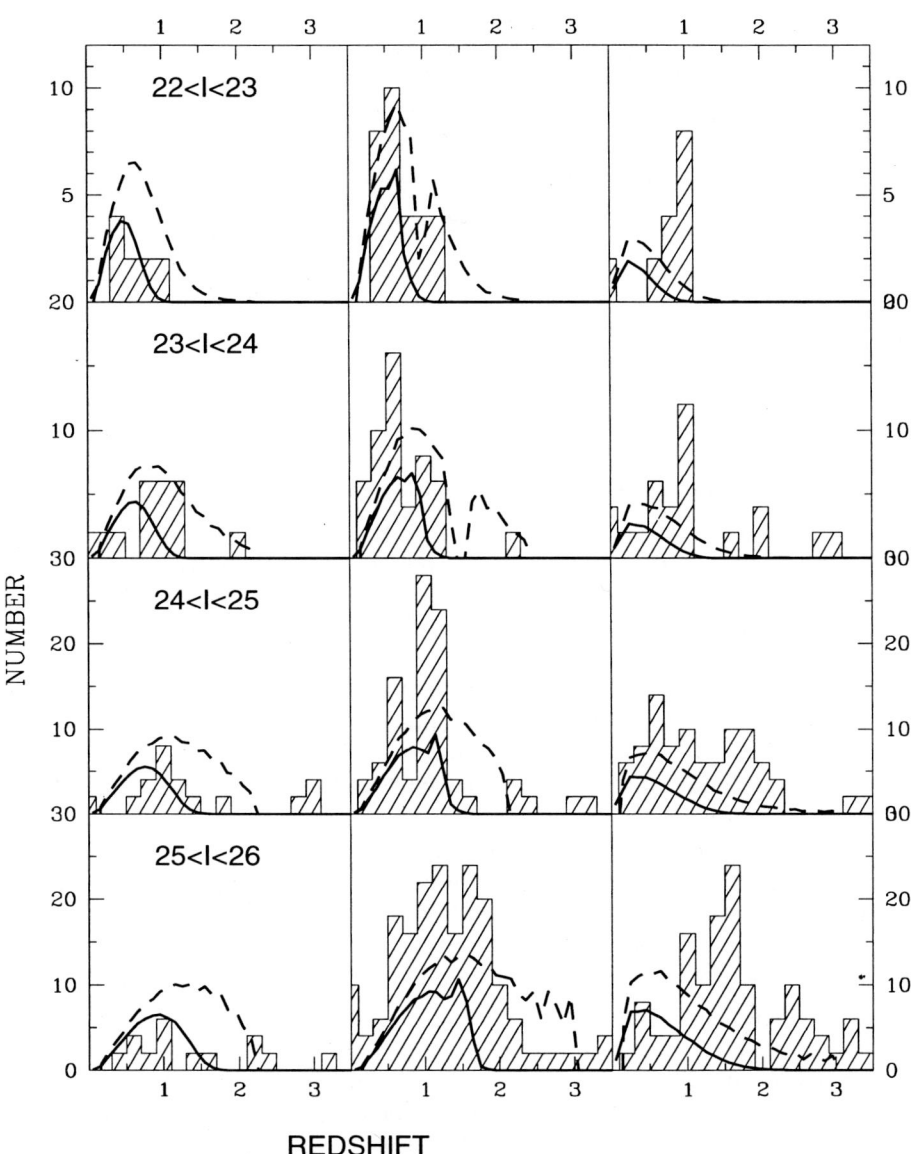

Fig. 2. Redshift distributions for: (a) ellipticals, (b) spirals and (c) irregulars) for progressively fainter magnitude intervals. The models lines show the zero- (solid) and maximal passive- (broken) evolution models based on a global renormalisation at $b_J = 18$.

required. A simpler interpretation is to simply adopt a lower normalisation for the elliptical models.

For the spirals we see that passive-evolution models match both the counts and the brightest N(z) distribution. However, disparity creeps into the plots at fainter magnitude intervals. In fact there is a repetition of the original faint blue galaxy problem within this sub-class in the sense that the observations agree in form to the zero-evolution models but are discrepant in amplitude. There are three obvious possibilities: (1) spiral galaxies undergo zero- evolution and exhibit a steeper local luminosity function than that derived from the CfA; (2) spiral galaxies have disassembled into a number of similar luminosity but less massive disk systems and (3) luminosity-dependent evolution. A frequent criticism leveled against morphological classification at faint magnitudes is the concern that regular galaxies simply appear more asymmetrical due to the K-correction *i.e.* objects viewed in the UV are intrinsically more irregular (e.g. Giavalisco *et al.* 1996). This is a valid concern but there are two points worth noting: Firstly, from Plate 1 the irregulars at $z > 1.5$ are *extremely* irregular and secondly, even with the potential for mis-classification we still see too many spirals at lower redshifts implying substantial number evolution ($\times 2$).

For the irregulars we saw from Figure 1, that even with the contentiously steep CfA local luminosity function, extremely strong evolution would be required before the model N(m,T=Sd/Irr) matches the observations. However the N(z,T=Sd/Irr)s of Figure 2 make this a moot point, and it is clear that the irregular class is irreconcilable with the observed N(z,T=Sd/Irr) distribution. The implication is that more than one population/process is contributing to the class of irregular galaxies: e.g. true irregulars — typically low-luminosity; and transient irregulars — peaking in the interval $z = 1 - 2$. Further work will be required to subdivide this population, however, before the nails are banged home in the dwarf-dominated coffin, we note that at no point does the N(z,T=Sd/Irr) model *strongly over predict* the observed distribution. Hence a steeply rising local luminosity function, akin to that seen in the CfA survey, is fully consistent and arguably favored by the low redshift end of the irregular N(z) distribution. However it is fair to say that the dwarf galaxy contribution to the faint galaxy counts is a minority component of the faint blue galaxy excess ($< 20\%$ to $I < 26$).

Finally we must also consider a more holistic interpretation allowing freedom of movement from one class to another. At some point all objects derive from some primordial density distribution and the distinction between morphological classes must eventually dissolve. Considering the extreme distances over which the HDF objects in this dataset are being seen this is clearly a serious consideration. One interesting coincidence from Figure 2, is that the high-z irregulars are typically observed, within each magnitude interval, at a redshift slightly higher to the spiral population. This begs the question as to whether this population represents the spiral progenitor population (non-relaxed spirals) ! In fact one might argue to simply combine the spiral and irregular sub-classes in which case the overall form of the N(m,T=disk) distributions agree well with the models and we are simply left with a disk normalisation problem.

4 What Next?

A few things prevent a definitive description and these are almost all due to uncertainties in our understanding of the local galaxy populations. The most important step forward over the next decade is to refocus our attention on redefining local galaxy samples and this has been recognised by the implementation of the 2dF and SLOAN surveys. However as has been shown here the inclusion of morphologies is crucial and both of these surveys need to find a way to incorporate such information. Accurate consensus of the local morphological luminosity functions and their precise normalisation would represent the single most important advancement. In fact it is the normalisation problem, not discussed here but see Shanks (1989), at $b_J = 18$, which stymies our attempts at a definitive explanation. The models shown here have been normalised uniformly to the total counts at $b_J = 18.0$, which in the absence of morphological data is an arbitrary decision taken to minimise the number of free parameters. If we allow ourselves the luxury of independent morphological renormalisations then we might adopt a renormalisation ratio of 1:2:3 for E/S0:Sabc:Sd/Irrs respectively. Such a normalisation goes a long way to simplifying the current interpretation outlined in the previous section. If the reason for the local normalisation problem is due to surface brightness selection effects as has been postulated then such a renormalisation would be favoured. Once again we see that we require local morphological information at *brighter* rather than fainter magnitudes (where morphological distinctions are expected to evaporate). How ironic it is then, that we now know more about the statistical properties of the distant galaxy population than that of the local population.

We acknowledge and thank the engineers, scientists and astronauts responsible for the ongoing success of the *Hubble Space Telescope*.

References

Driver S.P., Windhorst R.A., Ostrander E.J., Keel W.C., Griffiths R.E., & Ratnatunga K.U., 1995, ApJL, **449**, L23
Driver S.P., Windhorst R.A., Phillipps S., & Bristow P., 1996, ApJ, **461**, 525
Ellis R.S., 1998, ARA&A, **35**, 389
Glazebrook K., Ellis R.S., Colless M., Broadhurst T., Allington-Smith J., & Tanvir N., 1995a, MNRAS, **273**, 157
Glazebrook K., Ellis R.S., Santiago B., & Griffiths R.E., 1995b, MNRAS, **275**, L19
Giavalisco M., Bohlin R.C., Macchetto F.D., & Stecher T.P., 1996, AJ, **112**, 369
Hogg D., et al., 1998, ApJ, in press
Koo D.C., & Kron R.G., 1992, ARA&A, **30**, 613
Lilly S.J., Le Fevre O., Hammer F., Crampton D., 1996, ApJ, **460**, 1
Marzke R.O., Geller M.J., Huchra J.P., & Corwin H.G., Jr., 1994, AJ, **108**, 437
Metcalfe N., Shanks T., Fong R., & Roche N., 1995, MNRAS, **273**, 257
Odewahn S.C., Windhorst R.A., Keel W.C., Driver S.P., 1996, ApJL, **472**, L13
Poggianti B.M., 1997, A&AS, **122**, 399
Shanks T., 1989, in proc *The Extragalactic Background Light*, eds Bowyer S.C., Leinert C., (Publ: Kluwer Academic Press), 269

Fig. 3. A colour montage of all galaxies in the Hubble Deep Field with $I < 26$. The galaxies are subdivided into redshift intervals as indicated and according to apparent magnitude within each redshift interval. Note the increase in irregularity towards both faint magnitudes (true irregulars) and high redshift (evolutionary irregulars). The increase in irregularity at $z > 1.5$ coincides with the widely discussed "Madau peak".

Discussion

Illingworth: Your HST mosaics of images displayed by z and L are beautiful and very useful. It seems to me that the changes from $z = 3$ to $z \sim 1$ galaxies are most likely to be a combination of merging, new bursts and fading (merging "Christmas tree model"). You will not be able to put constraints on merging by looking at object-to-object correlation functions in visible/optical since many merged objects will be too faint or still just neutral gas. Need SKAI!

Driver: I agree a lack of signature in the angular correlation function only rules against some merger models e.g. a high merger rate between well formed galactic clumps of similar luminosity.

Illingworth: In your Marzke model fit to the HDF counts separated by type (E/S0/Spiral/Irr) you mentioned evolution of 1.8 mags by $z = 1$. This seems high – where did this value come from?

Driver: The 1.8 mags of evolution by $z = 1.0$ is that REQUIRED to reconcile the Sd/Irr sub-class to their number-counts if we adopt a "steep" Marzke LF. The required evolution for the giant classes is consistent with passive evolution.

Boyle: Is your preferred model of zero or passive evolution consistent with strong evolution in the UV luminosity density seen at $z < 1$ in the CFRS?

Driver: The preferred model of zero to passive evolution is only for the giants (i.e. ellipticals and spirals). It is only the Sd/Irr class which requires more dramatic evolution, to be consistent with the CFRS this presumably implies that it is also this class which must be responsible for the strong UV-evolution seen in the CFRS.

Mould: One of the things we can hope to do with the 8-metre telescopes is to study galaxy evolution by constructing the distribution function in the fundamental plane. In that diagnostic diagram – which should include spirals – there are no morphological decisions to be made and you can even plot the model predictions too!

Driver: Yes, ideally with high signal-to-noise high-resolution spectroscopy we can trace the variation in well known local structural properties. This is an obvious endeavour for the giant galaxies, unfortunately though it is the irregulars which appear to dominate the faint counts and for this class no local physical relations are known ... yet.

Rocca-Volmerange: The strong evolution derived from CFRS data in a "so called" Madau diagram, is not compatible with any model, even with evolution. The reason could be that the tracer of SF used to convert the CFRS data into a SFR rate could be not so significant or would need some correction function.

Driver: The Madau diagram contains no morphological information. The simplest explanation which would be consistent with our results and the Madau diagram is that ellipticals and spirals behave as expected while it is the irregulars which require dramatic evolution at $z = 1.5$. Alternatively as you suggest the conversion of the CFRS observables to SF rates may need further consideration.

Renzini: How secure are your photometric redshifts for, say, $1 < z < 2.5$?

Driver: In the range $1.5 < z < 2$ we have no spectroscopic redshifts with which to verify our photometric redshifts, this will eventually be rectified but for the moment is of some concern. Outside of this range however the agreement is very good.

An Extremely Blue Population in Multispectral Galaxy Surveys

Brigitte Rocca-Volmerange, Michel Fioc

Institut d'Astrophysique de Paris, 98 bis Bd Arago F-75014 Paris, France

Abstract. The nature of the blue galaxy excess, discovered in faint counts is tentatively analyzed in the frame of a multispectral approach including far-UV counts. The far-UV (2000Å) counts observed from the balloon experiment FOCA20000 bring conclusive constraints on star-forming populations. The evolution is followed with the help of our new model of galaxy evolution PEGASE. From Fioc and Rocca-Volmerange, 1997b (FRV97b), a population of dwarfs, periodically bursting, is needed to interpret the 2000Å data, in agreement with the optical and near-IR faint counts. However, this near UV-bright population is unable to solve the debated question of the high-z blue galaxy excess in a flat universe.

1 Introduction

The recent identification of a blue population in faint galaxy counts, confirmed in the deep Hubble Deep Field, hereafter HDF (Williams et al, 1996), requires a refined analysis with the help of evolutionary models. If the excess is due to nearby dwarf galaxies located at the faint end of the luminosity function, it will constrain the baryonic mass in the universe. If it is due to more distant galaxies, clues on the star formation history in the universe are likely waited for. The interpretation needs to take into account the distance effect as well as evolution effects of periodic starbursts and of evolved populations. At last the cosmological parameters (density parameter, cosmological and Hubble constants) play a crucial role at high redshifts, fundamental to interpret faint galaxy counts. In such a multispectral approach, the far-UV counts observed with the FOCA2000 balloon experiment are crucial constraints on the present star formation rate. A preliminary interpretation by Armand and Milliard, 1994 with a classical luminosity function, showed a deficiency of blue star forming galaxies. In FRV97b, we propose a new interpretation of the data with episodic starbursts added to a standard population, simultaneously followed in the optical and the near-IR. The total contribution of that UV population to the UV number counts does not exceed a factor 2 of the underlying population emission, while faint galaxy counts and color distributions in the optical and the near-infrared are significantly predicted.

2 Multispectral Surveys from the Far-UV to the Near-IR

In the visible and the near-infrared, faint galaxy counts $N(m)$ and color distributions $N(C)$ are complete down to V =29 (HDF) and K=24 (Moustakas et al,

1996), allowing to follow galaxy evolution on a long cosmic time-scale. In the best interpretations derived from simultaneous blue to red fits, red counts are partly arising the degeneracy resulting from only blue counts. Both colors are needed to coherently follow the young and evolved populations. Redshift surveys N(z) bring a third dimension till $z \simeq 1$ while the new telescope-time saving technique of photometric redshifts (Steidel, 1996) is useful in deeper surveys. Far-UV (2000Å) bright galaxy counts were observed with the balloon experiment FOCA2000 by Armand and Milliard, 1994 down to the magnitude $m_{2000\text{Å}}$ = 18. The preliminary interpretation by the authors with current luminosity functions concluded to the need of a star-forming population to conciliate the data with observations.

3 The Model of Galaxy Evolution PEGASE

The new spectrophotometric model of galaxy evolution PEGASE (Fioc & Rocca-Volmerange, 1997a) is built to interpret instantaneous starbursts and evolved galaxies from the near-infrared to the far-UV. Extinction and nebular contributions are coherently derived from the current radiation field. Codes and atlases are available for a friendly use by internet at http://www.iap.fr/users/fioc or /rocca. Its main characteristics are listed below: i) various sets of isochrones (Genova or Padova), both are extended to the thermal pulses of AGB and post-AGB phases ii) several optical stellar libraries; the near-infrared spectral library (Lançon & Rocca-Volmerange, 1992) with a 3Å resolution, extended through the J, H and K bands, will soon be included iii) the choice of star formation parameters : rates (gas depending, exponentials, others) and initial mass functions at choice. iv) the bolometric corrections are derived from the stellar SED, to avoid any inconsistency from the conversion of theoretical luminosities $L/L_\odot$ to magnitudes. v) the nebular emission corresponds to a typical HII region vi) the extinction correction depends on a star-dust relation. The metallicity effects will be published in a new version. Standard scenarios are fitted on reference templates at z=0.

4 The Modelling of a Far-UV (2000Å) Dwarf Population

The first interpretation of the far-UV counts (Armand and Milliard, 1994) with a standard luminosity function gave a deficiency by a factor 2, leading the authors to conclude in favour of a missing blue population. These results are confirmed with the recent model PEGASE, using new luminosity functions. Moreover, the colors of the missing population are extremely blue (B-I $\leq$ 1.6 and $m_{2000\text{Å}}$ - B $\leq$ 1.5). According to FRV97b, only an episodic star formation rate during its active phase is able to reproduce such blue colors. A luminosity function of these starbursting galaxies is built on the basis of evolving star formation rates. When compared to a Schechter function, $M_* = -17$, the faint end slope is α =1.3 and Φ_* is fitted on observations. Because these dwarfs are near, results

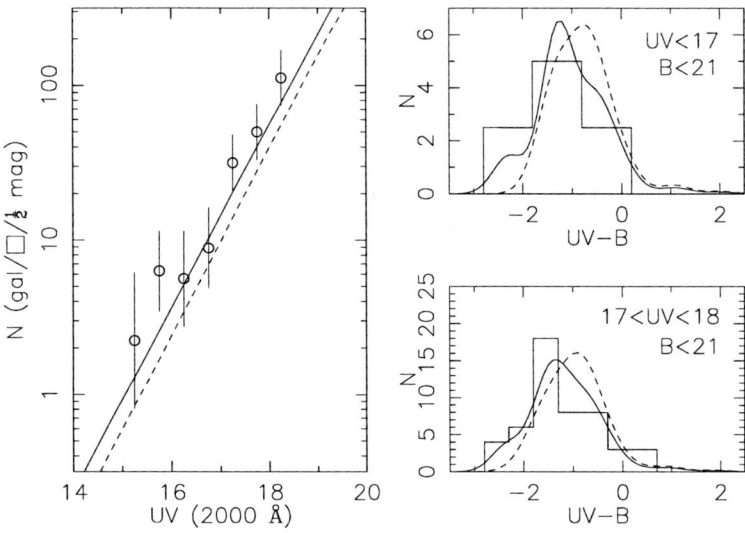

Fig. 1. Number counts and colour distributions predicted with Heyl et al. (1997) LF, without inclusion of bursting galaxies (dashed), and with them (solid), compared to the observations of Armand & Milliard (1994) (circles and histograms). Colour distributions are normalized to the area of the histograms.

are independent on the adopted cosmology. That blue population is added to the classical luminosity functions of the underlying stellar population. Fig. 1 present the comparison of galaxy counts and of $m_{2000\text{Å}}$ - B color distributions with the modelling with or without the starburst blue population. Figure 2 shows the deepest multispectral counts, including the HDF, fitted for various cosmologies with or not the bursting blue population. Its contribution is less than 10% to the U-band and optical filters and about null in the near-infrared. Clearly this population cannot explain the blue excess of galaxies in a flat universe without cosmological constant but is compatible with an open cosmology.

References

Armand C., Milliard B., 1994, *Astron.& Astrophys.* **282**, 1;
Fioc M., Rocca-Volmerange B., 1997a, *Astron.& Astrophys* **326**, 950;
Fioc M., Rocca-Volmerange B., 1997b, astro-ph/9709064 (FRV97b);
Heyl J., Colless M., Ellis, R. S., Broadhurst, T., 1997, *Month. Not. Royal. astro. Society* **285**, 613;
Lançon A., Rocca-Volmerange B.,*Astron & Astrophys. Sup.*, bf 96, 593;
Moustakas L.A., Davis, M., Graham, J.R.,Silk, J., Peterson, B.A., Yoshii, Y., 1996,*Astrophys. J.* **475**, 445;
Steidel C s, Giovalisco G., Pettini M, Dickinson M, Adelberger K L, *Astrophys. J.*, **462**, L17;
Williams R.E., et al., 1996, *Astro. J.* bf 112, 1335

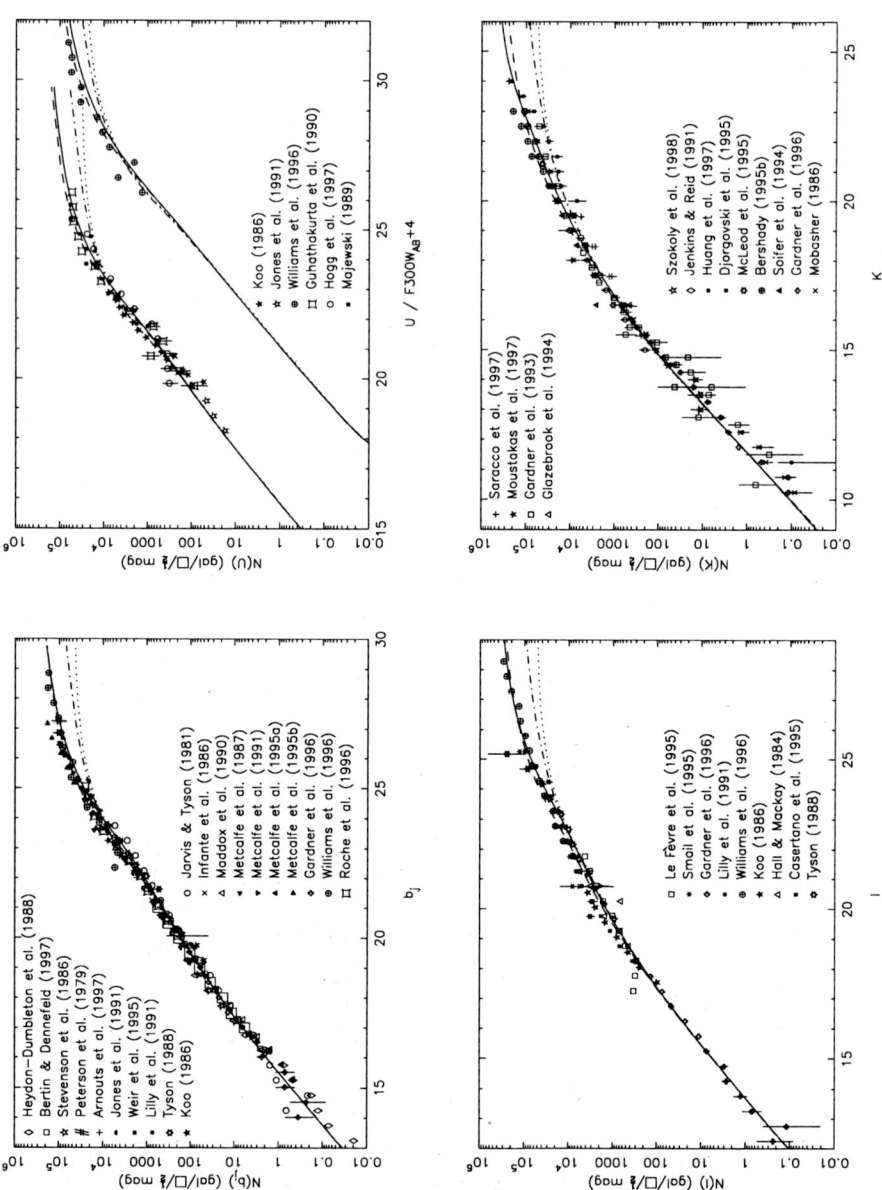

Fig. 2. Number magnitude counts (from FRV97b) in b_j, U (and $F300W$), I and K. Predictions including bursting galaxies are plotted for an open Universe ($\Omega_0=0.1$, $\Lambda_0=0$, $H_0=65$: solid), a flat Universe ($\Omega_0=1$, $\Lambda_0=0$, $H_0=50$: dot-dashed) and a Λ-dominated flat Universe ($\Omega_0=0.3$, $\Lambda_0=0.7$, $H_0=70$: dashed). The predictions without bursting galaxies are also plotted as a dotted line in the case of a flat ($\Omega_0=1$, $\Lambda_0=0$) Universe.

Discussion

Dickinson: What else is known (observationally) about the UV-excess galaxies in the FOCA sample? Have people measured redshifts, optical luminosities, etc?

Rocca-Volmerange: A redshift follow-up is in process by Dilliard and M.A. Treyer to derive a luminosity of the total sample. In recent proceedings a first estimate of star formation rate has been derived (Treyer, 1996). The sample of extremely-blue (B–I < -1.6) galaxies that I presented here, is only a subsample of the far-UV data from FOCA 2000.

Heisler: Does your PEGASE program fit the CO band head features in the near-IR?

Rocca-Volmerange: The CO band features were analyzed in our spectral analysis of starbursts in the near-IR (Lancon and Rocca-Volmerange 1996, New Astronomy, 1st Volume). To include such detailed analysis, we need reference spectra in the near-IR for any type of galaxies. I hope in a near future with the increasing samples of data from the IR camera arrays, we shall access to better spectral resolution spectra in the near-IR.

Spectro-Photometric Constraints on Distant Galaxies

Stéphane Charlot

Institut d'Astrophysique de Paris, CNRS, 98 bis Bvd Arago, 75014 Paris, France

Abstract. We describe primary spectro-photometric constraints to be gathered on distant galaxies in order to study their history of star formation. Fundamental degeneracies of spectral evolution models imply that the most straightforward way to constrain the history of star formation in a galaxy is in terms of its ongoing star formation rate and mass. In this context, we briefly review the reliability of the most commonly used star formation rate and mass estimators of distant galaxies.

1 Constraints on Nearby Stellar Populations

The most natural way to foresee what we might learn from future, detailed spectro-photometric observations of distant galaxies is to extrapolate on what we have learnt so far on the history of star formation of more nearby galaxies by scrutinizing their light. A necessary step to the interpretation of the spectral properties of galaxies in terms of current and past history of star formation is the modelling of the spectral energy distribution emitted by evolving populations of stars, or stellar population synthesis (e.g., Tinsley 1980). The most direct way to summarize the achievements of such models is to recall that any stellar population is expandable in a series of instantaneous starbursts. Therefore, our ability to constrain the star formation history of galaxies from their spectral properties relies ultimately on the accuracy to which the spectral evolution of a single, instantaneous-burst stellar population can be traced. Two main features of the spectral evolution of an instantaneous-burst population have fundamental implications for population synthesis studies of galaxies. First, the integrated ultraviolet luminosity of an instantaneous-burst population fades considerably during the first $\sim$ Gyr, as the most massive stars evolve off the main sequence. Second, the shape of the spectral energy distribution from the ultraviolet to the near-infrared does not evolve much after $\sim 4\,\mathrm{Gyr}$ because all low-mass stars evolve along roughly similar tracks in the Hertzsprung-Russell diagram. A detailed description of these properties can be found, for example, in Charlot (1996a).

These two fundamental properties of the spectral evolution of a burst stellar population imply a degeneracy in the interpretation of a galaxy's light in terms of the star formation history. In fact, a basic conclusion from population synthesis models is that the spectro-photometric properties of a galaxy depend essentially on the ratio $\psi(\mathrm{present})/\langle\psi\rangle$ of the present to past-averaged star formation rates, that controls the ratio of young (blue) to old (red) stars (e.g., Kennicutt, Tamblyn & Congdon 1994, and references therein). Dust and changes in metallicity introduce further degeneracies in the interpretation of the spectro-photometric properties because their effects are roughly parallel to that of age in a two-color diagram (Charlot 1996a). It is worth pointing out that within the framework of a specific population synthesis model, finer spectro-photometric diagnostics can be built to constrain the past history of star formation in a galaxy. In practice, however, the changes in photometric colors or line-index strength on which such finer diagnostics rely are more subtle than the typical discrepancies between the predictions of different models for the properties of stellar populations with fixed parameters (Charlot et al. 1996; Charlot 1996b).

The parameter $\psi(\text{present})/\langle\psi\rangle$ that can be most accurately constrained using population synthesis models does not provide us with any information on absolute ages nor on details of the past history of star formation in a galaxy. A direct implication of this degeneracy is that the spectral evolution linking galaxies at different redshifts is non-unique. In particular, there can be several ways to relate blue and red, bright and faint galaxies at various redshift, depending, for example, on whether star formation evolves smoothly or in more stochastic bursts. Perhaps one of the best-known illustrations of this degeneracy is the problem of the origin of early-type galaxies. Despite the wealth of observational constraints at redshifts up to $z \sim 1$, early-type galaxies can still be modeled either by assuming that they all were assembled early and evolved passively since a redshift of two or higher, or by assuming that they were assembled on average from more recent mergers (Kauffmann & Charlot 1998a, and references therein).

Hence, from the point of view of model interpretation, the most promising information to collect about distant galaxies is the ratio $\psi(\text{present})/\langle\psi\rangle$.

2 Star Formation Rate Estimators

We first investigate the accuracy to which $\psi(\text{present})$ can be estimated in galaxies. The most commonly used estimators of star formation are hydrogen recombination lines (Hα, Hβ), collisionally excited oxygen lines ([OII], [OIII]), the ultraviolet continuum from young massive stars (L_{1500} at 1500 Å, L_{2800} at 2800 Å), and the far-infrared radiation of dust heated by young stars (e.g., Kennicutt 1992; Meurer et al. 1995; Lilly et al. 1996; Madau et al. 1996). We do not discuss here Lyα emission as a star formation estimator because resonant scattering of Lyα photons by neutral atomic hydrogen dramatically affects their relation to the star formation rate in a galaxy (see Charlot & Fall 1993 for a review). Also, since we are interested in measurements of the star formation properties of galaxies from optical/near-infrared surveys, we do not discuss here the far-infrared radiation by dust grains (see, e.g., Meurer et al. 1995 and references therein). The star formation estimators listed above are optimal in different redshift ranges in the context of optical observations. In nearby galaxies, Hα and Hβ are the most commonly used estimators. In galaxies at redshifts up to $z \sim 0.8$, Hβ remains a viable estimator while [OII] can be used up to $z \sim 1.5$. Recent observations of extremely distant galaxies have led to routine use of L_{2800} (at $z \gtrsim 0.3$) and L_{1500} (at $z \gtrsim 1.3$) as star formation estimators up to $z \sim 4$. To reconstruct the history of star formation in galaxies from $z \sim 4$ all the way down to the present epoch, however, it is essential to calibrate consistently these various star formation estimators with respect to one another and in an absolute way.

Observed correlations linking the strengths of different emission lines and the ultraviolet and far-infrared fluxes from galaxies can be used to estimate and compare the relative sensitivities of the various indicators to star formation (e.g., Kennicutt 1992; Meurer et al. 1995). However, to ultimately relate observed fluxes to actual star formation rates — expressed in units of $M_\odot$/yr — requires to invoke models of the stellar and nebular emission from galaxies. Stellar emission corresponding to a given star formation history can be described using a population synthesis model (§1), while the reprocessing of stellar radiation by nebular gas is traditionally described using a photoionization model (e.g., Ferland et al. 1995 and references therein). Inevitably, this heavy modelling approach introduces a large number of free parameters (e.g., Stasińska & Leitherer 1996). The main free parameters of stellar population synthesis models are the stellar evolution prescription (from stellar evolution theory), initial mass function, metallicity and star formation history. The main free parameters of photoionization models are the density, filling factor, metallicity and dust contents of nebular gas inside HII regions. This parameter space is slightly reduced by the fact that combinations resulting in the same value of the ionization parameter lead to similar ionization structures of the gas and, for densities low enough to prevent important collisional deexcitation ($\lesssim 500\,\text{cm}^{-3}$), to similar line strength ratios.

There is strong observational evidence that the values of the critical input parameters of population synthesis and photoionization models which are relevant to studies of external galaxies vary on a case by case basis (e.g., Kennicutt 1998, and references therein). One of the most critical factors is the topology of the interstellar medium. For example, patchiness of the dust will weight reddening estimates, and hence star formation tracers, to the least extinguished regions. Determining the full set of model parameters appropriate for a specific object can be achieved, at least to some extent, in nearby galaxies with resolved HII regions and for which substantial observational constraints have been gathered on the metallicity, density and dust content of the gas as well as on stellar emission (e.g., García-Vargas et al. 1997). This approach, however, does not appear to be viable for systematic studies of more distant, coarsely resolved galaxies with fewer observational constraints.

Alternatively, one can illustrate the uncertainties affecting the star formation estimators commonly used to study distant galaxies by computing model predictions for a reasonably wide range of input parameters spanning the range observed in nearby galaxies. The dispersion in the response of a given estimator to the input star formation rate when sampling this range of parameters then gives an indication of its absolute reliability. Charlot & Longhetti (1998) have recently performed such an analysis, and we refer to their paper for detail.

As expected, H-recombination lines stand out as the most reliable star formation estimators in galaxies, although the relation between Hα or Hβ emission and star formation rate is uncertain by a factor of about 3 (or more if arbitrary extinction is allowed). The 1500 Å radiation from young stars would be an excellent tracer of star formation in dust-free galaxies. Extinction by dust, however, affects this estimator most dramatically. Also, the ultraviolet continuum at 2800 Å can be severely contaminated by older stars in galaxies with declining star formation rates (leading to substantial overestimates of the star formation rate). Finally, collisionally excited oxygen lines are highly sensitive to metallicity in star-forming galaxies, as has been noted before (e.g., Kennicutt 1992, and references therein).

The prevalence of H-recombination lines as star formation estimators implies that one must appeal to infrared spectroscopy to best constrain the star formation properties of distant ($z \gtrsim 1-3$) galaxies. Several programs are already underway to achieve this goal (e.g., Dickinson, this volume).

3 Mass Estimators

The other primary quantity to be determined about distant galaxies is their past-averaged star formation rate $\langle \psi \rangle$, which can be re-expressed as the mass in old stars divided by the age. Since old stars dominate the near-infrared light in a galaxy, mass estimates are often derived from the observed near-infrared luminosity. In fact, population synthesis models indicate that the mass-to-infrared luminosity ratio, M/L_K, is almost insensitive to details of the past star formation history in a galaxy (see for example Fig. 10 of Charlot 1996a). This is all the more remarkable in that different types of stars dominate the infrared light at different ages in an evolving stellar population (supergiant, asymptotic giant, and red giant stars) and conspire to maintain a roughly constant near-infrared luminosity. It is important to note, however, that since very low-mass stars on the lower main sequence are very faint, they can contribute significantly to the mass but not to the integrated light of a stellar population. Therefore, the lower part of the initial mass function is a key uncertain parameter in determinations of mass-to-light ratios.

In high-redshift galaxies, the observed K-band samples rest-frame emission at shorter wavelengths, and the mass-to-observed K-light ratio becomes more sensitive to the past history of star formation. The reason for this is that the mass-to-rest frame optical-light ratio increases as a stellar population ages, the optical luminosity declining steadily with decreasing turnoff mass (e.g., Fig. 9 of Charlot 1996a). Thus, at high redshift, the strength of the near-infrared light as a mass estimator is weakened by the uncertain k-correction to be applied to a galaxy. There are two distinct consequences from this. First, up to redshifts of $z \sim 1-3$, the dispersion in the observed M/L_K is still tolerable, and the K-band light still roughly traces mass (Fig. 1 of Kauffmann & Charlot 1998b). However, the absolute scale of the conversion between observed K light and mass changes significantly because of the k-correction (by a factor $2-3$ downward at $z \approx 1$).

Second, at redshifts $z \gtrsim 3$, where the K band samples rest-frame light blueward of the V band, relating the observed K magnitude to a mass becomes highly uncertain if the k-correction is not known.

Hence, changes in scaling at modest redshifts and high uncertainties at high redshifts must be kept in mind when estimating the mass of distant galaxies through their observed near-infrared light. Observations at longer wavelengths would appear more appropriate, but unfortunately the sky background already becomes untractable in the L band. The $1-5\,\mu$m infrared capabilities of the *Next Generation Space Telescope* will represent a breakthrough in this area. Finally, we note that another promising way of estimating masses of distant galaxies is to measure velocity dispersions by means of high-resolution spectroscopy. So far, this technique has mostly been applied to galaxies at relatively modest redshifts (e.g., Koo et al. 1997).

4 Conclusion

Fundamental degeneracies of spectral evolution models imply that the most straightforward way to constrain the history of star formation in a galaxy is in terms of its ongoing star formation rate and mass. Commonly used star formation rate and mass estimators, however, suffer from substantial uncertainties and become more difficult to handle when applied to high redshift galaxies. These uncertainties can be significantly reduced by appealing to near-infrared imaging and spectroscopy. It is worth noting that even if star formation rates and masses are ultimately known for all galaxies, evolutionary links between galaxy populations at different redshifts will not necessarily be readily identifiable (e.g., possibility of vanishing populations, top-heavy initial mass function, etc.). Nevertheless, the constraints will be invaluable to discriminate between competing models.

Acknowledgments

We gratefully acknowledge financial support from the organizers of this meeting.

References

Charlot, S. 1996a, in From Stars to Galaxies, ed. C. Leitherer, U. Fritze, & J. Huchra (ASP Conf. Series 98), 275
Charlot, S. 1996b, in the Universe at High Redshift, ed. E. Martínez-Gonzalez & J.L. Sanz (Springer: Lecture Notes in Physics 470), 53
Charlot, S., & Fall, S.M. 1993, ApJ, 415, 580
Charlot, S., & Longhetti, M. 1998, in preparation
Charlot, S., Worthey, G., & Bressan, A. 1996, ApJ, 457, 625
Ferland, G.J. et al. 1995, in Analysis of Emission Lines, ed. R.E. Williams & M. Livio (Cambridge: Cambridge University Press), 83
García-Vargas, M.L. et al. 1997, ApJ, 478, 112

Kauffmann, G., & Charlot, S. 1998a, MNRAS, in press
Kauffmann, G., & Charlot, S. 1998b, MNRAS, submitted
Kennicutt, R.C. 1992, ApJ, 388, 310
Kennicutt, R.C. 1998, ARAA, in press
Kennicutt, R.C., Tamblyn, P., Congdon, C.W. 1994, ApJ, 435, 22
Koo, D.C., Guzman, R., Gallego, J., & Wirth, G.D. 1997, ApJ, 478L
Lilly, S.J., Le Fèvre, O., Hammer, F., & Crampton, D. 1996, ApJ, 460, 1L
Madau, P. et al. 1996, MNRAS, 283,, 1388
Meurer, G.R., Heckman, T.M., Leitherer, C., Kinney, A., Robert, C. & Garnett, D.R. 1995, AJ, 110, 2665
Stasińska, G., & Leitherer, C. 1996, ApJS, 107, 661
Tinsley, B.M. 1980, Fundamentals of Cosmic Physics, 5, 287

Discussion

Bland-Hawthorn: Have you tried calibrating I(1500) from redshift zero galaxies?

Charlot: Your question is a bit unclear to me. What I presented was a set of model predictions of various SFR indicators, in which unlike Balmer line emission the L_{1500} luminosity was found to be very sensitive to dust. You would like to relate L_{1500} to an actual SFR (M.yr^{-1}) and for this you need to make models that depend on a large number of free parameters. In this sense, "calibration" is difficult and I did not attempt to achieve it. My goal, if you wish, was to show that it probably is difficult to calibrate (because of the strong dependence on extinction).

Glazebrook: We have measured Hα in 15 $z \sim 1$ CFRS galaxies and are finding "broad" agreement (within a factor of 2) between SFR(Hα) and SFR(L_{2800}) — with some oddballs. This is work in progress but we are encouraged that it agrees to a factor of 2 and is not out by a factor of 10.

Charlot: That is interesting.

Lahav: Could you comment on differences between various models and codes for galaxy spectra.

Charlot: We have investigated the differences in great detail in a recent paper (Charlot, Worthey, Bressan, 1996 ApJ). That paper was for models of "old" (≥ 1 Gyr old) stellar populations. A similar comparison for young (≤ 1 Gyr old) stellar populations can be found in Charlot (1996, in "From Stars to Galaxies", ed. C. Leitherer & V. Fritze (ASP conference series)).

Illingworth: Presumably it is best to use the rest-frame wavelength models for the infra-red (JHK) observations for estimating M/L instead of applying k-corrections.

Charlot: This is the point of the figure I showed. Unfortunately, at redshifts greater than 1, the rest-frame wavelength you probe is not a good mass estimator (V-band) if you have no good constraint on the galaxy spectral type.

Illingworth: Do you think it is okay to use Lyα to see lower limits to the star-formation rate?

Charlot: Certainly more okay than to use it to "estimate" the star-formation rate!

Semi-Analytic Models and Background Hydrogen-Ionizing Flux

Julien E.G. Devriendt, Bruno Guiderdoni, and Shiv K. Sethi

Institut d'Astrophysique de Paris, CNRS, 98bis Boulevard Arago, F-75014 Paris, France

Abstract. We estimate the contribution of galaxies to the cosmic background flux at 912Å by means of an extended semi-analytic model of galaxy formation and evolution which takes into account the absorption of Lyman-limit photons by HI and dust in the interstellar medium (ISM) of the galaxies. We find that, though the background Lyman-limit flux escaping from galaxies is negligible compared to the flux from quasars at high redshifts, these two contributions become comparable at $z \simeq 0$.

1 Modelling the Emission of Galaxies

1.1 Semi-Analytic Models

Using the framework of semi-analytic models of galaxy formation and evolution enables one to work explicitly in a cosmological context where one can track Cold Dark Matter halos and describe the dissipative processes the baryons will undergo in such halos (see Guiderdoni et al. (1998) and references therein for details). These astrophysical processes are still crudely modelled. Nevertheless, as shown in Devriendt et al. (1998) (hereafter DSGN) the models are able to reproduce — with a reasonably small number of free parameters — the global features of the universe such as the evolution of cosmic comoving emissivities in the optical, ultraviolet, near infrared and far infrared bands. Furthermore, assuming that damped Lyman–α systems (DLA) at high redshifts are the progenitors of present day galaxies, they also enable one to match the evolution of the neutral hydrogen content and average metallicity of such systems.

1.2 Hydrogen-Ionizing Flux

Using such models, we derived limits on the fraction of hydrogen-ionizing photons that can escape from galaxies. It is clear that this fraction will crucially depend on the relative distribution of stars, dust and neutral hydrogen in the ISM. In Fig. 1, we show the estimates for a fully ionized and non-dusty ISM, for a fully ionized but dusty ISM and finally for a non-ionized and dusty ISM where all the important elements (stars, dust and gas) are homogeneously mixed. For a more detailed discussion on the geometry as well as on the prescription we use to compute the optical depth of the intergalactic medium the reader is referred to DSGN.

2 Results

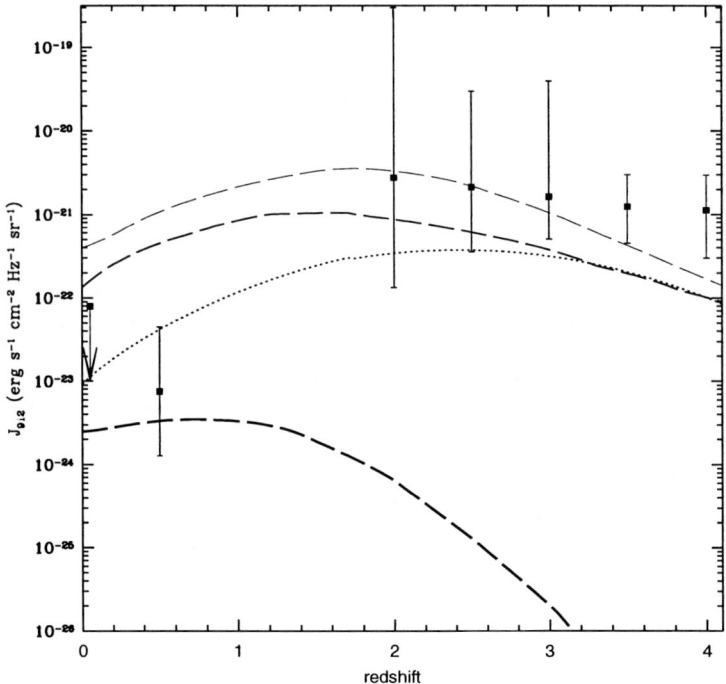

Fig. 1. The evolution of Lyman-limit background flux is shown for the various cases discussed in the text. The *dashed* lines give the predictions of the model. The lines of increasing thickness correspond respectively to models without dust and HI absorption, with dust but without HI absorption, and with both dust and HI absorption. The *dotted* line shows the evolution of background Lyman-limit flux from quasars. For details on the plotted data see DSGN.

From the curves shown in Fig. 1 we conclude that star-forming galaxies are very unlikely to give the major contribution to the background hydrogen-ionizing flux at high redshift, unless (i) most of the hydrogen they contain is ionized, **and** (ii) the DLA systems are not associated with star forming regions. On the other hand, this is not the case at low redshift where galaxies are likely to have a contribution comparable to the quasars'. We show in DSGN that this conclusion seems to remain valid with different distributions of the stars and gas.

References

Devriendt, J. E. G., Sethi, S. K., Guiderdoni, B., Nath, B. B. (1998): MNRAS *submitted*
Guiderdoni, B., Hivon, E., Bouchet, F. R., Maffei, B. (1998): MNRAS *in press*

Detection and Evolution of High-z Galaxies

Tom Broadhurst, Rychard Bouwens, Brenda Frye

Department of Astronomy, University of California, Berkeley, USA

Abstract. We present some results of a redshift survey designed to detect very distant galaxies behind massive lensing clusters. These objects are found to span the range $3.5 < z < 5.0$, they are intrinsically small ($< 1 kpc$) and display the spectral signature of gas outflow, at speeds of $\sim 200 - 300$km/s. Their space density, luminosity and size evolution are derived using a new model-independent method for comparing field galaxy populations at different redshifts. We find that distant galaxies observed in the HDF, $z > 2.0$, are far more luminous and compact than the low-z field galaxy population, for any reasonable choice of Cosmology. We also use this method to establish the relative evolution between the distant U,B and V-band selected galaxies, by cloning the HDF U_{300} dropouts to higher redshift. Little evolution in space-density or size is found over a wide redshift range, $2 < z < 6$, motivating higher redshift searches.

1 Cloning Low-z and Dropout Galaxies

Concrete statements regarding the evolution of faint field galaxies detected in deep HST images must take account of the considerable observational limitations affecting their detection. This includes surface-brightness dimming, k-corrections, sky noise, the instrumental resolution and sampling. To deal with this we have devised a model-independent 'cloning' technique for comparing resolved images of field galaxies at different redshifts. Complete samples of galaxies are extrapolated in a volume-limited way to higher redshift, in proportion to their individual space-densities, $1/V_{max}$, with a separate k-correction for every pixel belonging to each image. Deep fields of arbitrary depth are generated and then degraded to the match the observational parameters, thereby establishing a model-independent 'no-evolution' benchmark against which the basic statistical properties of higher redshift galaxies can be measured (Bouwens, Broadhurst & Silk 1998,1999).

Using the HDF images we quantify the relative evolution between high and low-z galaxies, in terms of space density, luminosity, size and morphology, over a wide range of redshifts $0.5 < z < 6.5$. Firstly we compare simulations based on the low redshift HDF population ($I < 22$) with a color selected sample of high redshift U_{300} dropouts. Figure 1 shows clearly that the number of galaxies with red U-B colours is underpredicted, even when Λ is included, enhancing the faint counts. The 'no-evolution' predictions lack the red 'plume', which comprises mainly $z > 2.0$ galaxies (See Dickinson these proceedings). The observed sizes and luminosities of the U_{300} dropout population, are found to be on average a factor 5-10 times higher surface brightness and 2-4 times smaller in scale length than the HDF field galaxy population at low redshift, depending on the choice of Cosmology.

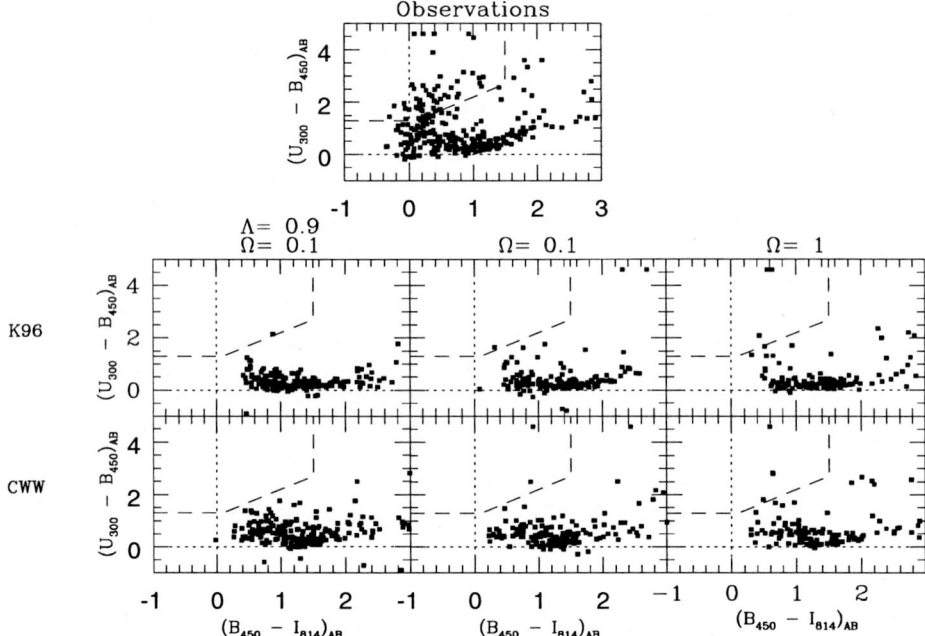

Fig. 1. Comparison of the observed UBI colour plane of the HDF, with our cloned 'no-evolution' prediction, based on the low redshift, $z \sim 0.5$ HDF field galaxy sample. Note the absence of the red 'plume' of high redshift, $z > 2$ galaxies, requiring strong evolution between $z = 0.5$ and $z = 2.5$

In the same way, we can 'clone' the U-band selected 'dropout' population to higher redshift, for comparison with faint B and V color selected Ly-depressed galaxies at higher redshift, allowing the relative evolution of these populations to be quantified over a wide redshift range, $2 < z < 6$, within the HDF. In the case of BVI color selection, spectroscopy shows this to be fairly robust way of detecting $z \sim 4$ galaxies (See Dickinson these proceedings). For objects detected in only V and I, most are very small, faint and of low surface brightness. Contamination by foreground early type galaxies is not likely for $V - I > 3$ (Figure 2), although our redshift survey of red galaxies finds a small contribution of dust reddened lower redshift objects (Frye et al 1998). Figure 2 shows the count comparison between these high-z populations, indicating no evidence of a decline in number with increasing redshift.

2 Redshift Survey of Magnified Red Galaxies

Lensing allows us to extend the limiting magnitude for detection of faint galaxies. The large magnification generated by clusters is responsible for the serendipitous discovery of some of the most distant galaxies known (Trager et al 1997,

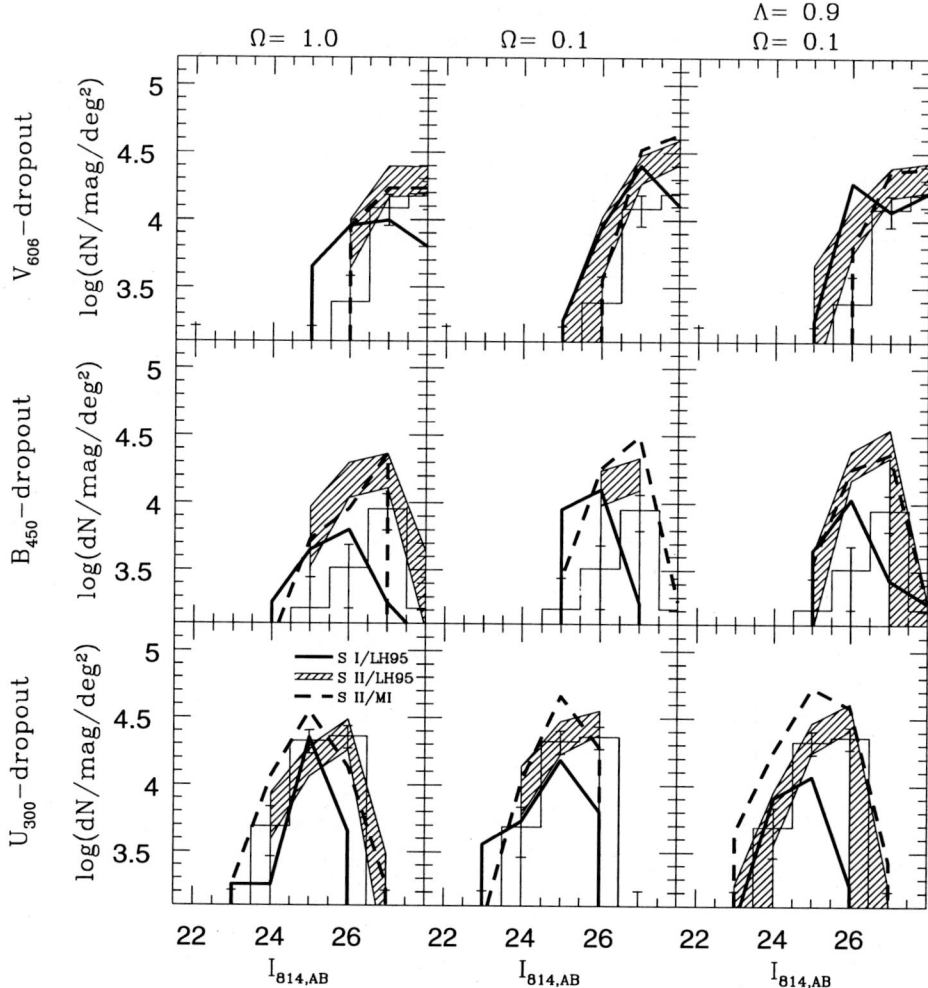

Fig. 2. Comparison of UBVI selected samples from the HDF with 'no-evolution' predictions cloned from the U_{300} sample (solid curves). No evidence of a decline in number with increasing redshift is found. Note the consistency of the U_{300} counts as required by our technique.

Franx et al 1997, Frye & Broadhurst 1998). We have carried out a systematic redshift survey of lensed galaxies with the aim of measuring the cluster mass by magnification and to obtain relatively detailed spectroscopy of very distant galaxies, $z > 3.5$.

Target selection is based only on color. Selecting all objects redder than the E/S0 sequence of the lensing cluster virtually guarantees the sample lies behind the cluster, where the k-correction of higher redshift early-type galaxies makes them appear redder. We do not restrict selection to arc shaped objects, since in general mass sub-structure can create apparently undistorted images, and secondly for intrinsically small sources the elongation due to lensing may still render the image too small to be well resolved (for this reason lensed QSO's do not appear arc shaped in optical images).

Our high redshift lensed galaxies all lie close to the critical curve of the cluster and span the range $3.5 < z < 5.0$. Figures 3 and 4 shows two interesting examples. The first is our brightest example of a distant galaxy, where 4 images are seen of a source at $z = 4.04$. This object is highly magnified in the tangential direction by the combined effect of the cluster potential and a massive elliptical galaxy lying close to the critical curve of the cluster.

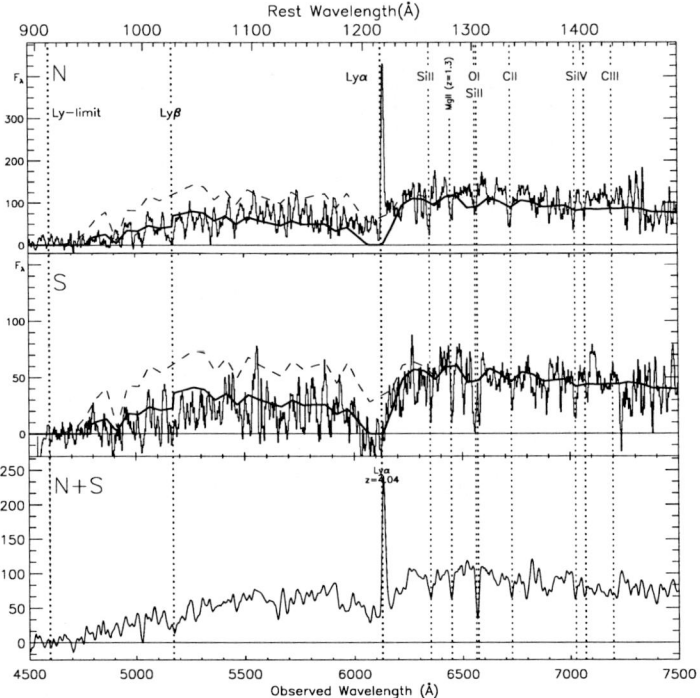

Fig. 3. Keck spectra of the arcs at z=4.04. The upper two panels show fluxed spectra of the Northern and Southern arcs taken with slits aligned normal to the long axes of these arcs. Overlayed is the best fitting B3-star continuum, with the opacity of the forest included, together with damped Lyα absorption at the galaxy redshift, but no reddening. The middle panel shows an independent observation of both these arcs with slits aligned along the length of the arcs, revealing metal lines at z=4.04, on which the redshift is based. These lines lie blueward of the centroid of the asymmetric Lyα emission by 300km/s.

Most objects do not show prominent Ly-α emission. Where it is seen with high S/N it appears redshifted with respect to the metal absorption lines, and asymmetric in shape (Figure 3). This behaviour is also seen in other sufficiently high quality spectra of distant galaxies most notably Franx, et al (1997), behaviour which has also been recently observed in local starburst galaxies, observed in the UV (Lagrand et al 1997). Lagrand et al (1997) explain the redward shift of the emission as arising from back-scattered Lyα photons from the HI around the expanding HII regions, so that the Doppler redshift of far-side emission is cleared of foreground absorption within the galaxy along the line of sight. The absence of Ly-α emission is therefore most likely due to a combination of high internal HI column and lower gas expansion rate, rather than dust absorption.

Our highest redshift galaxy to date lies at $z = 4.9$ (Figure 4) behind A1689 (Frye et al 1998). Its spectrum is very similar to that of Franx et al (1997) at this same redshift. A sharp flux decrement is observed shortward of Lyman-α due to the general forest absorption with a distinctive step like bump longward of Ly-β as predicted at the redshift of the galaxy (See Madau 1995), leading to a very red colour, V-I=4.2, redder than any passively evolving elliptical at lower redshift. This object is intrinsically very small (0.1–1Kpc), depending on the degree of magnification, barely resolved in HST images exposures, despite its close proximity to the critical curve of the cluster and hence large tangential magnification ($10 < \mu < 100$).

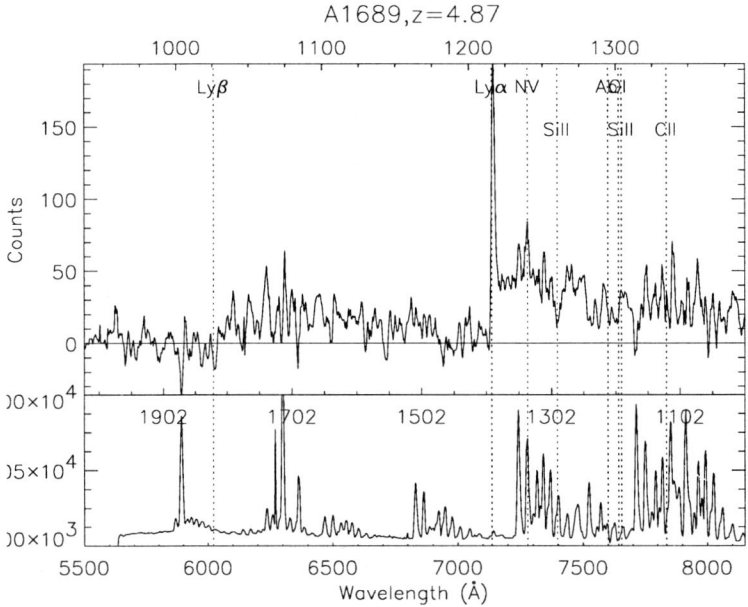

Fig. 4. This object is a highly magnified $I = 24$ galaxy at $z = 4.87$ behind the massive lensing cluster A1689. Despite the large magnification this object is barely resolved in HST images, indicating a small intrinsic size of only ~ 0.1Kpc and sub-L^* luminosity.

3 Discussion

The small sizes and high space-density inferred for these distant galaxies bears a qualitative resemblance to the early evolution expected in the hierarchical models for structure formation, in which increasingly more massive bound objects collapse and merge over time. However it appears that the formation of gaseous disks is a major problem in the context of this model. Detailed numerical simulations show that gas collapse is too efficient in a standard dark-matter halo to produce large rotationally supported disks (Steinmetz & Muller 1998), motivating discussions of 'feedback'. Perhaps the observed high velocity gas flows generate the required heating, but more plausibly they may help explain the widespread enrichment of the IGM at high redshift, requiring significant gas loss from the halos.

The most relevant observational information related to disk formation may lie at lower redshift, in the observationally difficult range $1 < z < 2$. Here we must expect some form of transition in the space density of dominant class of galaxy, between the luminous compact young galaxies at higher redshift and the large disk and E/SO galaxies abundant at lower redshift, $z < 1$. In this intermediate redshift range the Lyman-limit may prove the most efficient way of measuring redshifts, requiring only low resolution UV grism spectroscopy from space to define large redshift complete surveys.

References

Broadhurst, T., Bouwens, R., Silk, J.(1998): *Ap.J. in Press*
Franx, M., Illingworth, G.D. ,Kelson, D., VanDokkum, P.G.,Pieter, G.,Tran, K. (1997): *Ap.J. 486, L75*
Frye, B. Broadhurst, (1998): *Ap.J.Lett 499, L115*
Frye, B., Broadhurst, T., Guhathakurta, R., Kaiser, N, Dahle, H.(1998): *Ap.J. in Prep*
Lagrand, F., Kunth, D., Mas-Hesse, J, Lequeux, J., (1997): *A&A 326,929*
Madau, P. (1995): *Ap.J. 441, 18*
Steidel, C. C., Giavalisco, M., Dickinson, M., & Adelberger, K. L (1996): *AJ. 112,352*
Steinmetz, M. Muller, E., (1995): *MNRAS 276, 549*
Trager, S. C., Faber, S. M., Dressler, A. & Oemler, A (1997): *Ap.J. 485, 92*

On the Nature of Red Galaxies in the Early Universe

P.J. Francis, B.E. Woodgate, A.C. Danks

[1] Mt Stromlo & Siding Spring Observatory and Dept. of Physics & Theoretical Physics, Australian National University, Canberra ACT 0200, Australia
[2] NASA Goddard Space Flight Center, Code 681, Greenbelt, MD 20771, USA.
[3] Raytheon STX

Abstract. Increasing numbers of extremely red galaxies are being found in the high redshift universe. We present accurate near-IR photometry and NICMOS imaging for one of these galaxies, and demonstrate that it is a merging pair of early type galaxies. The merger is triggering only very moderate rates of star formation: even after many such mergers, the dominant stellar population will be old. This suggests that hierarchical formation models for elliptical galaxies can work.

1 Red Galaxies

Most currently known high redshift galaxies have the blue colours expected of new-born galaxies. A steady stream of much redder galaxies are, however, being found (e.g. Graham & Dey 1996). These could be dusty starbursts, or composed of old stellar populations. Either way, they could be an important part of the early universe. In this poster, we present new data on two of these red galaxies, 2139−4434 B1 and B4, at redshift $z = 2.38$, discovered by Francis, Woodgate & Danks (1997).

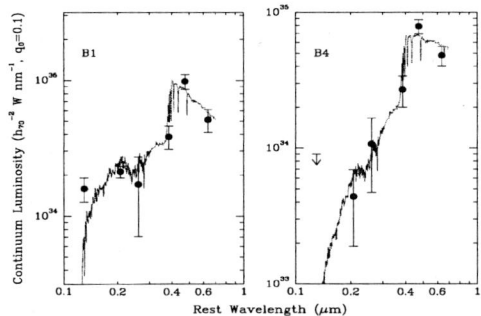

Fig. 1. Continuum photometry of the two red galaxies. Note the clear redshifted 400.0 nm breaks. The dotted lines are old stellar population model fits, using the spectral synthesis models of Bruzual & Charlot (1998, in press). Emission-line fluxes have been removed from the broad-band photometry.

Our deep near-IR and optical photometry (Fig 1) clearly shows that these two galaxies have large 400.0 nm breaks. The colours are inconsistent with dusty starburst models: we can only get acceptable fits if the continuum emission is dominated by an old (> 0.5 Gyr), massive ($> 5 \times 10^{11}$ solar masses) stellar population.

2 Hierarchical Formation

How can galaxies at $z = 2.38$ be this old and massive? Our photometry suggests that they formed before redshift 5: theories such as CDM are not consistent with the formation of bound objects so massive that early.

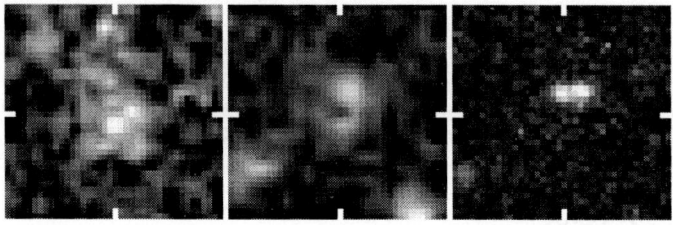

Fig. 2. Co-aligned images of B1. Narrow-band Lyα in the left panel, R-band in the centre, NICMOS F160W (H-band) in the right panel. All panels are $8''$ on a side, and registered to better than 1 pixel accuracy.

We obtained high resolution imaging of one of the red galaxies, B1 (Fig 2). It consists of two very red compact sources (either ellipticals or spiral bulges) separated by ~ 3 projected kpc. They seem to be interacting: a knot of rest-frame UV light (observed-frame R-band) comes from midway between them, and a UV-bright tail (a tidal tail?) extends away from them, copiously emitting Lyα. We therefore hypothesise that we have caught two massive elliptical galaxies in the act of merging.

Kauffmann (1996) suggested that giant ellipticals could form by the hierarchical merger of smaller ellipticals. If this could happen *without triggering significant star formation,* the stellar population of a red galaxy would be older than the galaxy itself, so no massive galaxies at $z > 5$ would be required. Our data seems consistent with this model: stars are being formed during the collision, but at only $\sim$1–5 solar mass per year: over the full merger timescale, newly formed stars will only account for 0.1% of the total stellar mass.

References

Graham, J. R., & Dey, A. 1996, ApJ 471, 720
Francis, P. J., Woodgate, B. E. & Danks, A. C. 1997, ApJL. 482, L25
Kauffmann, G. 1996, MNRAS, 281, 487

On the Evolution of X-Ray Clusters

Piero Rosati

European Southern Observatory, 85748 Garching b. München, Germany

Abstract. I summarize recent progress in the determination of the redshift evolution of the cluster abundance from large area, deep X-ray surveys with the ROSAT satellite. Serendipitous searches for galaxy clusters in deep PSPC pointings (such as the ROSAT Deep Cluster Survey) have revealed a sizeable population of clusters at $z > 0.5$ and are sensitive enough to detect clusters beyond redshift one. These surveys have significantly improved our knowledge of cluster evolution from an observational standpoint, but the interpretation of the new data within the framework of theories of structure formation is still a matter of debate.

1 Introduction

The redshift evolution of the abundance of galaxy clusters has long served as a valuable tool with which to test models of structure formation and set constraints on fundamental cosmological parameters, such as Ω_0. To this end, a considerable observational effort has been devoted over the last decade to the construction of homogeneous samples of clusters over a large redshift baseline. Until a few years ago, the difficulty of high redshift cluster searches in deep optical images and the limited sensitivity of early X-ray surveys had resulted only in a handful of spectroscopically confirmed clusters at $z > 0.5$. As a result, the evolution of the space densities of clusters, even at moderate look-back times, has until recently remained a controversial issue (Couch et al. 1991, Henry et al. 1992).

2 Searches for X-Ray Clusters

With the advent of X-ray imaging in the 80's, it was soon recognized that X-ray searches for galaxy clusters have the advantages of revealing physically-bound systems out to cosmologically interesting redshifts and lead to flux-limited samples with well-understood selection functions. Pioneering work in this field was carried out by Gioia et al. (1990) and Henry et al. (1992) based on the Einstein Medium Sensitivity Survey (EMSS). The ROSAT-PSPC detector, with its unprecedented sensitivity and spatial resolution, has allowed for a significant leap forward. ROSAT data have provided the means to carry out large contiguous area surveys of nearby clusters with the ROSAT All-Sky Survey (RASS), as well as much deeper serendipitous searches based on single pointings. On-going X-ray surveys of distant galaxy clusters include those utilizing PSPC archival data [RDCS (Rosati et al. 1995), SHARC (Collins et al. 1997), WARPS (Scharf et al. 1997), CfA (Vikhlinin et al. 1998)] and a survey in the North Ecliptic Pole (Henry et al. 1997), the deepest area scanned by the RASS.

The ROSAT Deep Cluster Survey (RDCS) (Rosati et al. 1995, 1998) has pushed these studies to the deepest X-ray fluxes yet and has amassed, to date, a large enough sample to allow a robust study of the evolutionary properties of the cluster population. A full discussion of the selection technique for the RDCS sample can be found in Rosati et al. (1995). Cluster candidates are selected from a serendipitous search for extended X-ray sources in deep pointed observations with flux $f_{-14} = f_X[0.5-2.0]\text{keV}/(10^{-14} \text{ erg cm}^{-2} \text{ s}^{-1}) > 1$. A search over ~ 50 deg^2 in 180 X-ray fields scattered across the two galactic caps and an extensive optical identification program have yielded about 130 clusters/groups to date. Spectroscopic observations have secured more than 100 cluster redshifts so far, spanning the range 0.045–0.83. A significant fraction of the newly discovered clusters lie at high redshift – about one-third at $z > 0.4$ and a quarter at $z > 0.5$ (Fig. 1). Several X-ray faint candidates have optical counterparts with V-I colors and magnitudes typical of clusters at $z \gtrsim 0.8$, suggesting that the highest redshift tail of the RDCS is yet to be discovered and will require near IR deep imaging and spectroscopy with 8m-class telescopes.

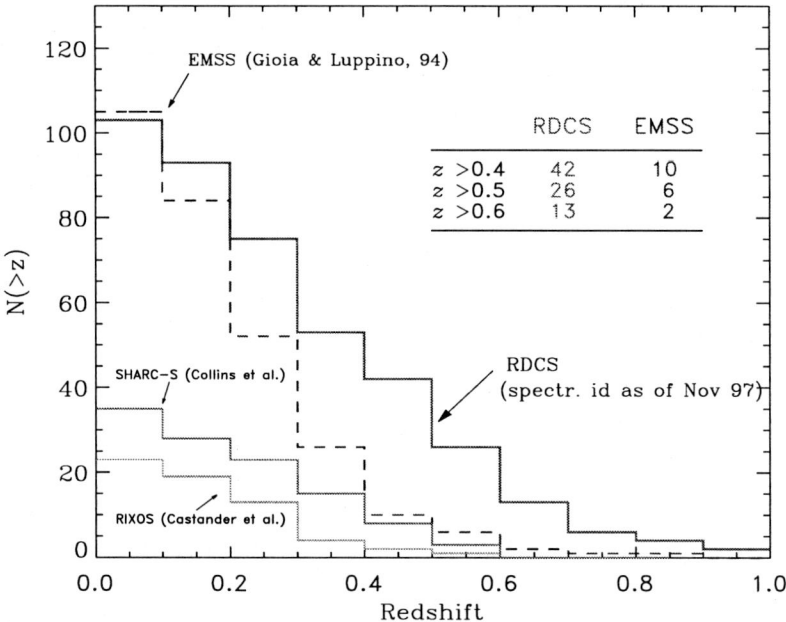

Fig. 1. Cumulative redshift distributions of *spectroscopically identified clusters* in several on-going ROSAT deep surveys compared with the EMSS sample (dashed line).

3 Evolution of the Cluster Abundance out to $z \simeq 0.8$

Using only the brightest half of the RDCS sample ($f_{-14} > 4$), a complete flux-limited sample of 70 clusters with measured redshifts can be defined and used to derive the X-ray Luminosity Function (XLF) as a function of redshift (Rosati et al. 1998). In Fig. 2 we show the RDCS XLF in 3 redshift bins, along with previous determinations of the local XLF. The XLF at higher redshifts from the Einstein Extended Medium Sensitivity Survey (EMSS) (Gioia et al. 1990, Henry et al. 1992) is also shown. The luminosities of RDCS clusters in the [0.5–2.0] keV rest-frame band range from $\sim 4 \times 10^{44}$ to 0.8×10^{42} erg s^{-1} and thus probe the region from moderately rich clusters (like Coma with $L_X \simeq L_X^*$) to poor groups. Given its relatively small surveyed area and volume (Fig. 3), the RDCS unfortunately does not provide a good probe of the XLF above $L_X \simeq 3 \times 10^{44}$ erg s^{-1}. This is the luminosity range where the EMSS found evidence of a negative evolution, i.e. a steepening of the high end of the XLF at $z \gtrsim 0.3$ (Gioia et al. 1990).

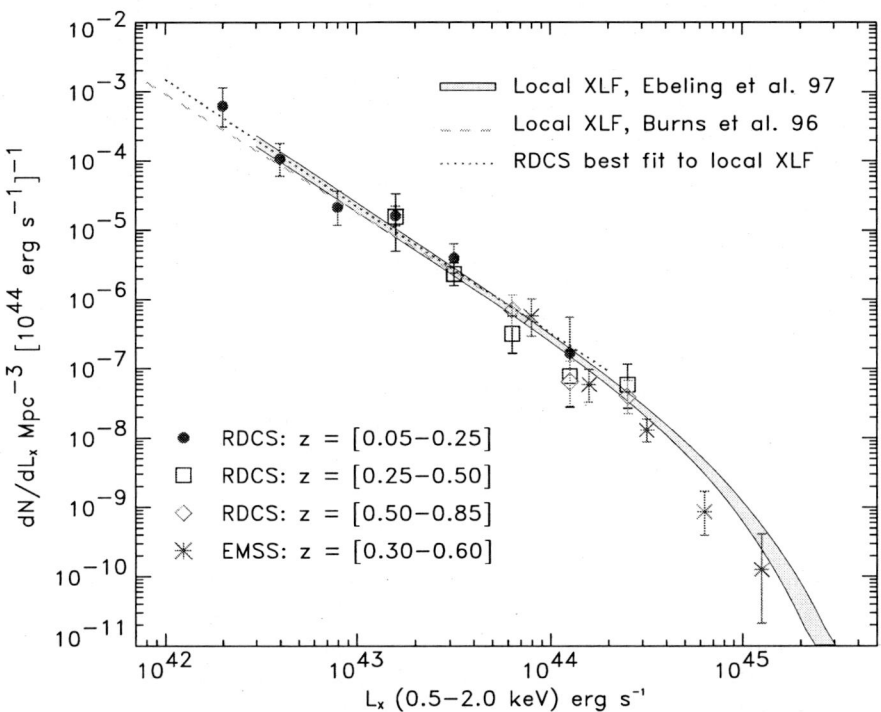

Fig. 2. The X-ray Luminosity Function for the RDCS sample in three redshift shells. Also plotted are two independent determinations of the local XLF from ROSAT All-Sky Survey data and the EMSS XLF at high redshifts (Rosati et al. 1998).

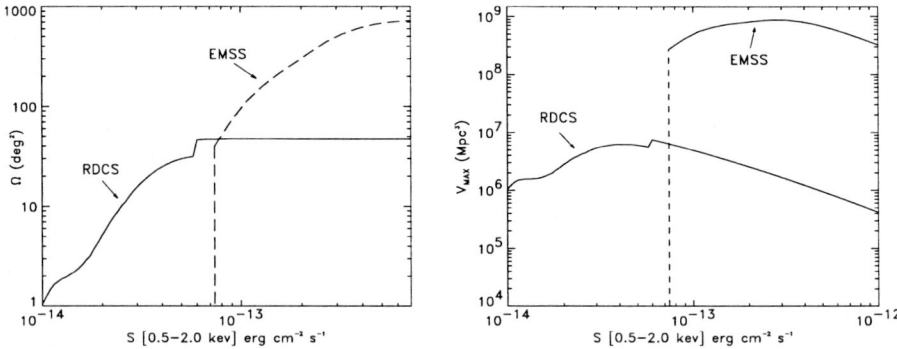

Fig. 3. *Left:* Effective sky coverage of the RDCS and EMSS surveys. *Right:* Corresponding volume explored by the two surveys for a typical L_X^* cluster.

Inspection of Fig. 2 shows several interesting features. Firstly, the local $(0.05 < z < 0.25)$ XLF of the RDCS is found to be in excellent agreement with two independent determinations from RASS data (Burns et al. 1996, Ebeling et al. 1997), with slopes in the range $1.75 - 1.85$ and consistent normalizations. This is quite remarkable, especially considering that these surveys used completely different selection techniques, with the RDCS being the only one whose selection is driven by X-ray properties alone. Thus, it would appear that the local cluster abundance is now well established. One should keep in mind that only two years ago, different surveys were finding faint end slopes in the range $1.1 - 2.2$, but the completeness levels and sky coverages of these surveys were not fully understood. A second point to note from Fig. 2 is that the RDCS XLF in the higher redshift bins *does not show any significant evolution*, at least out to $z \simeq 0.8$, over the probed luminosity range ($10^{43} \lesssim L_X(\text{erg s}^{-1}) \lesssim 3 \times 10^{44}$).

In Fig. 4, we show the cluster number counts down to $f_{-14} = 1$ derived from the whole RDCS sample which includes 130 clusters with $f_{-14} > 2$ (90% of which have been identified from optical imaging data) and 160 with $f_{-14} > 1$. The larger uncertainty associated with the faintest RDCS data point, reflects the incompleteness of the optical identification at $f_{-14} < 2$. Independent measurements from other sources are also plotted, the brightest counts being derived from RASS data (De Grandi, 1996; Ebeling et al. 1998). An agreement at the 2σ level is found among independent determinations spanning 3 decades in flux. The accurate determination of the local XLF allows us to compare the observed number counts with predictions based on a non-evolving XLF. The $\log N - \log S$ at low fluxes is very sensitive to the faint end slope of the XLF, as well as its evolution, which is not directly probed by the RDCS XLF in the two highest redshift bins. The agreement between the observed counts and the no evolution predictions (for any value of $0 \leq q_0 \leq 0.5$) further strengthens the evidence that the XLF does not evolve, within the present uncertainties, over a wide range of luminosities ($2 \times 10^{42} \lesssim L_X(\text{erg s}^{-1}) \lesssim 3 \times 10^{44}$). An independent study by

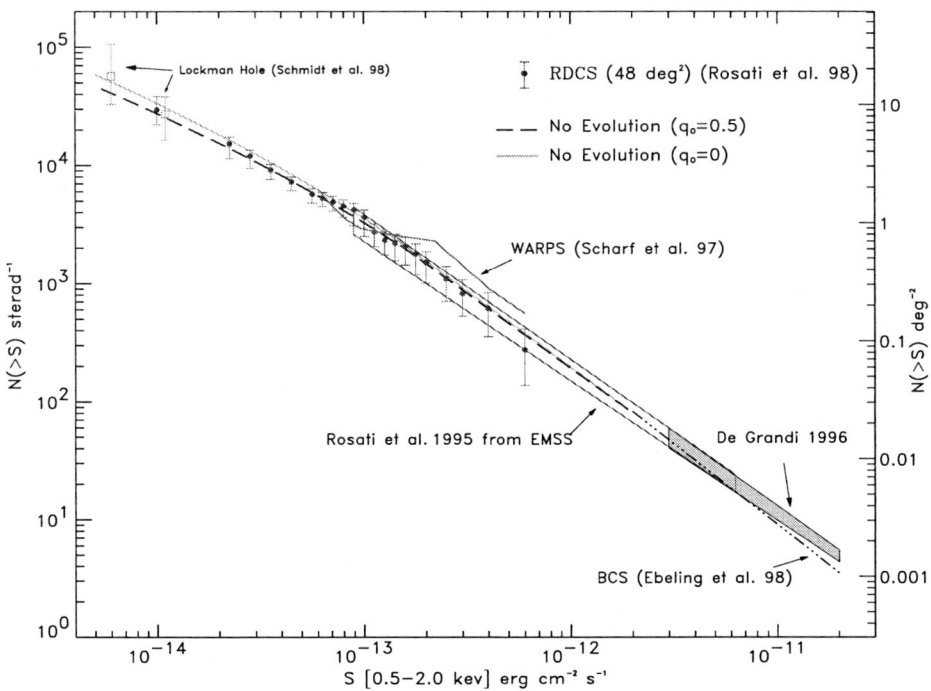

Fig. 4. Observed cluster cumulative number counts for the RDCS sample (Rosati et al. 1998) and other independent determinations. The two faint data points (open squares) have been derived from a sample of X-ray sources in the Lockman field with complete optical identification down to $f_x > 5.5 \times 10^{-15}$ erg cm^{-2} s^{-1} (Schmidt et al. 1998).

Burke et al. (1997), part of the SHARC-South collaboration, has also found no evidence of significant evolution of the cluster abundance at intermediate X-ray luminosities.

It should be stressed that these results complement those of the EMSS which samples a much large volume at moderate redshifts. Evolution (in density and/or luminosity) of the most luminous, presumably most massive systems at $z \gtrsim 0.4$ may very well occur. A moderately deep survey ($F_X > 5 \times 10^{-14}$ erg cm^{-2} s^{-1}) covering several hundreds square degrees would be required to further study this issue.

4 Discussion

Although X-ray surveys have shed much light on the evolution of the abundance of clusters at a given X-ray luminosity, the theoretical interpretation of these findings is still ambiguous. The implications that the RDCS $\log N$–$\log S$ have for CDM models of cluster formation have been discussed by Kitayama & Suto

(1997), Mathiesen & Evrard (1998) and Cavaliere et al. (1998). Borgani et al. (1998) have fully explored the space of cosmological parameters and evolutionary parameters of the intra-cluster medium by matching models with the observed distributions of the RDCS ($dN(z)/dL_X$, $N(z)$, $N(>S)$). This analysis has shown that without additional observational inputs from the temperatures of high-z clusters or a better understanding of the physics governing the evolution of their gaseous component, it is difficult to draw firm conclusions on the value of the density parameter Ω_0, although low-Ω universes are favoured.

In order to make further progress in understanding the evolutionary history of clusters and the background cosmology, a detailed study of a sizeable sample of high-z clusters is needed. This study should include the mass distribution of $z \gtrsim 0.6$ clusters via dynamical studies (velocity dispersion), as well as via the weak lensing technique (see G.Squires, this volume). Thus far, a census of $z > 1$ clusters has been difficult to assemble and represents one of the most challenging observational tasks for the next decade.

References

Borgani, S., Rosati, P., Tozzi, P., Norman, C. (1998), ApJ, submitted
Burke, D.J, Collins, C.A., Sharples, R.M., Romer, A.K., Holden, B.P., Nichol, R.C. (1997): ApJ, **488**, L83
Burns, J.O. et al. (1996): ApJ, **467**, L49
Castander, F.J. et al. (1995): Nat, **377**, 39
Cavaliere, A., Menci, N., Tozzi, P. (1998): ApJ, in press
Collins, C.A., Burke, D.J., Romer, A.K., Sharples, R.M., Nichol, R.C. (1997): ApJ, **479**, L117
Couch, W.J., Ellis, R.S, Malin, D.F., MacLaren, I. (1991): MNRAS, **249**, 606
De Grandi, S. (1996): Proceedings of Röntgenstrahlung from the Universe, MPE Report **263**, 577, ed. Zimmermann H.U., Trümper, J., Yorke, H.
Ebeling H., Edge, A.C, Fabian, A.C., Allen, S.W., Crawford, C.S., Böhringer, H. (1997): ApJ, **479**, L101
Ebeling, H. et al. (1998): MNRAS, submitted
Gioia, I.M., Henry, J.P., Maccacaro, T., Morris, S.L., Stocke, J.T., Wolter, A. (1990): ApJ, **356**, L35, 1990
Gioia I.M. & Luppino, G.A. (1994): ApJ Suppl. **94**, 583
Henry, J.P., Gioia, I.M., Maccacaro, T., Morris, S.L., Stocke, J.T., Wolter, A. (1992): ApJ, **386**, 408
Henry, J.P. (1997): ApJ, **489**, L1
Kitayama, T. & Suto, S. (1997): ApJ, **490**, 557
Mathiesen, B. & Evrard, A.E. (1998): MNRAS, **295**, 769
Rosati, P., Della Ceca, R., Burg, R., Norman, C., Giacconi, R. (1995): ApJ, **445**, L11
Rosati, P., Della Ceca, R., Norman, C., Giacconi, R. (1998): ApJ, **492**, L21
Scharf, C.A., Jones, L.R., Ebeling H., Perlman, E., Malkan, M., Wegner, G. (1997): ApJ, **477**, 79
Schmidt, M. et al (1998): A&A, **329**, 495
Vikhlinin et al. (1998): ApJ, in press

Discussion

Dickinson: Of the four $z = 0.8$ clusters you and Isabella Gioia have shown at this meeting, three have this elongated, filamentary appearance. Do clusters with similarly large L_X in the local universe ever show similar structure?

Rosati: Not enough statistics to provide a firm answer.

Bland-Hawthorn: Do we need X-rays? If your ultimate goal is sigma, and you get a weak lensing signature in a few minutes for your X-ray cluster, why not take hundreds of images and select clusters with weak lensing?

Rosati: There are roughly ~ 1 cluster/deg^2 with $L_X > 10^{44}$ erg/s out to $z \sim 1$, too few for a serendipitous search with a small field of view like IRIS (25 arcmin^2). You would need ~ 200 exposures to detect just one cluster. Moreover, 10 min exposure is enough to detect the weak lensing signal in a cluster at $z = 0.3$, but ~ 2 hrs are needed for a cluster at $z = 0.8$.

Rich Clusters of Galaxies at Low to Intermediate Redshift

Eileen O'Hely[1], Warrick J. Couch[1], Ian Smail[2], Ann Zabludoff[3], and Alastair Edge[4]

[1] Department of Astrophysics and Optics, School of Physics,
    University of New South Wales, Sydney, Australia, 2052
[2] Department of Physics, University of Durham, South Rd, Durham, UK, DH1 3LE
[3] UCO/Lick Observatory and Board of Astronomy and Astrophysics,
    University of California at Santa Cruz, Santa Cruz, USA, CA 95064
[4] Institute of Astronomy, Madingley Rd, Cambridge, UK, CB3 OHA

Abstract. Rich clusters of galaxies are fundamental components of the large-scale structure seen throughout the universe. Their large populations of co-distant galaxies provide a very efficient means of studying volume-limited samples of galaxies. Comparisons of clusters at high z to those at low z show that they have undergone significant evolution, both dynamically and in their galaxy populations, at recent times. There is, however, a $\simeq 2$ Gyr evolutionary gap in cluster observations over the important transitory epoch between these two extremes. The LARC (LCO/AAT Rich Cluster) Survey will bridge this gap by using a combination of high resolution X-ray imaging, optical imaging and spectroscopic data of clusters in the redshift range $0.07 \leq z \leq 0.16$. From such data we can perform detailed analysis of clusters' dynamics and the evolutionary histories of their constituent galaxies.

1 Motivation

Evolutionary studies have compared clusters at high redshift to their low redshift counterparts. There are remarkable distinctions in the galaxy populations of clusters at these two extremes of epoch. Clusters at $z > 0.4$ have a bluer core galaxy population than their present day counterparts (Butcher and Oemler, 1978). This blue galaxy fraction increases with redshift from $z = 0.2$ for rich, compact clusters (Couch, 1981; Butcher and Oemler, 1984) and is commonly known as the Butcher-Oemler effect. Tracing remnants of this excess blue population at intermediate z is one of the key issues addressed by this survey.

The usefulness of a model of galaxy evolution depends on how well it reproduces the observed characteristics of galaxies at particular epochs. It is necessary to have a broad observational comparison on hand to test such models. We are compiling a large data base of images and spectra of galaxies in rich clusters in the range $0.07 \leq z \leq 0.16$. This will serve as an observational standard for testing models of galaxy evolution in the rich cluster environment. Our survey will allow the characterisation of clusters at low and intermediate z, providing continuous observations of the cluster evolutionary sequence to the lower bounds of surveys at higher z.

2 Survey Design and Status

We adopt the technique used by Zabludoff and Zaritsky (1995) in their study of Abell 754 and extend it to a sub-set of clusters chosen from the X-ray Brightest, Abell-type Cluster Survey catalogue (Ebeling et al., 1996). The analysis of X-ray and optical images and spectroscopy of Abell 754 revealed the presence of substructure within the cluster. By measuring the frequency of subclustering at recent epochs, we can constrain the cluster relaxation timescale and Ω_0, the mean mass density parameter; a high frequency of substructure at a late epoch implying a high value of Ω_0 (Evrard et al., 1993).

Assuming an infall radius of 5 Mpc, clusters at this epoch are 2 degrees on the sky. The first step is to take photometric, multi-colour images of each cluster in order to derive basic morphological information and map the galaxy distribution. We have obtained broad-band R and B CCD images, to a completeness limit of 21 and 22.5 respectively, of a 2-degree field around 17 clusters in our sample. These data were taken on the 40 inch Swope telescope at Las Campanas Observatory in Chile. The CCD camera has a 23×23 arcmin^2 field of view, so a 5×5 mosaic was necessary to attain the spatial coverage. The data is being photometrically calibrated and preliminary results show a slight variation in colour for increasing radius. However, before drawing any strong conclusions we need spectral data to distinguish cluster galaxies from line-of-sight interlopers.

To this end we are performing multi-fibre spectroscopy with the new 2dF facility at the Anglo-Australian Observatory which will also provide information on the galaxies' starformation histories. The field of view of this instrument exactly matches the area imaged of clusters in our survey, and 2dF's 400 multiplex power makes our observations very economical; 2 configurations per cluster are expected to yield $\simeq 350 - 400$ spectra of cluster galaxies.

To see what role, if any, environment plays in galaxy evolution we need to combine spectral data with high resolution optical and X-ray imaging. To map the gas content of the cluster we are using both archived data and our own X-ray observations from the ROSAT HRI camera. The spatial resolution of the HRI ($20h_{100}^{-1}$kpc) enables us to look at structure in the intra-cluster medium on galactic scales. By identifying the preferred environment of galaxies of fixed morphology we can draw conclusions on the environmental factors responsible for their evolution. Results will be forthcoming in late 1998.

References

Butcher, H., Oemler, A. (1978): Ap. J. **219**, 18
Butcher, H., Oemler, A. (1984): Ap. J. **285**, 426
Couch, W. J. (1981): *PhD Thesis* Australian National University.
Ebeling, H., Voges, W., Bohringer, H., Edge, A. C., Huchra, J. P., Briel, U. G. (1996): MNRAS **281**, 799
Evrard, A. E., Mohr, J. J., Fabricant, D. G., Geller, M. J. (1993): Ap. J. L. **419**, L9
Zabludoff, A. I., Zaritsky, D. (1995): Ap. J. L. **447**, 21

Tunable Filter-Selected Hα Emission in the Rich Cluster A 3665 (AC 106)

D. Heath Jones[1] and Joss Bland-Hawthorn[2]

[1] Mount Stromlo and Siding Spring Observatories, Private Bag, Weston Creek PO, Weston Creek, ACT 2611, Australia
[2] Anglo-Australian Observatory, PO Box 296, Epping, NSW 2121, Australia

Abstract. We present work in progress on an Hα emission-selected sample of galaxies in the rich $z = 0.237$ cluster A 3665 (AC 106; Couch and Newell 1984). The sample was derived by scanning the monochromatic field ($\delta\lambda = 15\text{Å}$) of the Taurus Tunable Filter (TTF) in search of redshifted Hα. The TTF sample represents a strict volume-limited selection for Hα emission through 5300 km s^{-1} in the rest frame of the cluster, encompassing cluster redshift. From such a sample we can derive Hα line luminosities and infer star-formation rates. Our average star-formation rates for A 3665 show little or no excess to that of the field population at this redshift.

Here we demonstrate the emission-line selection techniques applied to the cluster for this project.

1 Introduction

Tunable filters are specially modified Fabry-Perot devices in which the spacing between the two glass plates is much narrower and adjustable over a much wider range than traditional instruments. This gives a tunable filter greater flexibility in wavelength coverage and spectral resolution, over a much wider dynamic range (Bland-Hawthorn and Jones 1998). Scanning tunable filter techniques offer significant advances over traditional spectroscopic surveys for emission-line galaxies: (1) the passband can be tuned to optimize the detection of faint emission-lines, (2) all galaxies in the field are sampled simultaneously, and (3) the sampling is strictly volume-limited.

We maintain a WWW site describing all aspects of TTF and its operation (http://msowww.anu.edu.au/~dhj/ttf.html).

2 Observations

Figure 1 demonstrates detection power of TTF using the emission-line galaxies of an A 3665 field. We observed seven 14 Å passbands (inset, *left*) for 20 min each. These slices were combined into three bands: $B1$ (slices 1 and 2), $B2$ (slices 3 and 4) and $B3$ (slices 6 and 7). The magnitude differences $M(B2) - M(B1)$ and $M(B3) - F(B2)$ are plotted in Fig. 1. To aid clarity, error bars have only been shown for those objects with colour terms greater than one from $(0,0)$. We expect objects containing redshifted Hα emission to lie along the diagonal in the lower right-hand quadrant. Indeed we find a clear population of such objects in Fig. 1. Deviations from the diagonal are likely to be due to contamination by [N II] for which no correction has been made. The excess of objects in the upper left-hand quadrant indicate an absence of flux from the central band. Given the Butcher-Oemler nature of A 3665 we suspect these are E+A galaxies. The population of objects along the line $M(B3) - M(B2) = 0$ are objects near the detection threshold, indicative of a higher signal-to-noise ratio in $B1$ than either $B2$ or $B3$. We conclude from Fig. 1 that our distribution of bands is sufficient to encompass most Hα features of galaxies within the cluster.

Figure 1(a) shows examples of three of the stronger emission-line galaxies from our sample as evident from single-frame analysis. The emission-line galaxy in row 2 is near the centre of the cluster and has been noted independently by Couch and Newell (1984) for its spectrum of strong emission lines, including [O II], [Ne III], [O III], He I, Hγ and Hβ. Figure 1(b) shows relative photometry of the same three galaxies from individual frames. Single frame photometry is adopted when redshift measurements are needed or when galaxy redshifts are not known *a priori*, such as in blind field searches.

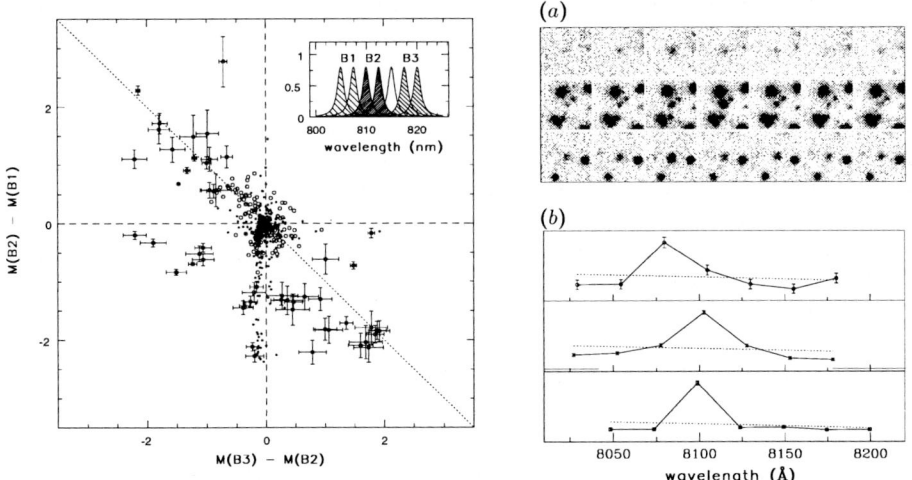

Fig. 1. Examples of combined (*left*) and single (*right*) frame detection of Hα emission-line galaxies in A 3665. Details in the text.

References

Bland-Hawthorn, J., Jones, D. H. (1998): in *Optical Astronomical Instrumentation*, Proc. SPIE **3355**, S. D'Odorico (ed.), in press

Couch, W. J., Newell, E. B. (1984): Distant Clusters of Galaxies. I. Uniform Photometry of 14 Rich Clusters, ApJS **56**, 143–192

Luminosity Function of Cluster Galaxies

E. Molinari, A. Moretti, G. Chincarini, S. De Grandi

[1] Osservatorio Astronomico di Brera, via Bianchi 46, I-23807 Merate (LC), Italy
[2] Università degli Studi di Milano, Milano, Italy
[3] Max-Planck-Institut für extraterrestrische Physik, Garching bei München, Germany

Abstract. We started a project for the study of the luminosity function (LF) in clusters of galaxies selected by their X-ray brightness. These low redshift clusters propose new puzzling features which challenge the interpretation of the cluster evolution.

1 The Sample

Despite of its importance only recently some attempts to characterise properly the luminosity function (LF) have been made. The work on Virgo cluster summarised in Binggeli et al (1988) showed the existence of different LFs for different morphological types. Bernstein et al (1995) pointed out the high number of faint galaxies in the Coma cluster while Biviano et al (1995) showed the presence of peculiar gaps.
We started in the beginning of 1995 a series of observations aiming to produce three-colours photometry of a set of 18 clusters of galaxies selected from the ROSAT All Sky Survey. From the list of candidate clusters we selected the most bright which had a very high probability to be real extended sources. All our clusters are of Bauts-Morgan type I or II, characterising their evolved dynamical status.

2 The Luminosity Function

Abell 496 is the first cluster to be completely analysed and served to tune our procedure. The field of view covered reaches more than 3 core radii. This allows us to use the farthest region as a background corrector. The distribution of magnitudes of this area will be scaled and subtracted from the central one. Number counts in the control area closely match those of Tyson (1988).
The spatial distribution of the galaxies in A496 shows immediately the concentration of redder objects in the center. Red (fiducial early-type) galaxies are the only ones to follow the potential well, as seen in X-ray images. Bluer (probably member spiral) galaxies distribute preferably in a intermediate shell just outside 1 core radius. We avoid this area for the purpose of measuring LF in this work. After estimating the number counts in the far area and smoothing the distribution to minimise the effects of low statistics we compute the LF for the core of A496, focussing in this way on the E/S0 galaxies in the potential well.
Left side of Fig. 1 plots the luminosity function for the three colours available

in our observations. On the faint end, where we easily reach the dwarf domain, the number count steepening seems to differ from one colour to another, with a flatter slope for bluer passband. This could be a selection effects, as our procedure of choosing the $R < 1R_C$ area and correcting background contamination using the $R > 2R_C$ counts leads to favour E/S0 galaxies with redder colours.

More incontestably the presence of a gap in the magnitude distribution again strikes out. The $\sim 2\sigma$ level in each single LF is straightened by the simultaneous presence in every filter. The typical magnitude of these gaps lies in the region where the dwarf elliptical begin to dominate the population mix. In the right side of Fig. 1 a tentative comparison of three type of clusters is shown. Upper: the LF of E/S0 and dE from the Binggeli et al (1988) compilation shows no gap when co-added. Middle: the gap for a BM-type II. Lower: A496, a cD dominated cluster, with a more pronounced gap at somewhat fainter magnitude. Are we looking at a real dynamical sequence?

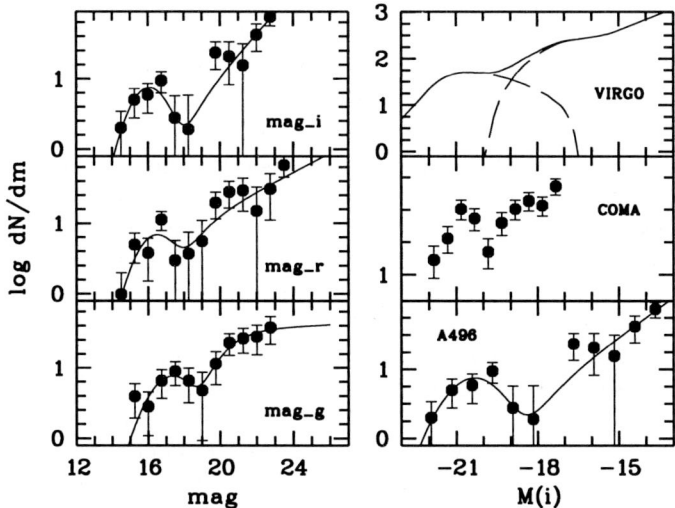

Fig. 1. Left: the LF for A496 in g, r and i. Error bars are poisson counting statistics and the solid line is the fit with the sum of a gaussian and a Schechter function. On the right the comparison with the Virgo early type LF (divided into E/S0 and dwarf E, Binggeli et al 1988), the Coma core LF (Biviano et al 1995) and A496. All LF are translated into the i passband.

References

Bernstein, Nichol, Tyson, Ulmer, Wittman 1995, AJ 110, 1507
Binggeli, Sandage, Tammann 1988, ARA&A 26, 509
Biviano, Durret, Gerbal, Le Fèvre, Lobo, Mazure, Slezak 1995, A&A 297, 610
Tyson 1988, AJ 96, 1

A Dwarf Galaxy Population – Density Relation

Steven Phillipps[1], Simon Driver[2], Warrick Couch[2] and Rodney Smith[3]

[1] University of Bristol, UK
[2] University of New South Wales, Australia
[3] University of Cardiff, U.K.

Abstract. Recent surveys of clusters of galaxies have revealed the presence of large numbers of dwarf (ie. low luminosity) galaxies. However, clusters do show variations in their luminosity functions, with some having larger numbers of dwarfs per giant galaxy than others. Based on our recent cluster surveys we suggest that the dwarf to giant ratio is environment dependent, with lower density clusters, or parts thereof, having the highest relative numbers of dwarfs.

1 Dwarf Galaxies in Clusters and the Galaxy Luminosity Function

Many recent surveys of local and moderate redshift clusters of galaxies (e.g. Smith et al. 1997, Wilson et al.1997, Trentham 1997) have revealed high numbers of dwarf (ie. low luminosity) members. In particular, faintwards of about $M_R = -19$ (for $H_0 = 50$ km s^{-1} Mpc^{-1}), the luminosity function appears to take a sharp upturn from its canonical $\alpha = -1$ slope, the faint end exhibiting slopes in the range $\alpha = -1.5$ to -2.

Although this behaviour is seen quite generally, the precise form of the luminosity function does vary from cluster to cluster when we consider clusters of different morphology (for instance Bautz-Morgan type). In particular, we have found that denser, more centrally concentrated clusters appear to have smaller ratios of dwarf galaxies to giants than do more diffuse lower density clusters.

In addition, it appears that the dwarf-to-giant ratio (DGR) varies with position within a cluster, the dwarfs showing a slower radial decline in number density than the giants.

2 Local Galaxy Density

Combining these results, the implication is that relatively more dwarfs are seen in lower density environments. This is illustrated in Figure 1. Notice in particular that the DGR is similar for the centres of loose clusters (closed symbols) and for the equal density regions on the outskirts of more compact clusters (open symbols). This dwarf population – local (projected) giant galaxy density relation is obviously reminiscent of the well known morphology – density relation, and the two together imply that dwarfs may be distributed rather like spiral galaxies, both being more common in low density regions (or, equivalently, tending to avoid very dense regions). The basic cause of the two relations may also be similar, arising out of the formation processes of galaxies in hierarchical merger models (see e.g. Baugh, Cole & Frenk 1996, Kauffman, Nusser & Steinmetz 1997).

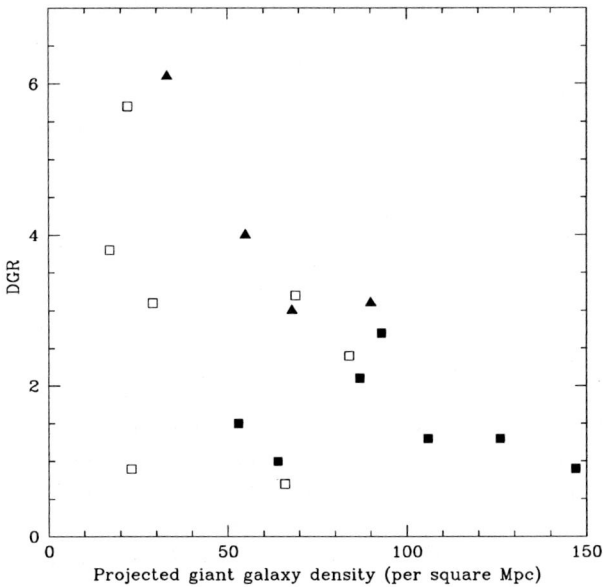

Fig. 1. Variation of dwarf-to-giant ratio with local density. Filled symbols are for the central 1 square Mpc area of our clusters, open symbols for regions further out. The triangles show in slightly more detail the radial variation of DGR across Abell 2554.

References

Baugh C., Cole S., Frenk C.S., 1996, MNRAS, 261, 1378
Kauffmann G., Nusser A., Steinmetz M., 1997, MNRAS, 286, 795
Phillipps S., Driver S.P., Couch W.J., Smith R.M., 1998, ApJL, in press
Smith R.M., Driver S.P., Phillipps S., 1997, MNRAS, 287, 415
Trentham N., 1997, MNRAS, 286, 133
Wilson G., Smail I., Ellis R.S., Couch W.J., 1997, MNRAS, 284, 915

Radio Survey of Merging Clusters in the Core of the Shapley Concentration

T. Venturi, S. Bardelli, R. Morganti, R.W. Hunstead

[1] Istituto di Radioastronomia del CNR, via Gobetti 101, 40129 Bologna, Italy
[2] Osservatorio Astronomico di Trieste, Via G.B. Tiepolo 11, 34131 Trieste, Italy
[3] School of Physics, University of Sydney, NSW 2006, Australia

Abstract. We present preliminary results of a radio survey in the chain of the interacting Abell clusters A3556-A3558-A3562, belonging to the central region of the Shapley Concentration. The survey was carried out at 22 cm with the Australia Telescope Compact Array. In our preliminary list of optical identifications we find 18 radio galaxies whose optical counterparts are members of the Shapley Concentration. Four of them have extended radio morphologies.

1 The Core Region of the Shapley Concentration

The Shapley Concentration is a supercluster of galaxies, located in the Southern Hemisphere at a redshift going from $z \sim 0.03$ to $z \sim 0.05$, and extending over several square degrees on the plane of the sky. It is characterised by a very unrelaxed dynamical state (Bardelli et al. 1994, Bardelli et al. 1998) and it is therefore an ideal laboratory to study the properties and the consequences of cluster merging. The central region of the Shapley Concentration is formed by a chain of three clusters, A3556, A3558 and A3562, located almost orthogonal to the line of sight, and containing mostly early-type galaxies (see Figure 1 in Venturi et al. 1997). A3558 is the central and most massive cluster in the chain (richness class R=4) and is dominated by a cD galaxy; A3562 is an intermediate richness (R=2) cluster dominated by a cD galaxy with a very extended asymmetric optical halo, and A3556 is the poorest cluster in the chain (R=0), with the most intriguing optical properties (see Bardelli et al. 1998 for a detailed discussion). The cores of these three main clusters are connected by a stream of poorer groups of galaxies.

The spectroscopic and X-ray properties of the whole A3558 complex make it very similar to cluster collision simulations carried out by Roettiger et al. (1993), suggesting that a major merging event may have just taken place between A3558 and another massive cluster, which has partly crossed the core of A3558 (A3562) and has been partly left behind (A3556) and is now merging with A3558.

2 Radio Observations and Results

We performed ATCA observations of small groups in the A3558 complex at 22 cm with the 1.5B and 6C configurations, for 12 hours in each configuration. We made three pointings located respectively (a) in the region between A3556 and A3558, south-west of the A3558 core; (b) between the core of A3558 and the group SC1329-313; (c) east of the A3562 core. Each mapped field covers a sky region ~ 1 square degree. All sources brighter than 1 mJy were searched for an optical counterpart on the Digitised Sky Survey. Adding the results obtained for A3556 (Venturi et al. 1997) to the present data, we found 64 identifications. 21 galaxies have known redshift, and 18 of them belong to the A3558 complex. If we consider that counterparts without redshift and brighter than $b_J \sim 17$ are very likely to be cluster members, the identified radio galaxies could rise to 27.

Four radio galaxies belonging to the Shapley Concentration have extended radio emission. Interestingly, they are all located at the periphery of the A3558 complex, in the most unrelaxed parts of the chain, i.e. two of them belong to A3556 (Venturi et al. 1997) and the remaining two belong to A3562. One is the head-tail radio galaxy J1333−3141, located at a projected distance of ~ 1' from the dominant cD, and the other is the classical double J1335−3153. In Figure 1 22 cm radio contours of these two radio galaxies are given, overlaid on the Digitised Sky Survey. These results give support to the idea that head-tail sources are preferentially found in merging environments (Bliton et al. 1998; Reid et al. 1998; Venturi et al., submitted to MNRAS).

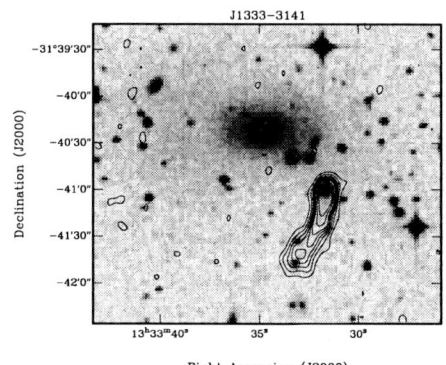
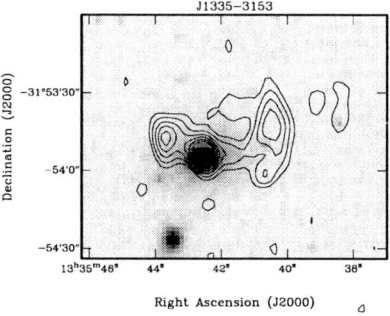

References

Bardelli, S., Zucca, E., Vettolani, G. et al. (1994): MNRAS **267**, 665
Bardelli, S., Zucca, E., Zamorani, G. et al. (1998): MNRAS, in press
Bliton, M., Rizza, E., Burns, J., et al. (1998): 191st AAS Bulletin, 106.08
Reid, A., Hunstead, R.W., Pierre, M. (1998): MNRAS, in press
Roettiger, K., Burns, J., Loken, C. (1993): ApJ Let. **407**, 470
Venturi, T., Bardelli, S., Morganti, R., Hunstead, R.W. (1997): MNRAS **285**, 898

Cosmological Parameters as Measured by Type Ia Supernovae

Bruno Leibundgut[1], Brian Schmidt[2], Jason Spyromilio[1], Mark Phillips[3]

[1] European Southern Observatory, Karl-Schwarzschild-Strasse 2, D-85748 Garching, Germany
[2] Mount Stromlo and Siding Springs Observatory, Private Bag, Weston Creek P.O. 2611 Australia
[3] Cerro Tololo Inter-American Observatory, Casilla 603, La Serena, Chile

Abstract. Supernovae are among the most important cosmological distance indicators. Through the application of distance independent correction methods to the peak luminosity, it has become possible to determine very accurate luminosity distances. Type Ia supernovae have been used extensively to measure the current Hubble constant, H_0. More recently, they have been employed to determine the deceleration of the universe. From a first sample of distant supernovae, strong indications of a non-vanishing cosmological constant Λ emerge. A careful analysis of the systematic effects does not change this result. The distant supernovae also indicate a dynamic age which is consistent with age determinations of the oldest stellar components of the Galaxy.

1 Introduction

The ability of Type Ia supernovae (SNe Ia) to measure cosmological distances is largely based on empirical evidence. Samples of nearby SNe Ia display a very narrow scatter around the expected linear expansion of the local universe (Kowal 1968, Tammann & Leibundgut 1990) and can be further improved by application of distance independent corrections like for light curve shape (Hamuy et al. 1996a, Riess et al. 1996). The dispersion of the nearby sample decreases to around 0.15 magnitude after an additional correction for absorption. Attempts to explain the empirical correlation between the peak luminosity and the decline rate of SNe Ia after maximum are, however, still controversial (Höflich et al. 1996, Eastman 1997).

With such a modified standard candle it is possible to derive accurate values of the few parameters which govern the expansion in a homogeneous, isotropic universe (e.g. Leibundgut & Pinto 1992, Goobar & Perlmutter 1995). Samples of SNe Ia out to redshifts of $z = 0.1$ combined with a measurement of the absolute luminosity yield values for the Hubble constant (Hamuy et al. 1996a, Riess et al. 1996, Reiss et al. 1998).

♣ We report results by the High-z Supernovae Search Team. Members of the team are P. Challis, A. Clocchiatti, A. Diercks, A. Filippenko, P. Garnavich, R. Gilliland, C. Hogan, S. Jha, R. Kirshner, D. Reiss, A. Riess, R. Schommer, C. Smith, C. Stubbs, N. Suntzeff, J. Tonry, and P. Woudt.

With a sample of distant supernovae ($z > 0.3$) we can measure the curvature of the universe and any possible contributions of a cosmological constant. This is a geometric measure of the universe. The deceleration can be determined by comparing the distant supernovae with the local sample (Goobar & Perlmutter 1995, Leibundgut 1998). Two experiments are trying to make use of SNe Ia to achieve a determination of the deceleration (Perlmutter et al. 1995, 1997, 1998; Leibundgut et al. 1996, Garnavich et al. 1998, Schmidt et al. 1998, Riess et al. 1998). The current status of the two searches is summarized in section 3.

Yet supernovae are not true standard candles. The light curves and spectral evolution show distinct variations among individual events (Phillips et al. 1987, 1992, Filippenko et al. 1992a, 1992b, Leibundgut et al. 1993). In particular the near-infrared light curves show a second maximum for many, but not all, SNe Ia (Elias et al. 1985, Frogel et al. 1987, Suntzeff 1996). The strength of the second maximum changes considerably and is also distinguishable in bolometric light curves (Vacca & Leibundgut 1996, Contardo et al. 1998).

2 Brief Overview of SNe Ia

A generally accepted theory of SN Ia explosions has not yet emerged. For more than a decade explosion models with a subsonic burning front have been favored (Nomoto et al. 1984, Woosley & Weaver 1986), but the physics which sustains such a burning front without turning into a detonation is not understood. New models which explode a white dwarf through He detonation at the surface (Arnett & Livne 1994, Woosley & Weaver 1994), rather than the direct explosive carbon burning inside, cannot be distinguished by the current observations (Branch et al. 1995). The exact trigger of the explosion and the reaction of the white dwarf is not constrained and many models have been proposed (e.g. Höflich & Khokhlov 1996). The radiation transport is largely unsolved. The supernova radiation is not thermalized in the ejecta and the observed pseudo-continuum is dominated by line radiation. A simple way to see this is the fact that the occurrence of maximum in the observed filter light curves does not indicate a cooling shell, but rather a very complex redistribution of the emission (Contardo et al. 1998). The bolometric light curves display the characteristic shoulder of the near-infrared light curves and the total energy changes from event to event.

With this theoretical background it is difficult to support the standard candle picture. The observations have shown a significant variety of appearance. Nonetheless, most variations appear to correlate strongly so that the light curve decline (Hamuy et al. 1996a, Riess et al. 1996), the strength of the Ca II and Si II lines (Nugent et al. 1995), as well as the colors near maximum (Riess et al. 1998, Phillips et al. 1998) provide distance independent indicators of the intrinsic luminosity of SNe Ia. In this sense, SNe Ia appear to be an ordered set and can empirically be used as modified standard candles.

3 The Hubble Constant as Derived from SNe Ia

The derivation of the Hubble constant from standard candles is in principle very simple. With knowledge of the absolute magnitude the linear relation of redshift and apparent magnitude provides the Hubble constant (Tammann & Leibundgut 1990, Leibundgut & Pinto 1992). The recent programs to determine distances to nearby galaxies which contained SNe Ia (Saha et al. 1997) provide a fairly accurate determination of the luminosity of SNe Ia. Differences arise in the treatment of the relative absorption of the Cepheid sample and the supernova, and corrections to the supernova light curves themselves. The small scatter of the measured peak luminosity of a sample of 7 nearby SNe Ia (Saha et al. 1997) is further indication of the standard behavior of these objects.

The distribution of SNe Ia in the Hubble diagram then yields a value of the Hubble constant. The controversy of the previous decades is finally subsiding and values of $55 < H_0 < 70$ km s^{-1} Mpc^{-1} are generally found (Hamuy et al. 1996a, Riess et al. 1996, Saha et al. 1997). This provides a measurement of H_0 to 15%. The differences are mainly in the treatment of the supernova light curves. It has to be noted that the light curve correction in general changes the value of H_0 by about 10% simply because the nearby SN sample with Cepheid distances and the set of SNe Ia in the Hubble flow have a zero-point difference of this order (Hamuy et al. 1996a, Leibundgut 1998). Other points of discussion are the exact selection of the nearby sample which, although distance limited, suffers from partially inadequate historical records of supernova photometry. These two effects combine to the differences in the reported values. With more Cepheid distances towards supernovae and better observed light curves the discrepancy will disappear.

4 Distant Supernovae: Ω_M and Ω_Λ

The determination of the cosmological energy density rests on the measurement of objects distant enough so that the curvature of the universe becomes significant. SNe Ia discovered at redshifts larger than about 0.3 are very well suited for such an investigation. Two groups aggressively pursue this experiment (Perlmutter et al. 1995, 1997, 1998, Leibundgut et al. 1996, Garnavich et al. 1998, Schmidt et al. 1998, Riess et al. 1998). The cosmological frame work has been laid out by Carroll et al. (1992) and Goobar & Perlmutter (1995).

Supernovae are now regularly found at large redshifts. So far, there is no indication that their light curves and spectra differ from nearby objects. Many SNe Ia have been classified and the light curves are being analyzed. The results very strongly favor a positive value for Ω_Λ (Riess et al. 1998).

Of fundamental importance are systematic effects which can alter the result. Several technical steps lead to the luminosity distances. Accurate photometry of faint sources on a spatially variable background, phase-dependent K-corrections, and light curve determinations have to be applied. All of these corrections have been tested thoroughly and independently in several applications (Schmidt et al. 1998, Riess et al. 1998).

Other uncertainties are introduced by the light curve shape correction, extinction, possible evolution between the distant and the nearby supernovae, sample selection, and (de-)magnification due to gravitational lensing. A careful analysis can test for all effects with the exception of gravitational lensing which is achromatic and static. The systematic influence of weak lensing, however, is expected to be very small (Wambsganss et al. 1997). Should the supernovae be lensed significantly the scatter in the Hubble diagram would be increased, which is not observed (Riess et al. 1998).

A luminosity shift between SNe Ia in early and late type galaxies has been observed (Hamuy et al. 1996b). This has been attributed to the different ages, masses, or metallicities of the progenitor population. The light curve corrections can successfully neutralize the offset and the luminosity difference between samples in elliptical and spiral galaxies can be reduced to 0.05 magnitudes (Schmidt et al. 1998). This is an indication of what has to be expected for a comparison between the nearby and the distant sample. A critical parameter appears to be the C/O composition of the progenitor white dwarf (Höflich et al. 1998). Extinction can be detected through the color of the supernovae. Very few distant objects appear reddened (Garnavich et al. 1998, Riess et al. 1998). There is a bias which favors discovery of unabsorbed objects due to the luminosity and also the selection of objects well separated from the host galaxy. The latter is not a true selection by the search, but rather is applied when decisions on the follow up observations are taken. Thus, the absorption in the distant sample is likely to be less on average than in the nearby sample. Another important issue is the concordance of the nearby and the distant sample. Any systematic shift between the two would result in a distortion of the derived cosmological parameters. An obvious systematic effect would be different average luminosities of the two samples. This can be tested by comparing the light curve shapes of the two samples. For our sample of 10 distant and 27 nearby supernovae the average Δ of the MLCS method (Riess et al. 1996) is -0.126 and -0.147 for the nearby and the distant sample, respectively. The systematic offset is thus 0.02 magnitudes, which is negligible. The same investigation for the analysis with Δm_{15} yields values which are slightly different. The mean Δm_{15} for the local sample is 1.223, while the distant sample has 1.173 on average. This difference translates into a zero-point offset of 0.04 (0.03) magnitudes for B (V) using the second order treatment proposed by Phillips et al. (1998). The weighted mean value becomes somewhat larger: 0.09 (0.08) magnitudes for B (V). The local sample is on average less luminous than the distant one. This is also exemplified by Fig. 10 in Riess et al. (1998). Figure 1 displays the magnitude difference of the observed supernovae from an Einstein-de Sitter Universe ($\Omega_M = 1, \Omega_\Lambda = 0$). Most objects clearly lie above the line for an empty universe (0,0) and thus indicate a contribution of the vacuum energy to the energy density of the universe.

The comparison of 14 distant $z>0.16$ supernovae has shown that they appear to be too dim for an empty universe. No solution for a positive matter density ($\Omega_M > 0$) is found if $\Lambda = 0$ at a $> 2.5\sigma$ significance (Riess et al. 1998). Due to the rather limited range of redshifts an exact value for Ω_M and Ω_Λ cannot

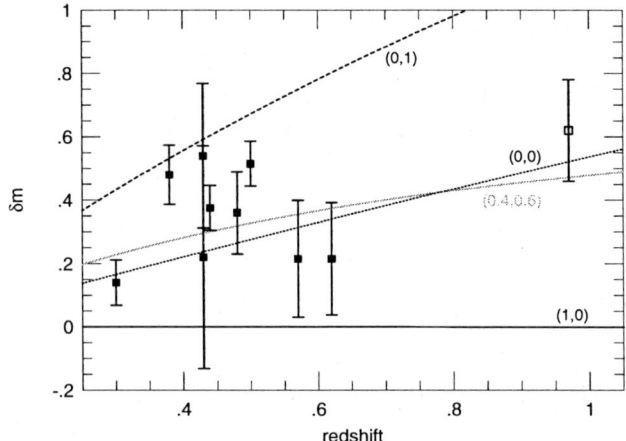

Fig. 1. Magnitude difference from an Einstein-de Sitter Universe ($\Omega_M = 1, \Omega_\Lambda = 0$). The empty case ($\Omega_M = 0, \Omega_\Lambda = 0$) and a mixed model with ($\Omega_M = 0.4, \Omega_\Lambda = 0.6$) are also indicated. The observed δm for the supernovae are based on B and V light curves. SN 1997ck is shown as an open symbol as no color information nor a spectrum for this supernova are available.

be derived yet without further constraints. This confirms the preliminary result presented by Garnavich et al. (1998) and is also in accordance with the one reported by Perlmutter et al. (1998).

5 Conclusions

The analysis of our distant set with two slightly different analysis methods resulted in a clear signature of a positive cosmological constant. A universe with negligible Λ is excluded at the 2.5 σ level, as no solution with a positive matter density is found. The effect is so strong that the value of the deceleration parameter q_0 is negative which indicates an accelerated expansion over the time span sampled by our observations (5 Gyr; Riess et al. 1998).

The dynamic age shows a dependency very similar to the uncertainty regions determined by the supernovae. It is thus possible to restrict the dynamic age of the universe quite accurately. Riess et al. (1998) find values in the range of 13 to 15 Gyr which is consistent with most age determination of the oldest stellar components.

Essential issues remain the systematic uncertainties of the experiment. Evolution, extinction and sample selection are the most important contributors. The only checks for evolution are extensive observations of distant objects to detect any differences with nearby supernovae. A spectroscopic sequence with good signal would allow us to investigate small differences in the ejecta. There are clear signatures in the R and I light curves of SNe Ia which can be observed for objects with $z < 0.4$.

References

Arnett, W. D., & Livne, E. 1994, ApJ, 427, 315
Branch, D., Livio, M., Yungelson, L. R., Boffi, F. R., & Baron, E. 1995, PASP, 107, 1019
Carroll, S. M., Press, W. H., & Turner, E. L. 1992, ARA&A, 30, 499
Contardo, G., Leibundgut, B., & Vacca, W. D. 1998, in preparation
Eastman, R. G. 1997, Thermonuclear Supernovae, eds. P. Ruiz-Lapuente, R. Canal, & J. Isern, Dordrecht: Kluwer, 571
Elias, J. H., Matthews, K., Neugebauer, G., & Persson, S. E. 1985, ApJ, 196, 379
Filippenko, A. V., et al. 1992a, ApJ, 384, L15
Filippenko, A. V., et al. 1992b, AJ, 104, 1534
Frogel, J. A., et al. 1987, ApJ, 315, L129
Garnavich, P., et al. 1998, ApJ, 493, L53
Goobar, A., & Perlmutter, S. 1995, ApJ, 450, 14
Höflich, P., & Khokhlov, A., 1996, ApJ, 457, 500
Höflich, P., Khokhlov, A., Wheeler, J. C., Phillips, M. M., Suntzeff, N. B., & Hamuy, M. 1996, ApJ, 472, L81
Höflich, P., Wheeler, J. C., & Thielemann, F.-K. 1998, ApJ, 495, 617
Hamuy, M., et al. 1996a, AJ, 112, 2398
Hamuy, M., et al. 1996b, AJ, 112, 2391
Kowal, C. T. 1968, AJ, 73, 1021
Leibundgut, B. 1998, Supernovae and Cosmology, eds. L. Labhardt, B. Binggeli, & R. Buser, Basel: University of Basel, in press (astro-ph/980169)
Leibundgut, B., & Pinto, P. A. 1992, ApJ, 401, 49
Leibundgut, B., et al. 1993, AJ, 105, 301
Leibundgut, B., et al. 1996, ApJ, 466, L21
Nomoto, K., Thielemann, F.-L., & Yokoi, K. 1984, ApJ, 286, 644
Nugent, P., Phillips, M. M., Baron, E., Branch, D., & Hauschildt, P. 1995, ApJ, 455, L147
Perlmutter, S., et al. 1995, ApJ, 440, L41
Perlmutter, S., et al. 1997, ApJ, 483, 565
Perlmutter, S., et al. 1998, Nature, 391, 51
Phillips, M. M., et al. 1987, PASP. 99, 592
Phillips, M. M., Wells, L. A., Suntzeff, N. B., Hamuy, M., Leibundgut, B., Kirshner, R. P., & Foltz, C. B. 1992, AJ, 103, 1632
Phillips, M. M., et al. 1998, in preparation
Reiss, D. J., Germany, L. M., Schmidt, B. P., & Stubbs, C. W. 1998, AJ, 115, 26
Riess, A. G., Press, W. M., & Kirshner, R. P. 1996, 473, 88
Riess, A. G., et al. 1998, AJ, submitted
Saha, A., Sandage, A., Labhardt, L., Tammann, G. A., Macchetto, F. D., & Panagia, N. 1997, ApJ, 486, 1
Schmidt, B. P., et al. 1998, ApJ, in press
Suntzeff, N. B. 1996, IAU Colloquium 145: Supernovae and Supernova Remnants, ed. R. McCray, (Cambridge: Cambridge University Press), 41
Tammann, G. A., & Leibundgut, B. 1990, A&A, 236, 9
Vacca, W. D., & Leibundgut, B., 1996, ApJ, 471, L37
Wambsganss, J., Cen, R., Xu, G., & Ostriker, J. P. 1997, ApJ, 475, L81
Woosley, S. E., & Weaver, T. A. 1986, ARA&A, 24, 205
Woosley, S. E., & Weaver, T. A. 1994, ApJ, 423, 371

Author Index

Allen, M.G. 248
Andreani, P. 216
Ashley, M.C.B. 201
Banks, G. 132
Bardelli, S. 326
Bernardeau, F. 59
Birkinshaw, M. 159
Bland-Hawthorn, J. **91**, 320
Bouwens, R. 303
Boyle, B. **16**, **275**, 278
Broadhurst, T. 70, **303**
Burton, M.G. **201**
Carilli, C. 246
Charlot, S. **294**
Chincarini, G. 322
Claeskens, J.-F. 89
Clay, R.W. **199**
Colless, M. **9**
Conconi, P. 157
Couch, W. 318, 324
Cram, L. 120
Cristiani, S. 216
Croom, S.M. 16
da Costa, L. **34**
Danks, A.C. 309
Danziger, I.J. **194**
Davies, J.I. 21
Dawson, B.R. 199
De Breuck, C. 236, **246**
De Grandi, S. 322
de Ruiter, R.H. 114
Devriendt, J.E.G. **301**
Dickens, R.J. 21
Dickinson, M. **262**
Dickson, R. 130
Drinkwater, M. **21**, 278
Driver, S. **280**, 324
Edge, A. 318
Ekers, R. 114, **146**
Fernández-Soto, A. **270**
Fioc, M. 289

Fluke, C. **66**
Fosbury, R.A.E. 248
Francis, P.J. **309**
Freudling, W. **29**
Frye, B. 303
Gallimore, J. 248
Georgakakis, A. 120
Gioia, I.M. **181**
Giommi, P. 187
Gregg, M.D. 21
Gregorini, L. 114
Guiderdoni, B. 301
Hadley, B. 78
Hook, I.M. **211**
Hopkins, A. **120**
Hunstead, R.W. **226**, 326
Illingworth, G. **250**
Jackson, C.A. **110**
Jean, C. **89**
Jones, D.H. **320**
Jones, L.R. 187
Juraszek, S. 23
Kilborn, V. **132**
Koribalski, B. 23, 132
Kraan-Korteweg, R.C. **23**
La Franca, F. **216**
Lahav, O. **42**, 112
Lanzetta, K.M. 270
Le Fèvre, O. 59
Leibundgut, B. **328**
Liang, H. **159**
Lineweaver, C.H. **167**
Liske, J. **234**
Loaring, N. 16
Maddox, S.J. 112
Magliocchetti, M. **112**
Malin, D. **78**
McMahon, R.G. 211
Mellier, Y. **59**
Miley, G. 236, 246
Miller, L. 16

Mobasher, B. 120
Molinari, E. **157**, **322**
Monai, S. 157
Moretti, A. 322
Morganti, R. **130**, 248, 326
Mortlock, D.J. **68**
Mould, J. **3**
Norris, R.P. **140**
O'Hely, E. **318**
Padovani, P. **187**
Parker, Q.A. 21, **83**
Parma, P. **114**
Perlman, E. 187
Phillipps, S. 21, 83, **324**
Phillips, M. 328
Prandoni, I. **114**
Pucillo, M. 157
Putman, M. **132**
Renzini, A. **34**
Reynolds, J. 187
Rocca-Volmerange, B. **289**
Rosati, P. 70, **311**
Röttgering, H. 236, 246
Sadler, E.M. 21, **103**
Sambruna, R. 187
Schmidt, B. 328
Schneider, P. **51**
Sealey, K. **278**
Sethi, S.K. 301
Shanks, T. 16
Shaver, P. **153**, 211

Silk, J. 70
Smail, I. 318
Smith, M.G. **72**
Smith, R. 324
Smith, R.J. 16
Smith, R.M. 21
Spyromilio, J. 328
Squires, G. **70**
Stanford, A. 236
Staveley-Smith, L. **132**
Storey, J.W.V. 201
Surdej, J. 89
Tadhunter, C. 130
Taylor, K. **257**
Tsvetanov, Z. **248**
Tzioumis, A. 187
van Breugel, W. **236**, 246
van Waerbeke, L. 59
Venturi, T. **326**
Vettolani, G. 114
Villar-Martin, M. 130
Wall, J.V. 110, 112
Webb, J. 234, 278
Webster, R. 66, 68, 132
Wielebinski, R. **125**
Wieringa, M.H. 114
Wisotzki, L. **221**
Woodgate, B.E. 309
Yahil, A. 270
Yamashita, K. **174**
Zabludoff, A. 318

ESO ASTROPHYSICS SYMPOSIA
European Southern Observatory

Series Editor: Philippe Crane

C. G. Tinney (Ed.), **The Bottom of the Main Sequence – And Beyond**
Proceedings, 1994. XVII, 309 pages. 1995.

G. Meylan (Ed.), **QSO Absorption Lines**
Proceedings, 1994. XXIII, 471 pages. 1995.

D. Minniti, H.-W. Rix (Eds.), **Spiral Galaxies in the Near-IR**
Proceedings, 1995. X, 350 pages. 1996.

H. U. Käufl, R. Siebenmorgen (Eds.), **The Role of Dust in the Formation of Stars**
Proceedings, 1995. XXII, 461 pages. 1996.

P. A. Shaver (Ed.), **Science with Large Millimetre Arrays**
Proceedings, 1995. XVII, 408 pages. 1996.

J. Bergeron (Ed.), **The Early Universe with the VLT**
Proceedings, 1996. XXII, 438 pages. 1997.

F. Paresce (Ed.), **Science with the VLT Interferometer**
Proceedings, 1996. XXII, 406 pages. 1997.

D. L. Clements, I. Pérez-Fournon (Eds.), **Quasar Hosts**
Proceedings, 1996. XVII, 336 pages. 1997.

L. N. da Costa, A. Renzini (Eds.), **Galaxy Scaling Relations: Origins, Evolution and Applications**
Proceedings, 1996. XX, 404 pages. 1997.

L. Kaper, A. W. Fullerton (Eds.), **Cyclical Variability in Stellar Winds**
Proceedings, 1997. XXII, 415 pages. 1998.

R. Morganti, W. J. Couch (Eds.), **Looking Deep in the Southern Sky**. Proceedings, 1997. XXIII, 336 pages. 1999.

Printing: Druckhaus Beltz, Hemsbach
Binding: Buchbinderei Schäffer, Grünstadt